Computer Processing of Remotely-Sensed Images

An Introduction

Second Edition

Paul M. Mather

School of Geography, The University of Nottingham, UK

JOHN WILEY & SONS
Chichester · New York · Weinheim · Brisbane · Singapore · Toronto

Other Wiley Editorial Offices

John Wiley & Sons, Inc., 605 Third Avenue,
New York, NY 10158-0012, USA

WILEY-VCH Verlag GmbH, Pappelallee 3,
D-69469 Weinheim, Germany

Jacaranda Wiley Ltd, 33 Park Road, Milton,
Queensland 4064, Australia

John Wiley & Sons (Asia) Pte Ltd, 2 Clementi Loop #02-01,
Jin Xing Distripark, Singapore 129809

John Wiley & Sons (Canada) Ltd, 22 Worcester Road,
Rexdale, Ontario M9W 1L1, Canada

Library of Congress Cataloging-in-Publication Data

Mather, Paul M.
    Computer processing of remotely sensed images: an introduction/
  Paul M. Mather. – 2nd ed.
      p.   cm.
    Includes bibliographical references and index.
    ISBN 0-471-98550-3 (alk. paper)
    1. Remote sensing – Data processing. I. Title.
  G70.4.M38 1999
  621.36′78 – dc21   98-37544
                                                        CIP

British Library Cataloguing in Publication Data

A catalogue record for this book is available from the British Library

ISBN 0 471 98550 3

Typeset in 9/11pt Times by Vision Typesetting, Manchester
Printed and bound in Great Britain by Bookcraft (Bath) Ltd, Midsomer Norton
This book is printed on acid-free paper responsibly manufactured from sustainable forestry,
in which at least two trees are planted for each one used for paper production.

03891874

'I hope that posterity will judge me kindly, not only as to the things which I have explained but also as to those which I have intentionally omitted so as to leave to others the pleasure of discovery.'

René Descartes

'I am none the wiser, but I am much better informed.'

Queen Victoria

Contents

Preface

Many things have changed since the first edition of this book was written, more than 10 years ago. The increasing emphasis on scientific rigour in remote sensing (or Earth observation by remote sensing, as it is now known), the rise of interest in global monitoring and large-scale climate modelling, the increasing number of satellite-borne sensors in orbit, the development of geographical information systems (GIS) technology and the expansion in the number of taught Masters courses in GIS and remote sensing are all noteworthy developments. Perhaps the most significant single change in the world of remote sensing over the past decade has been the rapid increase in and the significantly reduced cost of computing power and software available to students and researchers alike, which allows them to deal with growing volumes of data and more sophisticated and demanding processing tools. In 1987 the level of computing power available to researchers was minute in comparison with that which is readily available today. I wrote the first edition of this book using a BBC Model B computer, which had 32 kb of memory, 100 kb diskettes and a processor that would barely run a modern refrigerator. Now I am using a 266 MHz Pentium II with 64 Mb of memory and a 2.1 Gb disk. It has a word processor that corrects my spelling mistakes (though its grammar checking can be infuriating). I can connect from my home to the University of Nottingham computers by optic fibre cable and run advanced software packages. The cost of this computer is about one per cent of that of the VAX 11/730 that is mentioned in the Preface to the first edition of this book.

Although the basic structure of the book remains largely unaltered, I have taken the opportunity to revise all of the chapters to bring them up to date, as well as to add some new material, to delete obsolescent and uninteresting paragraphs, and to revise some infelicitous and unintelligible passages. For example, Chapter 4 now contains new sections covering sensor calibration, plus radiometric and topographic correction. The use of artificial neural networks in image classification has grown considerably in the years since 1987, and a new section on this topic is added to Chapter 8, which also covers other recent developments in pattern recognition and methods of estimating Earth-surface properties. Chapter 3, which provides a survey of computer hardware and software, has been almost completely rewritten. In Chapter 2 I have tried to give a brief overview of a range of present and past sensor systems but have not attempted to give a full summary of every sensor, because details of new developments are now readily available via the World Wide Web. I doubt whether anyone would read this book simply because of its coverage of details of individual sensors.

Other chapters are less significantly affected by recent research as they are concerned with the basics of image processing (filtering, enhancement and image transforms), details of which have not changed much since 1987, though I have added new references and attempted to improve the presentation. I have, however, resisted the temptation to write a new chapter on GIS, largely because there are several good books on this topic that are widely accessible – for example, Bonham-Carter (1994) and Maguire *et al.* (1991) – but also because I feel that this book is primarily about image processing. The addition of a chapter on GIS would neither do justice to that subject nor enhance the reader's understanding of digital processing techniques. However, I have made reference to GIS and spatial databases at a number of appropriate points in the text. My omission of a survey of GIS techniques does not imply that I consider digital image processing to be a 'stand-alone' topic. Clearly, there are significant benefits to be derived from the use of spatial data of all kinds within an integrated environment, and this point is emphasised in a number of places in this book. I have added a significant number of new references to each of the chapters, in the hope that the reader might be encouraged to enjoy the comforts of his or her local library.

I have also added some 'self-assessment' questions at the end of each chapter. These questions are not intended to constitute a sample examination paper, nor do they provide a checklist of 'important' topics (the implication being that the other topics covered in the book are unimportant). They are simply a random set of questions – if you can answer them, then you probably understand the contents of the chapter. Readers should use the MIPS software (see below) described in Appendices A and B to try out the methods mentioned in these questions. Data sets are also available on the accompanying CD, and are described in Appendix D.

Perhaps the most significant innovation that this book offers is the provision of a CD containing software and images. I am not a mathematician, and so I learn by trying out ideas rather than exclusively by reading or listening. I learn new methods by writing computer programs and applying them to various data sets. I am including a small selection of the many programs that I have produced over the past 30 years, in the hope that others may find them useful. These programs are described in Appendix B. I have been teaching a course on remote sensing for the past 14 years. When this course began there were no software packages available, so I wrote my own (my students will remember NIPS, the Nottingham Image Processing System, with varying degrees of hostility). I have completely rewritten and extended NIPS so that it now runs under Microsoft Windows 95. I have renamed it Mather's Image Processing System (MIPS), which is rather an unimaginative name, but is nevertheless pithy. It is described in Appendix A. Many of the procedures described in this book are implemented in MIPS, and I encourage readers to try out the methods discussed in each chapter. It is only by experimenting with these methods, using a range of images, that you will learn how they work in practice. MIPS was developed on an old 486-based machine with 12 Mb of RAM and a 200 Mb disk, so it should run on most PCs available in today's impoverished universities and colleges. MIPS is not a commercial system, and should be used only for familiarisation before the reader moves on to the software behemoths that are so readily available for both PCs and UNIX workstations. Comments and suggestions for improving MIPS are welcome (preferably by e-mail), though I warn readers that I cannot offer an advisory service or assist in research planning!

Appendix D contains a number of Landsat, SPOT, AVHRR and RADARSAT images, mainly extracts of size 512×512 pixels. I am grateful to the copyright owners for permission to use these data sets. The images can be used by the reader to gain practical knowledge and experience of image processing operations. Many university libraries contain map collections, and I have given sufficient details of each image to allow the reader to locate appropriate maps and other back-up material that will help in the interpretation of the features shown on the images.

The audience for this book is seen to be advanced undergraduate and Masters students, as was the case in 1987. It is very easy to forget that today's student of remote sensing and image processing is starting from the same level of background knowledge as his or her predecessors in the 1980s. Consequently, I have tried to restrain myself from including details of every technique that is mentioned in the literature. This is not a research monograph or a literature survey, nor is it primarily an exercise in self-indulgence, and so some restriction on the level and scope of the coverage provided is essential if the reader is not to be overwhelmed with detail and thus discouraged from investigating further. Nevertheless, I have tried to provide references on more advanced subjects for the interested reader to follow up. The volume of published material in the field of remote sensing is now very considerable, and a full survey of the literature of the last 20 years or so would be both unrewarding and tedious. In any case, online searches of library catalogues and databases are now available from networked computers. Readers should, however, note that this book provides them only with a background introduction – successful project work will require a few visits to the library to peruse recent publications, as well as practical experience of image processing.

I am most grateful for comments from readers, a number of whom have written to me, mainly to offer useful suggestions. The new edition has, I hope, benefited from these ideas. Over the past years, I have been fortunate enough to act as supervisor to a number of postgraduate research students from various countries around the world. Their enthusiasm and commitment to research have always been a factor in maintaining my own level of interest, and I take this opportunity to express my gratitude to all of them. My friends and colleagues in the Remote Sensing Society, especially Jim Young, Robin Vaughan, Arthur Cracknell, Don Hardy and Karen Korzeniewski, have always been helpful and supportive. Discussions with many people, including Mike Barnsley, Giles Foody and Robert Gurney, have added to my knowledge and awareness of key issues in remote sensing. I also acknowledge the assistance given by Dr Magaly Koch, Remote Sensing Center, Boston University, who has tested several of the procedures reported in this book and included on the CD. Her careful and thoughtful advice, support and encouragement have kept me from straying too far from reality on many occasions. My colleagues in the School of Geography in the University of Nottingham continue to provide a friendly and productive working environment, and have been known occasionally to laugh at some of my jokes. Thanks are due especially to Chris Lewis and Elaine Watts for helping to sort out the diagrams for the new edition, and to Dee Omar for his patient assistance and support. Michael McCullagh has been very helpful, and has provided a lot of invaluable assistance. The staff of John Wiley & Sons have been extremely supportive, as always. Finally, my wife Rosalind deserves considerable credit for the production of this book, as she has quietly undertaken many of the tasks that, in fairness, I should have carried out during the many evenings and weekends that I have spent in front of the computer. Moreover, she has never complained about the chaotic state of our dining room, or about the intrusive sound of Wagner's music dramas. There are many people, in many places, who have helped or assisted me; it is impossible to name all of them, but I am nevertheless grateful. Naturally, I take full responsibility for all errors and omissions.

Paul M. Mather   *June, 1998.*
Nottingham

paul.mather@nottingham.ac.uk

Preface to the First Edition

Environmental remote sensing is the measurement, from a distance, of the spectral features of the Earth's surface and atmosphere. These measurements are normally made by instruments carried by satellites or aircraft, and are used to infer the nature and characteristics of the land or sea surface, or of the atmosphere, at the time of observation. The successful application of remote sensing techniques to particular problems, whether they be geographical, geological, oceanographic or cartographic, requires knowledge and skills drawn from several areas of science. An understanding of the way in which remotely-sensed data are acquired by a sensor mounted on board an aircraft or satellite needs a basic knowledge of the physics involved, in particular environmental physics and optics. The use of remotely-sensed data, which are inherently digital, demands a degree of mathematical and statistical skill plus some familiarity with digital computers and their operation. A high level of competence in the field in which the remotely-sensed data are to be used is essential if full use of the information contained in those data is to be made. The term 'remote-sensing specialist' is thus, apparently, a contradiction in terms for a remote-sensing scientist must possess a broad range of expertise across a variety of disciplines. While it is, of course, possible to specialise in some particular aspect of remote sensing, it is difficult to cut oneself off from the essential multidisciplinary nature of the subject.

This book is concerned with one specialised area of remote sensing, that of digital image processing of remotely-sensed data but, as we have just seen, this topic cannot be treated in isolation and so Chapter 1 covers in an introductory fashion the physical principles of remote sensing. Satellite platforms currently or recently in use, as well as those proposed for the near future, are described in Chapter 2, which also contains a description of the nature and sources of remotely-sensed data. The characteristics of digital computers as they relate to the processing of remotely-sensed image data is the subject of Chapter 3. The remaining five chapters cover particular topics within the general field of the processing of remotely-sensed data in the form of digital images, and their application to a range of problems drawn from the Earth and environmental sciences. Chapters 1–3 can be considered to form the introduction to the material treated in later chapters.

The audience for this book is perceived as consisting of undergraduates taking advanced options in remote sensing in universities and colleges as part of a first degree course in geography, geology, botany, environmental science, civil engineering or agricultural science, together with postgraduate students following taught Masters courses in remote sensing. In addition, postgraduate research students and other research workers whose studies involve the use of remotely-sensed images can use this book as an introduction to the digital processing of such data. Readers whose main scientific interests lie elsewhere might find here a general survey of this relatively new and rapidly developing area of science and technology. The nature of the intended audience requires that the formal presentation is kept to a level that is intelligible to those who do not have the benefit of a degree in mathematics, physics, computer science or engineering. This is not a research monograph, complete in every detail and pushing out to the frontiers of knowledge. Rather, it is a relatively gentle introduction to a subject which can, at first sight, appear to be overwhelming to those lacking mathematical sophistication, statistical cunning or computational genius. As such it relies to some extent on verbal rather than numerical expression of ideas and concepts. The author's intention is to provide the foundations upon which readers may build their knowledge of the more complex and detailed aspects of the use of remote-sensing techniques in their own subject rather than add to the already extensive literature which caters for a mathematically orientated readership. Because of the multidisciplinary nature of the intended audience, and since the book is primarily concerned with techniques, the examples have been kept simple, and do not assume any specialist knowledge of geology, ecology, oceanography, or other branch of Earth science. It is expected that the reader is capable of working out potential applications in his or her own field, or of following up the references given here.

It is assumed that most readers will have access to a digital image processing system, either within their own department or institution, or at a regional or national remote sensing centre. Such processors normally have a built-in software package containing programs to carry out most, if not all, of the operations described in this book. It is hoped that the material presented here will provide such readers with the background necessary to make sensible use of the facilities at their disposal. Enough detail is provided, however, to allow interested readers to develop computer programs to implement their own ideas or to modify the operation of a standard package. Such ventures are to be encouraged, for software skills are an important part of the remote sensing scientist's training. Furthermore, the development and testing of individual ideas and conjectures provide the opportunity for experiment and innovation, which is to be preferred to the routine use of available software. It is my contention that solutions to problems are not always to be found in the user manual of a standard software package.

I owe a great deal to many people who have helped or encouraged me in the writing of this book. Michael Coombes of John Wiley and Sons, Ltd. took the risk of asking me to embark upon the venture, and has proved a reliable and sympathetic source of guidance as well as a model of patience. The Geography Department, University of Nottingham, kindly allowed me to use the facilities of the Remote Sensing Unit. I am grateful also to Jim Cahill for many helpful comments on early drafts, to Michael Steven for reading part of Chapter 1 and for providing advice on some diagrams, and to Sally Ash-

ford for giving a student's view and to George Korybut-Daszkiewicz for his assistance with some of the photography. An anonymous referee made many useful suggestions. My children deserve a mention (my evenings on the word-processor robbed them of the chance to play their favourite computer games) as does my wife for tolerating me. The contribution of the University of Nottingham and the Shell Exploration and Development Company to the replacement of an ageing PDP11 computer by a VAX 11/730-based image processing system allowed the continuation of remote sensing activities in the university and, consequently, the successful completion of this book. Many of the ideas presented here are the result of the development of the image processing software system now in use at the University of Nottingham and the teaching of advanced undergraduate and Masters degree courses. I am also grateful to Mr J. Winn, Chief Technician, Mr C. Lewis and Miss E. Watts, Cartographic Unit, and Mr M.A. Evans, Photographic Unit, Geography Department, University of Nottingham, for their invaluable and always friendly assistance in the production of the photographs and diagrams. None of those mentioned can be held responsible for any errors or misrepresentations that might be present in this book; it is the author's prerogative to accept liability for these.

Paul M. Mather,   *March, 1987.*
Remote Sensing Unit,
Department of Geography,
The University,
Nottingham NG7 2RD

1

Remote Sensing: Basic Principles

'Electromagnetic radiation is just basically mysterious.'

B.K. Ridley, 1984

1.1 INTRODUCTION

The science of remote sensing consists of the interpretation of measurements of electromagnetic energy reflected from or emitted by a target from a vantage-point that is distant from the target. *Earth observation* by remote sensing (EO) is the interpretation and understanding of measurements of electromagnetic energy that is reflected from or emitted by the Earth's surface or atmosphere, and the establishment of relationships between these measurements and the nature and distribution of phenomena on the Earth's surface or within the atmosphere. Sensors mounted on aircraft or satellite platforms record the amounts of energy reflected from or emitted by the Earth's surface. These measurements are made at a large number of points distributed either along a one-dimensional profile on the ground below the platform or over a two-dimensional area below or to

one side of the ground track of the platform. Figure 1.1 shows an image being collected by a nadir-looking sensor.

Data in the form of profiles will not be considered in this book, which is concerned with the processing of two-dimensional (spatial) data collected by imaging sensors. Such sensors are either *nadir-looking* or *side-looking*. In the former case, the ground area to either side of the satellite or aircraft platform is imaged, while in the latter case an area of the Earth's surface lying to one side of the satellite track is imaged. The most familiar kinds of images, such as those collected by the Thematic Mapper instrument carried by Landsat-4 and -5, or by the High Resolution Visible (HRV) instrument on-board the SPOT satellites, are scanned line by line. Scanners use the forward motion of the satellite or aircraft to build up an image of the Earth's surface by the collection of successive scan lines. Electro-mechanical scanners have

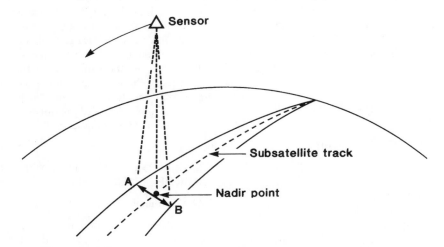

Figure 1.1 Sensor on-board satellite scans along line AB. As the satellite platform moves forward, an image of the swath region is built up. The subsatellite track is an imaginary line on the ground traced beneath the satellite. The nadir point is the ground point on the subsatellite track immediately beneath the satellite at a given instant in time.

a small number of detectors and use an oscillating mirror to direct reflected or emitted electromagnetic energy on to these detectors. Push-broom scanners simultaneously record an entire scan line by using an array of charge-coupled devices (CCDs). At a given moment in time, each detector is observing a small area of the Earth's surface along the given scan line. This ground area is called a *pixel*. A remotely-sensed image is made up of a rectangular matrix of measurements on individual pixels, each representing the magnitude of upwelling electromagnetic energy for a small ground area. This is a rather oversimplified account of image acquisition, and more detail is provided in Chapter 2.

Because image data collected in this way are digital in nature, computers are used to display, enhance and manipulate them, and the main part of this book deals with techniques used in these types of processing. Spatial patterns evident in remotely-sensed images can be interpreted in terms of geographical variations in the nature of the material forming the surface of the Earth. Such Earth-surface materials may be vegetation, exposed soil and rock, or water surfaces. Notice that the characteristics of these materials are not themselves detected directly by remote sensing. Their nature is inferred from the properties of the incident electromagnetic radiation that is reflected, scattered, or emitted by these materials on the Earth's surface and recorded by the sensor. Another characteristic of digital image data is that they can be calibrated in order to provide estimates of physical measurements of properties of the target such as radiance, reflection or albedo. These values are used in modelling of climate or crop growth. Examples of the uses of remotely-sensed image data in Earth science and environmental management can be found in Calder (1991). A number of World Wide Web sites provide access to image libraries for those with computers connected to the Internet. A good starting point is my own page of links at

http://www.geog.nottingham.ac.uk/~mather/useful__links.html

The NASA tutorial by Nicholas Short at

http://rst.gsfc.nasa.gov/TofC/Coverpage.html

is strongly recommended.

Aerial photography is a familiar form of Earth observation by remote sensing. Air photographs differ from digital images in that they are analogue rather than digital in nature. *Analogue* means: using some alternative physical representation to display some property of interest. For example, a photographic film displays different light levels in terms of the differential response of silver halide particles in the film emulsion. Analogue images cannot be processed by computer unless they are converted to digital form, using a scanning device. Computer scanners operate much in the same way as those carried by satellites in that they view a small area of the photograph and convert the grey level to a number, usually in the range 0 (black) to 255 (white). Conventional photography using a hand-held camera has been, until recently, almost entirely analogue. Digital cameras are now readily available. Images produced by these cameras are similar in nature to those produced by the push-broom type of sensor mentioned above. Instead of a film, a digital camera has a two-dimensional array of CCDs. The amount of light from the scene that impinges on an individual CCD is recorded as a number in the range 0 (no light) to 255 (detector saturated). A two-dimensional set of CCD measurements produces a grey-scale image. Three sets of CCDs are used to produce a colour image, just as three layers of emulsion are used to generate an analogue colour photograph. The three sets of CCDs measure the amounts of red, green and blue light that reach the camera. Nowadays, digital imagery is relatively easily available from digital cameras, from scanned analogue photographs, as well as from sensors carried by aircraft and satellites.

The nature and properties of electromagnetic radiation are considered in section 1.2 and are those which concern its interaction with the atmosphere, through which the electromagnetic radiation passes on its route from the Sun (or from another source such as a microwave radar) to the Earth's surface and back to the sensor mounted on-board an aircraft or satellite. The interactions between electromagnetic radiation and Earth-surface materials are summarised in section 1.3. It is by studying these interactions that the types and properties of the material forming the Earth's surface are inferred. See Bird (1991a, b), Elachi (1987) and Rees (1990) for more extended treatments of the material discussed in the following sections.

1.2 ELECTROMAGNETIC RADIATION AND ITS PROPERTIES

1.2.1 Terminology

The terminology used in remote sensing is often understood only imprecisely, and is therefore occasionally used loosely. A brief guide is therefore given in this section. It is neither complete nor comprehensive, and is meant only to introduce some basic ideas. The subject is dealt with more thoroughly by Bird (1991a, b), Chapman (1995), Elachi (1987), Rees (1990), Slater (1980) and Schowengerdt (1997). A good summary, available via the World Wide Web, is

http://www.jpl.nasa.gov/basics/bsf6-1.htm

Note that the term *electromagnetic radiation* is abbreviated to *EMR* in the following sections.

Energy is the capacity to do work. It is expressed in *joules* (J), named after James Prescott Joule, an English brewer whose hobby was physics. *Radiant energy* is the energy associated with EMR. The rate of transfer of energy from one place to another (for example, from the Sun to the Earth) is termed the *flux* of energy, the term being derived from the Latin word meaning 'flow'. It is measured in *watts* (W). The interaction between electromagnetic radiation and surfaces such as that of the Earth can be understood more clearly if the concept of *radiant flux density* is introduced. Radiant flux is the rate of transfer of radiant (electromagnetic) energy. Density implies variability over the two-dimensional surface on which the radiant energy falls. Hence radiant flux density is the magnitude of the radiant flux that is incident upon or, conversely, is emitted by a surface of unit area (measured in $W\,m^{-2}$). The topic of emission of EMR by the Earth's surface in the form of heat is considered at a later stage. If radiant energy falls (is incident) upon a surface then the term *irradiance* is used in place of radiant flux density. If the energy flow is away from the surface, as in the case of thermal energy emitted by the Earth or solar energy that is reflected by the Earth, then the term *radiant exitance* or *radiant emittance* (measured in units of $W\,m^{-2}$) is appropriate.

The term *radiance* is used to mean the radiant flux density transmitted from a unit area on the Earth's surface as viewed through a unit solid (three-dimensional) angle. This solid angle is measured in *steradians*, the three-dimensional equivalent of the familiar radian (defined as the angle subtended at the centre of a circle by a sector which cuts out a section of the circumference that is equal in length to the radius of the circle). If, for the moment, we consider that the irradiance reaching the surface is backscattered in all upward directions (Figure 1.2(a)), then a proportion of the radiant flux would be measured per unit solid viewing angle. This proportion is the radiance (Figure 1.3). It is measured in watts per square metre per steradian ($W\,m^{-2}\,sr^{-1}$). The concepts of the radian and steradian are illustrated in Figure 1.4.

Reflectance, ρ, is the ratio between the irradiance and the radiant emittance of an object. The reflectance of a given object is independent of irradiance, as it is a ratio.

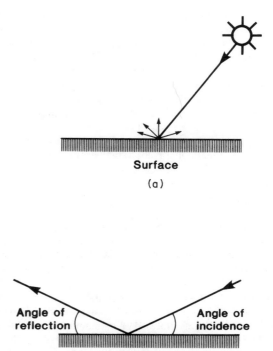

Figure 1.2   (a) Diffuse reflectance: radiance incident upon the surface is backscattered in all upward directions. (b) Specular reflectance: angles of incidence and reflection are equal and no scattering occurs at the surface.

When remotely-sensed images collected over a time period are to be compared it is common practice to convert the radiance values recorded by the sensor into reflectance factors in order to eliminate the effects of variable irradiance over the seasons of the year. This topic is considered further in section 4.6.

The quantities described above can be used to refer to particular wavebands rather than to the whole electromagnetic spectrum (section 1.2.3). The terms are then preceded by the adjective *spectral*; for example, the spectral radiance for a given waveband is the radiant flux density in that waveband (i.e. spectral radiant flux density) per unit solid angle. Terms such as *spectral irradiance*, *spectral reflectance* and *spectral exitance* are defined in a similar fashion.

1.2.2   Nature of electromagnetic radiation

One important point of controversy in physics over the past 250 years has concerned the nature of EMR.

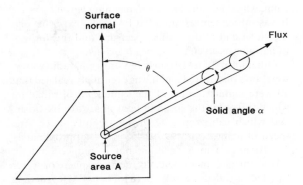

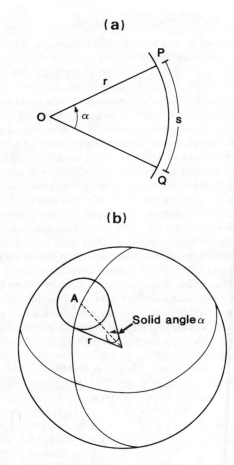

Figure 1.3 Radiant flux leaving source A in direction θ per solid angle α is the radiance in watts per square metre per steradian ($W\,m^{-2}\,sr^{-1}$). Based on *Manual of Remote Sensing*, 2nd edition, ed. R.N. Colwell, 1983, figure 2.10; reproduced by permission of the American Society for Photogrammetry and Remote Sensing.

Newton, while not explicitly rejecting the idea that light is a wave-like form of energy (the wave theory), inclined to the view that it is formed of a stream of particles (the corpuscular theory). The wave–corpuscle dichotomy was not to be resolved until the early years of the twentieth century with the work of Planck and Einstein. The importance to remote sensing of the nature of EMR is considerable, for we need to consider radiation both as a waveform and as a stream of particles. The wave-like characteristics of EMR allow the distinction to be made between different manifestations of such radiation (for example, microwave and infrared radiation) while, in order to understand the interactions between EMR and the Earth's atmosphere and surface, the idea that EMR consists of a stream of particles or *photons* is most easily used. Einstein showed that EMR could travel or propagate through a vacuum; thus, the need for an all-pervading ether was avoided. Building on the work of Planck, Einstein proposed in 1905 that light consists of particles called photons which, in most respects, were similar to other sub-atomic particles such as protons and neutrons. It was found that, at the sub-atomic level, both wave-like and particle-like properties were exhibited, and that phenomena at this level appear to be both waves and particles. Schrödinger, in *Science, Theory and Man*, wrote as follows:

'In the new setting of ideas, the distinction [between particles and waves] has vanished, because it was discovered that all particles have also wave properties, and vice-versa. Neither of the concepts must be discarded, they must be amalgamated. Which aspect

Figure 1.4 (a) The angle formed when arc length s equals r, the radius of the circle, is equal to 1 radian. Thus, angle $\alpha = s/r$ radians. There are 2π radians (360 degrees) in a circle. (b) A steradian is the solid (three-dimensional) angle formed when the area A delimited on the surface of the sphere is equal to the square of the radius of the sphere. A need not refer to a uniform shape. Thus angle $\alpha = A/r^2$ steradians (sr). There are 4π radians in a sphere.

obtrudes itself depends not on the physical object but on the experimental device set up to examine it.'

Thus, from the point of view of quantum mechanics, EMR is both a wave and a stream of particles. Whichever view is taken will depend on the particular situation. In section 1.2.5, the particle theory is best suited to explain the manner in which incident EMR interacts with the atoms, molecules and other particles that form the Earth's atmosphere.

1.2.3   The electromagnetic spectrum

The Sun's light is the form of EMR that is most familiar to human beings. Sunlight that is reflected by physical objects travels in most situations in a straight line to the observer's eye. On reaching the retina, it generates electrical signals which are transmitted to the brain by the optic nerve. These signals are used by the brain to construct an image of the viewer's surroundings. This is the process of vision, which is closely analogous to the process of remote sensing; indeed, vision is a form of remote sensing (Greenfield, 1997). A discussion of the human visual process can be found in Chapter 5.

Visible light is so called because it is detected by the eye, whereas other forms of EMR are invisible to the unaided eye. Newton investigated the nature of white light, and came to the conclusion that it is made up of differently coloured components which he saw by passing white light through a prism to form a rainbow-like spectrum. This is the *visible spectrum*, which ranges from red through orange, yellow and green, to blue, indigo and violet. Later, the astronomer Herschel demonstrated the existence of EMR beyond the visible spectrum, which he called *infrared*, meaning *beyond the red*. It was subsequently found that EMR also exists beyond the violet end of the visible spectrum, and this form of radiation was given the name *ultraviolet*.

Other forms of EMR, such as X-rays and radio waves, were soon discovered and it was eventually realised that all were manifestations of the same kind of radiation which travels at the speed of light in a wave-like form, and which can propagate through empty space. The speed of light (c_0) is $299\,792\,458\,\mathrm{m\,s^{-1}}$ (approximately $3 \times 10^8\,\mathrm{m\,s^{-1}}$) in a vacuum, but is reduced if light travels through media such as the atmosphere or water by a factor called the *index of refraction*. EMR reaching the Earth comes mainly from the Sun and is produced by thermonuclear reactions in the Sun's core. The set of all electromagnetic waves is called the *electromagnetic spectrum*, which includes the range from the long radio waves, through the microwave and infrared wavelengths to visible light waves, and beyond to the ultraviolet and to the short-wave X-rays and gamma-rays (Figure 1.5).

Symmetric waves can be described in terms of their *frequency (f)*, which is the number of waveforms passing

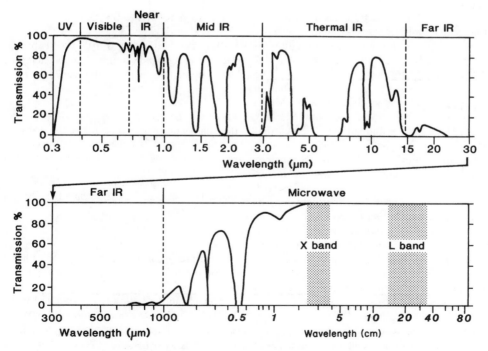

Figure 1.5   Electromagnetic spectrum showing nomenclature associated with spectral bands. The curve shows spectral transmission in per cent (see section 1.2.5). Reprinted with permission from Goetz, A.F.H. and Rowan, L.C., 1981, Geologic remote sensing. *Science*, **211**, 781–791 (figure 1). © 1981 American Association for the Advancement of Science.

a fixed point in unit time. This quantity used to be known as *cycles per second* (cps) but the preferred term is *hertz* (Hz), after Heinrich Hertz, who discovered radio waves in 1888. Alternatively, the concept of *wavelength* can be used (Figure 1.6). The wavelength is the distance between successive peaks (or successive troughs) of a waveform, and is normally measured in metres or fractions of a metre (Table 1.1). Both measures – frequency and wavelength – convey the same information and are often used interchangeably. Another measure of the nature of a waveform is its period (*T*). This is the time, in seconds, needed for one full wave to pass a fixed point. The relationships between wavelength, frequency and period are given by:

$$f = c/\lambda$$
$$\lambda = c/f$$
$$T = 1/f = \lambda/c$$

In these expressions, *c* is the speed of light. The velocity of propagation (*v*) is the product of wave frequency and wavelength, i.e.

$$v = \lambda f$$

The *amplitude* (*A*) of a wave is the maximum distance attained by the wave from its mean position (Figure 1.6). The amount of energy, or intensity, of the waveform is proportional to the square of the amplitude. Using the relationships specified earlier we can compute the frequency given the wavelength, and vice versa. If, for example, wavelength λ is 0.6 µm or 6×10^{-7} m, then, since velocity *v* equals the product of wavelength and frequency *f*, it follows that

$$v = 6 \times 10^{-7} f$$

so that

$$f = \frac{c_0}{v} = \frac{3 \times 10^8}{6 \times 10^{-7}} \text{Hz}$$

i.e.

$$f = 0.5 \times 10^{15} \text{Hz} = 0.5 \text{PHz}$$

1 PHz (petahertz) equals 10^{15} Hz (Table 1.1). Thus, electromagnetic radiation with a wavelength of 0.6 µm has a frequency of 0.5×10^{15} Hz. The *period* is the

Table 1.1   Terms and symbols used in measurement.

Factor	Prefix	Symbol	Factor	Prefix	Symbol
10^{-18}	atto	a			
10^{-15}	femto	f	10^{15}	peta	P
10^{-12}	pico	p	10^{12}	tera	T
10^{-9}	nano	n	10^{9}	giga	G
10^{-6}	micro	µ	10^{6}	mega	M
10^{-3}	milli	m	10^{3}	kilo	k

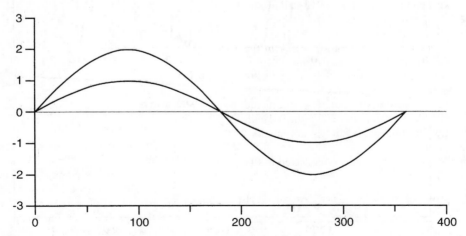

Figure 1.6   Two curves (waveforms) have the same wavelength (360 units, *x*-axis). However, one curve has an amplitude of two units (*y*-axis) while the other has an amplitude of one unit. If we imagine that the two curves are moving to the right, like traces on an oscilloscope, then the frequency is the number of waveforms (0–360 units) that pass a fixed point in unit time (usually measured in cycles per second or hertz, Hz). The period of the waveform is the time taken for one full waveform to pass a fixed point. These two waveforms have the same wavelength, frequency and period and differ only in terms of their amplitude.

reciprocal of the frequency, so one wave of this frequency will pass a fixed point in 2×10^{-15} seconds. The amount of energy carried by the waveform, or the squared amplitude of the wave, is defined for a single photon by the relationship:

$$E = hf$$

where E is energy, h is a constant known as Planck's constant (6.625×10^{-34} J s) and f is frequency. Energy thus increases with frequency, so that high-frequency, short-wavelength electromagnetic radiation such as X-rays carries more energy than visible light or radio waves.

While electromagnetic radiation with particular temporal and spatial properties is used in remote sensing to convey information about a target, it is interesting to note that both time and space are defined in terms of specific characteristics of electromagnetic radiation. A second is the duration of 9 192 631 770 periods of the caesium radiation (in other words, that number of wavelengths or cycles are emitted by caesium radiation in one second; its frequency is approximately 9 GHz or a wavelength of around 0.03 m). A metre is defined as 1 650 764.73 vacuum wavelengths of the orange-red light emitted by krypton-86.

Visible light is electromagnetic radiation having wavelengths between 0.4 μm and 0.7 μm. We call the shorter-wavelength (0.4 μm) end of the visible spectrum 'blue' and the longer-wavelength end (0.7 μm) 'red' (Table 1.2). The eye is not uniformly sensitive to light within this range, and has its peak sensitivity at around 0.55 μm, which lies in the green part of the visible spectrum (Figure 1.7). This peak in the response function of the human eye corresponds closely to the peak in the Sun's radiation emittance distribution (section 1.2.4).

The process of atmospheric scattering, discussed in section 1.2.5 below, deflects light rays from a straight path and thus causes blurring or haziness. It affects the blue end of the visible spectrum more than the red end, and consequently the blue waveband is not used in many remote-sensing systems.

Figure 1.8(a)–(c) shows three images collected in the

Table 1.2 Wavebands corresponding to perceived colours of visible light.

Colour	Waveband (μm)	Colour	Waveband (μm)
red	0.780–0.622	green	0.588–0.492
orange	0.622–0.597	blue	0.492–0.455
yellow	0.597–0.588	violet	0.455–0.390

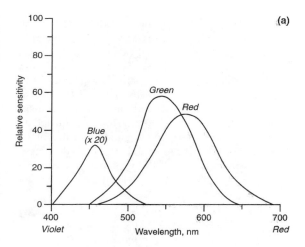

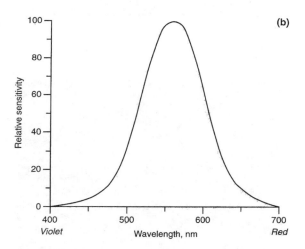

Figure 1.7 (a) Response function of the red-, green- and blue-sensitive cones on the retina of the human eye. (b) Overall response of the human eye. Peak sensitivity occurs at 0.55 μm.

blue-green, green and red wavebands respectively by a sensor called the *Thematic Mapper* (TM) carried by the American Landsat-5 satellite (Chapter 2). The different land cover types reflect energy in the visible spectrum in a differential manner, although the clouds and cloud shadows in the upper centre of the image are clearly visible in all three images. Various crops in the fields round the village of Littleport (north of Cambridge in eastern England) can be discriminated, and the river Ouse can also be seen as it flows northwards in the right-hand side of the image area. It is dangerous to rely on visual interpretation of images such as these. Since

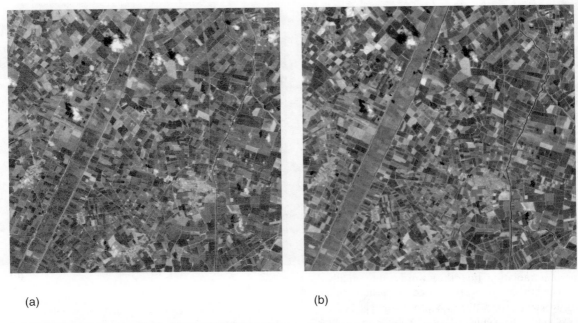

(a)

(b)

(c)

Figure 1.8 Images collected in the (a) blue-green, (b) green and (c) red wavebands of the optical spectrum by the Thematic Mapper sensor carried by the Landsat satellite. These images are the Landsat Thematic Mapper bands 1, 2 and 3, respectively. The area shown is near the town of Littleport in Cambridgeshire, eastern England. Original data © ESA 1994, distributed by Eurimage.

they are digital in nature they can be manipulated by computer – that is what this book is about.

Electromagnetic radiation with wavelengths shorter than those of visible light (less than 0.4 µm) is divided into three spectral regions, which are called gamma rays, X-rays and ultraviolet radiation. Because of the effects of atmospheric scattering and absorption, none of these wavebands is used in satellite remote sensing, although gamma-ray emissions from radioactive materials in the Earth's crust can be detected by low-flying aircraft. Since radiation in these wavebands is dangerous to life, the fact that it is mostly absorbed or scattered by the atmosphere allows life to exist on Earth. In terms of the discussion of the wave–particle duality in section 1.2.2, it should be noted that gamma radiation has the highest energy levels and is the most 'particle-like' of all electromagnetic radiation, whereas radiofrequency radiation is most 'wave-like' and has the lowest energy levels.

Wavelengths that are longer than the visible red are subdivided into the infrared (IR), microwave and radiofrequency wavebands. The *infrared* waveband, extending from 0.7 µm to 1 mm, is not a uniform region. Short-wavelength or near-infrared energy, with wavelengths between 0.7 and 0.9 µm, behaves like visible light and can be detected by special photographic film. Infrared radiation with a wavelength of up to 3.0 µm is primarily of solar origin and, like visible light, is reflected by the surface of the Earth. Hence, these wavebands are often known as the *optical* bands. Figure 1.9(a) shows an

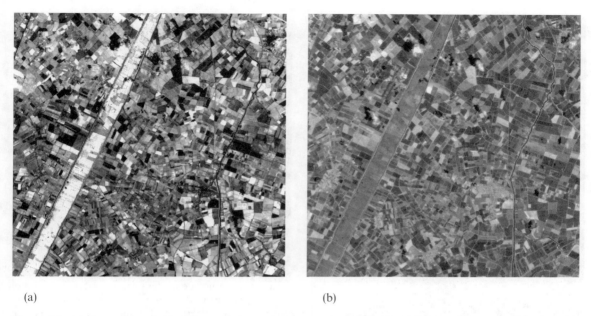

(a)   (b)

Figure 1.9 (a) Image of ground surface reflectance in the 0.75–0.90 μm band (near-infrared) and (b) middle-infrared (2.08–2.35 μm) image of the same area as that shown in Figure 1.8. These images are Landsat Thematic Mapper bands 4 and 7. Original data © ESA 1994, distributed by Eurimage.

image collected by the Landsat Thematic Mapper sensor in the near-infrared region of the spectrum (0.75–0.90 μm) of the area shown in Figure 1.8. This image is considerably clearer than the visible spectrum images shown in Figure 1.8. We will see in section 1.3.2 that the differences in reflection between vegetation, water and soil are probably greatest in this near-infrared band. An image of surface reflection in the Landsat Thematic Mapper mid-infrared waveband (2.08–2.35 μm) of the same area is shown in Figure 1.9(b).

Beyond a wavelength of around 3 μm, IR radiation emitted by the Earth's surface can be sensed in the form of heat. The amount and wavelength of this radiation depend on the temperature of the source (section 1.2.4). Because these longer IR wavebands are sensed as heat they are called the *thermal infrared* wavebands. Much of the emitted radiation is absorbed by, and consequently heats, the atmosphere, thus making life possible on the Earth (Figure 1.5). There is, however, a 'window' between 8 and 14 μm which allows a satellite sensor above the atmosphere to detect the thermal emittance from the Earth, which has its peak wavelength at 9.6 μm. Note, though, that the presence of ozone in the atmosphere creates a narrow absorption band within this window, centred at 9.5 μm. Absorption of longer-wave radiation

by the atmosphere has the effect of warming the atmosphere (the natural greenhouse effect). Water vapour and carbon dioxide are the main absorbing agents, together with ozone. The increase in the carbon dioxide content of the atmosphere over the past century, due to the burning of fossil fuels, is thought to enhance the greenhouse effect and to raise the temperature of the atmosphere above its natural level. This could have long-term climatic consequences. A thermal image of part of Western Europe acquired by the Advanced Very High Resolution Radiometer (AVHRR) carried by the US NOAA-14 satellite is shown in Figure 1.10. The different grey scales show different levels of emitted thermal radia-tion in the 11.5–12.5 μm waveband. Before these grey levels can be interpreted in terms of temperatures, the effects of the atmosphere as well as the nature of the sensor calibration must be considered. Both these topics are covered in Chapter 4. For comparison, a visible band image of Europe and North Africa produced by the Meteosat-6 satellite is shown in Figure 1.11. Both images were collected by the NERC-funded satellite receiving station at Dundee University, Scotland.

That region of the spectrum composed of electromagnetic radiation with wavelengths between 1 mm and 300 cm is called the *microwave band*. Radiation at these wavelengths can penetrate cloud, and the microwave

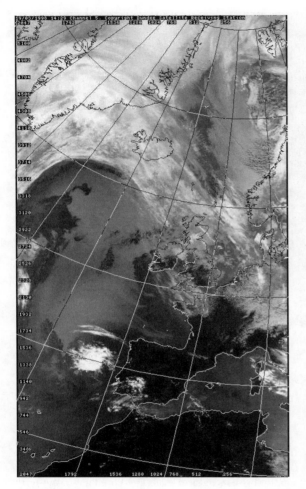

Figure 1.10 NOAA-14 AVHRR band 5 (thermal infrared, 11.5–12.5 μm) image of Western Europe and NW Africa taken at 1420 on 19 March 1998. The image was downloaded by the receiving station located at Dundee University, Scotland, where the image was geometrically rectified (Chapter 4). A grid and a digitised coastline were added at the receiving station. Dark areas indicate greater thermal emissions. The position of a high-pressure area (anticyclone) can be inferred from cloud patterns, which are white as cloud-top temperatures are low. The satellite took just over 15 minutes to travel from the south to the north of the area shown. © Dundee Satellite Receiving Station, Dundee University.

band is thus a valuable region for remote sensing in temperate and tropical areas where cloud cover restricts the collection of optical and thermal infrared images. Some microwave sensors can detect the small amounts of radiation at these wavelengths that are emitted by the Earth. Such sensors are called *passive* because they de-

tect EMR that is generated externally, for example, by emittance by or reflectance from a target. Passive microwave radiometers such as the SMMR (Scanning Multichannel Microwave Radiometer) produce imagery with a low spatial resolution (section 2.2.1) which is used to provide measurements of sea-surface temperature and wind speed, and also to detect sea ice.

Because the level of microwave energy emitted by the Earth is very low, a high-resolution imaging microwave sensor generates its own source of radiation, and detects the strength of the return from the ground. Such devices are called *active* sensors to distinguish them from the *passive* sensors that are used to detect and record radiation of solar or terrestrial origin in the visible, infrared and microwave wavebands. An active microwave sensor generates energy, which is directed towards and scattered by a target. It also receives and records radiation that is backscattered by the target in the direction of the sensor. Thus, active microwave instruments are not dependent on an external source of radiation such as the Sun or, in the case of thermal emittance, the Earth. Active microwave sensors can operate independently by day or by night. An analogy that is often used is that of a camera. In normal daylight reflected radiation from the target enters the camera lens and exposes the film. Where illumination conditions are poor, the photographer employs a flash-gun, which generates radiation in visible wavebands, and the film is exposed by light from the flash-gun that is reflected by the target. The microwave instrument produces pulses of energy, usually at centimetre wavelengths, that are transmitted by an antenna. The same antenna picks up the reflection of these energy pulses as they return from the target.

Microwave imaging sensors are called *imaging radars* (the word *radar* is an acronym, derived from RAdio Detection And Ranging). The spatial resolution (section 2.2.1) of an imaging radar is a function of antenna length. If a conventional ('brute-force') radar is used then antenna lengths become considerable. Schreier (1993b, p. 107) notes that if the Seasat satellite had used a brute-force approach then its 10 m long antenna would have generated images with a spatial resolution of 20 km. A different approach, using several views of the target as the satellite approaches, reaches and passes the target, provides a means of achieving high resolution without the need for excessive antenna sizes. This approach uses *synthetic aperture radar* (SAR), and all of the satellite-borne radar systems have used the SAR principle. The main advantage of SAR is that it is an all-weather, day–night, high-spatial-resolution instrument, which can operate independently of weather con-

Figure 1.11 Portion of a Meteosat-6 visible channel image of Europe and North Africa taken at 1800 on 17 March 1998, when the lights were going on across Europe. Image received by Dundee University, Scotland. © Dundee Satellite Receiving Station, Dundee University.

ditions or solar illumination. This makes it an ideal instrument for observing areas of the world such as the temperate and tropical regions, which are often cloud-covered and therefore inaccessible to optical and infrared imaging sensors.

Imaging radars are side-looking rather than nadir-looking instruments, and their geometry is complicated by foreshortening (the top of a mountain appearing closer to the sensor than does the foot of the mountain) and shadow, caused by the 'far side' of a mountain or hill being invisible to the side-looking radar sensor – see Schreier (1993a) for a thorough description of synthetic aperture radar. Furthermore, the interaction between microwave radiation and the ground surface generates a phenomenon called *speckle*, which is the result of interference arising from the coherent integration of the contributions of all the scatterers in the pixel area (Quegan and Rhodes, 1994). Speckle magnitude is proportional to the magnitude of the backscattered signal, and is rather more difficult to remove from the image than is additive noise. Filtering, which is used to remove unwanted features such as speckle, is considered in detail in Chapter 7.

Radar wavelength bands are described by a code such as 'L band' or 'C band' that was developed during World War II for security purposes. Unfortunately, several versions of the code were used, which confused

Allied scientists as much as the enemy. Table 1.3 shows one commonly accepted delimitation of the radar wavelengths.

Although a radar signal does not detect colour information (which is gained from optical wavelength sensors) or temperature information (derived from thermal infrared sensors) it detects surface roughness and electrical conductivity information (which can be related to soil moisture conditions). Because radar is an active instrument the characteristics of the transmitted signal can be controlled. In particular the wavelength, depression angle and the polarisation of the signal are important properties. Radar wavelength determines the observed roughness of the surface; any surface that has a

Table 1.3 Radar wavebands and nomenclature.

Band designation	Frequency (MHz)	Wavelength (cm)
P	300–1000	30–100
L	1000–2000	15–30
S	2000–4000	7.5–15
C	4000–8000	3.75–7.5
X	8000–12 000	2.5–3.75
K_u	12 000–18 000	1.667–2.5
K	18 000–27 000	1.111–1.667
K_a	27 000–40 000	0.75–1.111

roughness that has a frequency less than that of the microwave radiation used by the radar is seen as smooth. Radar wavelength also has a bearing on the degree of penetration of the surface material that is achieved by the microwave pulses. At L-band wavelengths, microwave radiation can penetrate the foliage of trees and, depending on the height of the tree, may reach the ground. Backscatter occurs from the leaves, branches, trunks and the ground surface. In areas of dry alluvial or sandy soils, L-band radar can penetrate for several metres. The same is true of glacier ice. The shorter-wavelength C-band radiation will penetrate the canopies of trees, and the upper layers of soil and ice. Shorter-wavelength X-band radar mainly 'sees' the top of the vegetation canopy and the soil and ice surface. An X-band image of the area around Perpignan in SW France is shown in Figure 1.12.

Beyond the microwave region is the *radio* band. Radio wavelengths are used in remote sensing, but not to detect Earth-surface phenomena. Commands sent to the satellite utilise radio wavelengths. Image data are transmitted to the ground receiving stations using wavelengths in the microwave region of the spectrum; these data are recorded on the ground by high-speed tape-recorders while the satellite is in direct line of sight of a ground receiving station. Image data for regions of the world that are not within range of ground receiving stations are recorded by on-board tape-recorders, and the data so recorded are subsequently transmitted together with currently scanned data when the satellite is within the reception range of a ground receiving station. The first three Landsat satellites (Chapter 2) used on-board tape-recorders to supplement data that were directly transmitted to the ground. Landsats-4 and -5 rely on the TDRS (Tracking and Data Relay Satellite), which allows direct broadcast of data from Earth resources satellites to one of a set of communications satellites located above the Equator in geostationary orbit (meaning that the satellite's orbital velocity is just sufficient to keep pace with the rotation of the Earth). The signal is relayed by the TDRS to a ground receiving station at White Sands, New Mexico.

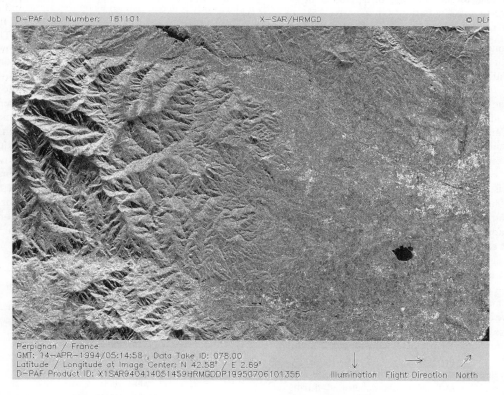

Figure 1.12   X-band radar image of part of SW France round Perpignan collected as part of the 1994 Shuttle Imaging Radar SIR -C/X-SAR experiment. © DLR/DFD 1995. Deutsches Zentrum für Luft- und Raumfahrt e.V.

1.2.4 Sources of electromagnetic radiation

All objects whose temperature is greater than absolute zero ($-273\,°C$, $0\,K$) emit radiation. However, the distribution of the amount of radiation at each wavelength across the spectrum is not uniform. Radiation is emitted by the stars and the planets; chief of these is the Sun, which provides the heat and light radiation needed for life on Earth. The Sun is an almost-spherical body with a diameter of $1.39 \times 10^6\,km$ and a mean distance from Earth of $150 \times 10^6\,km$. Its chief constituents are hydrogen and helium. The conversion of hydrogen to helium in the Sun's core provides the energy that is radiated from the outer layers. At the edge of the Earth's atmosphere the power received from the Sun, measured over the surface area of the Earth, is approximately $3.9 \times 10^{22}\,MW$ which, if it were distributed evenly over the Earth, would give an incident radiant flux density of $1367\,W\,m^{-2}$. This value is known as the *solar constant*, even though it varies throughout the year by about $\pm 3.5\%$, depending on the distance of the Earth from the Sun. Bonhomme (1993) provides a useful summary of a number of aspects relating to solar radiation. On average, 35% of the incident radiant flux is reflected from the Earth (including clouds and atmosphere), 17% is absorbed by the atmosphere and 47% is absorbed by Earth-surface materials. From the Stefan–Boltzmann law (see below) it can be shown that the Sun's temperature is $5777\,K$ if the solar constant is $1367\,W\,m^{-2}$. Other estimates of the Sun's temperature range from $5500\,K$ to $6200\,K$. The importance of establishing the surface temperature of the Sun lies in the fact that the distribution of energy emitted in the different regions of the electromagnetic spectrum depends upon the temperature of the source.

If the Sun were a perfect emitter it would be an example of a theoretical ideal, called a *blackbody*. A blackbody transforms heat energy into radiant energy at the maximum rate that is consistent with the laws of thermodynamics (Suits, 1983). *Planck's law* describes the spectral exitance (i.e. the distribution of radiant flux density with wavelength, section 1.2.1) of a blackbody as:

$$M_\lambda = \frac{c_1}{\lambda^5 [\exp(c_2/\lambda T) - 1]}$$

in which $c_1 = 3.742 \times 10^{-16}\,W\,m^{-2}$, $c_2 = 1.4388 \times 10^{-2}\,m\,K$, $\lambda = $ wavelength (m), $T = $ temperature (in kelvins) and $M_\lambda = $ spectral exitance per unit wavelength.

Curves showing the spectral exitance for blackbodies at temperatures of $1000\,K$, $1600\,K$ and $2000\,K$ are shown in Figure 1.13. The total radiant energy emitted by a blackbody is dependent on its temperature, and as temperature increases so the wavelength at which the maximum spectral exitance is achieved is reduced. The dotted line in Figure 1.13 joins the peaks of the spectral exitance curves. It is described by *Wien's displacement law*, which gives the wavelength of maximum spectral exitance (λ_m) in terms of temperature:

$$\lambda_m = \frac{c_3}{T}$$

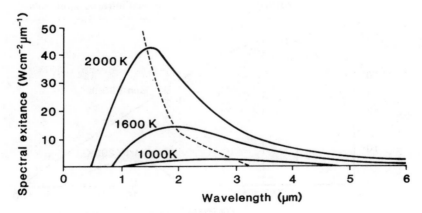

Figure 1.13 Spectral exitance curves for blackbodies at temperatures of 1000, 1600 and 2000 K. The dotted line joins the emittance peaks of the curves and is described by Wien's displacement law (see text).

and

$$c_3 = 2.898 \times 10^{-3}\,\text{m K}$$

The total spectral exitance of a blackbody at temperature T is given by the *Stefan–Boltzmann law* as:

$$M = \sigma T^4$$

where

$$\sigma = 5.6697 \times 10^{-8}\,\text{W m}^{-2}\,\text{K}^{-4}$$

The distribution of the spectral exitance for a blackbody at 5900 K closely approximates the Sun's spectral exitance curve, while the Earth can be considered to act like a blackbody with a temperature of 290 K (Figure 1.14). The solar radiation maximum occurs in the visible spectrum, with maximum irradiance at 0.47 µm. About 46% of the total energy transmitted by the Sun falls into the visible waveband (0.4 to 0.76 µm).

Wavelength-dependent mechanisms of atmospheric absorption alter the actual amounts of solar irradiance that reach the surface of the Earth. Figure 1.15 shows the spectral irradiance from the Sun at the edge of the atmosphere (solid curve) and at the Earth's surface (broken curve). Further discussion of absorption and scattering can be found in section 1.2.5. The spectral distribution of radiant energy emitted by the Earth (Figure 1.14) peaks in the thermal infrared wavebands at 9.7 µm. The amount of terrestrial emission is low in comparison

with solar irradiance. However, the solar radiation absorbed by the atmosphere is balanced by terrestrial emission in the thermal infrared, keeping the temperature of the atmosphere approximately constant. Furthermore, terrestrial thermal infrared emission provides sufficient energy for remote sensing from orbital altitudes. The characteristics of the radiation sources used in remote sensing impose some limitations on the range of wavebands available for use. In general, remote-sensing instruments that measure the spectral reflectance of solar radiation from the Earth's surface are restricted to the wavelengths shorter than 2.5 µm. Instruments to detect terrestrial radiant exitance operate in the spectral region between 3 and 14 µm. Because of atmospheric absorption by carbon dioxide, ozone and water vapour, only the 3–5 µm and 8–14 µm regions of the thermal infrared band are useful in remote sensing. An absorption band is also present in the 9–10 µm region. As noted earlier, the Earth's emittance peak occurs at 9.7 µm so satellite-borne thermal sensors normally operate in the 10.5–12.5 µm spectral region. The 3–5 µm spectral window can be used to detect targets that are hotter than their surroundings, for example, forest fires. Since the 3–5 µm region also contains some reflected solar radiation it can only be used for temperature sensing at night.

Wien's displacement law (Figure 1.13) shows that the radiant power peak moves to shorter wavelengths as temperature increases, so that a forest fire will have a radiant energy peak at a wavelength shorter than 9.7 µm. Since targets such as forest fires are sporadic in

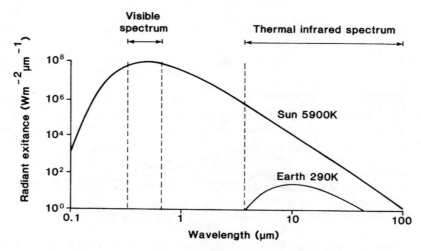

Figure 1.14   Spectral exitance curves for blackbodies at 290 and 5900 K, the approximate temperatures of the Earth and the Sun.

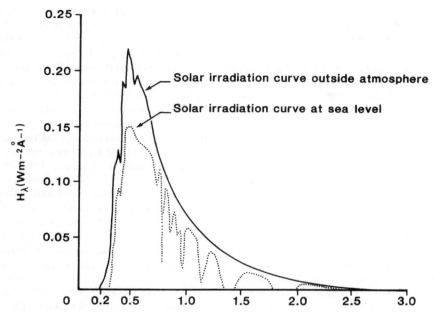

Figure 1.15 Solar irradiance at the top of the atmosphere (solid line) and at sea level (dotted line). Differences are due to atmospheric effects as discussed in the text. See also Figure 1.5. Based on *Manual of Remote Sensing*, 2nd edition, ed. R.N. Colwell, 1983, figure 5.5; reproduced by permission of the American Society for Photogrammetry and Remote Sensing.

nature and require high-resolution imagery, the 3–5 μm spectral region is used by aircraft-mounted thermal detectors. This is a difficult region for remote sensing because it contains a mixture of reflected and emitted radiation, the effects of which are not easy to separate.

The selection of wavebands for use in remote sensing is therefore seen to be limited by several factors, primarily (i) the characteristics of the radiation source, as discussed in this section, (ii) the effects of atmospheric absorption and scattering (section 1.2.5) and (iii) the nature of the target. This last point is considered in section 1.3.

1.2.5 Interactions with the Earth's atmosphere

In later chapters we consider measurements of radiance from the Earth's surface made by satellites such as Landsat and SPOT, which carry sensors that operate in the optical wavebands, composed of those parts of the electromagnetic spectrum with properties similar to visible light. The Sun is the source of radiation, and EMR from the Sun that is reflected by the Earth (the amount reflected depending on the reflectivity or albedo of the surface) and detected by the satellite or aircraft-borne sensor must pass through the atmosphere twice – once on its journey from the Sun to the Earth, and once after

being reflected by the surface of the Earth back to the sensor. During its passage through the atmosphere EMR interacts with particulate matter suspended in the atmosphere and with the molecules of the constituent gases. This interaction is usually described in terms of two processes. One, called *scattering*, deflects the radiation while the second process, *absorption*, converts the energy present in electromagnetic radiation into the internal energy of the absorbing molecule. Both absorption and scattering vary in their effect from one part of the spectrum to another. Remote sensing is impossible in those parts of the spectrum that are seriously affected by scattering and/or absorption, for these mechanisms effectively render the atmosphere opaque to incoming or outgoing radiation. As far as remote sensing of the Earth's surface is concerned, the atmosphere

'. . . appears no other thing to me but a foul and pestilential congregation of vapours'

Hamlet, Act 2, Scene 2

Regions of the spectrum that are relatively free from the effects of scattering and absorption are called *atmospheric windows*; electromagnetic radiation in these regions passes through the atmosphere with less change than does the radiation at other wavelengths. This effect can be compared to the way in which the bony tissues of the human body are opaque to X-rays, whereas the soft muscle tissue and blood are transparent. Similarly, glass is opaque to ultraviolet radiation but is transparent at the visible wavelengths. Figure 1.5 shows a plot of wavelength against the percentage of incoming radiation transmitted through the atmosphere; the window regions are those with a high transmittance.

The effect of the processes of scattering and absorption is to add a degree of haze to the image, that is, to reduce the contrast of the image, and to reduce the amount of radiation returning to the sensor from the Earth's surface. A certain amount of radiation that is reflected from the neighbourhood of each target may also be recorded by the sensor as originating from the target. This is because scattering deflects the path taken by electromagnetic radiation as it travels through the atmosphere, while absorption involves the interception of photons or particles of radiation. Our eyes operate in the visible part of the spectrum by observing the light reflected by an object. The position of the object is deduced from the assumption that this light has travelled in a straight line between the object and our eyes. If some of the light reflected towards our eyes from the object is diverted from a straight path then the object will appear less bright. If light from other objects has been deflected so that it is apparently coming to our eyes from the direction of the first object then that first object will become blurred. Taken further, this scattering process will make it appear to our eyes that light is travelling from all target objects in a random fashion, and no objects will be distinguishable. Absorption reduces the amount of light that reaches our eyes, making a scene relatively duller. Both scattering and absorption, therefore, limit the usefulness of some portions of the electromagnetic spectrum for remote-sensing purposes. They are known collectively as *attenuation* or *extinction*.

Scattering is the result of interactions between electromagnetic radiation and particles or gas molecules that are present in the atmosphere. These particles and molecules range in size from the microscopic (such that they have radii approximately equal to the wavelength of the electromagnetic radiation) to raindrop size (100 μm and larger). The effect of scattering is to redirect the incident radiation, or to deflect it from its path. The atmospheric gases that primarily cause scattering include oxygen, nitrogen and ozone. Their molecules have radii of less than 1 μm and affect electromagnetic radiation with wavelengths of 1 μm or less. Other types of

particles reach the atmosphere both by natural causes (such as salt particles from oceanic evaporation or dust entrained by aeolian processes) and as a result of human activity (for instance, dust from soil erosion caused by poor land management practices, and smoke particles from industrial and domestic pollution). Some particles are generated as a result of photochemical reactions involving trace gases such as sulphur dioxide or hydrogen sulphide. The former may reach the atmosphere from car exhausts or from the combustion of fossil fuels. Another type of particle is the raindrop, which tends to be larger in size than the other kinds of particles mentioned previously (10–100 μm compared to 0.1–10 μm radius). The concentration of particulate matter varies both in time and over space. Human activities, particularly agriculture and industry, are not evenly spread throughout the world, nor are natural processes such as wind erosion or volcanic activity. Meteorological factors will cause variations in atmospheric turbidity over time, as well as over space. Thus, the effects of scattering will be uneven spatially (the degree of variation depending on weather conditions) and will vary from time to time. Remotely-sensed images of a particular area will thus be subjected to different degrees of atmospheric scattering on each occasion that they are produced. Differences in atmospheric conditions over time are the cause of considerable difficulty in the quantitative analysis of time sequences of remotely-sensed images.

The mechanisms of scattering are complex, and are beyond the scope of this book. However, it is possible to make a simple distinction between selective and nonselective scattering. Selective scattering affects specific wavelengths of electromagnetic radiation, whilst nonselective scattering is wavelength-independent. Very small particles and molecules, with radii far less than the wavelength of the electromagnetic radiation of interest, are responsible for *Rayleigh scattering*. The effect of this type of scattering is inversely proportional to the fourth power of the wavelength, which implies that shorter wavelengths are much more seriously affected than longer wavelengths. Blue light (wavelength 0.4–0.5 μm) is thus more powerfully scattered than red light (0.6–0.7 μm). This is why the sky seems blue, for incoming blue light is so scattered by the atmosphere that it seems to reach our eyes from all directions, whereas at the red end of the visible spectrum scattering is much less significant so that red light maintains its directional properties. The sky appears to be much darker blue when seen from a high altitude, such as from the top of a mountain or from an aeroplane, because the degree of scattering is reduced due to the reduction in the length of the path traversed through the atmosphere by the incoming solar radiation. Scattered light reaching the Earth's surface is termed *diffuse* (as opposed to *direct*) irradiance or, more simply, *skylight*. Radiation that has been scattered within the atmosphere and reaches the sensor without having made contact with the Earth's surface is called the *atmospheric path radiance*.

Mie scattering is caused by particles that have radii between 0.1 and 10 μm, that is, approximately the same magnitude as the wavelengths of electromagnetic radiation in the visible, near-infrared and thermal infrared regions of the spectrum. Particles of smoke, dust and salt have radii of these dimensions. The intensity of Mie scattering is inversely proportional to wavelength, as in the case of Rayleigh scattering. However, the exponent ranges from -0.7 to -2 rather than the -4 of Rayleigh scattering. Mie scattering affects shorter wavelengths more than longer wavelengths, but the disparity is not so great as in the case of Rayleigh scattering.

Non-selective scattering is wavelength-independent. It is produced by particles whose radii exceed 10 μm. Such particles include water droplets and ice fragments present in clouds. All visible wavelengths are scattered by such particles. We cannot see through clouds because all visible wavelengths are non-selectively scattered by the water droplets of which the cloud is formed. The effect of scattering is, as mentioned earlier, to increase the haze level or reduce the contrast in an image. If contrast is defined as the ratio between the brightest and darkest areas of an image, and if brightness is measured on a scale running from 0 (darkest) to 100 (brightest), then a given image with a brightest area of 90 and a darkest area of 10 will have a contrast of 90:10 or 9. If scattering has the effect of adding a component of upwelling radiation of 10 units then the contrast becomes 100:20 or 5. This reduction in contrast will result in a decrease in the detectability of features present in the image. Figure 1.16 shows relative scatter as a function of wavelength for the 0.3–1.0 μm region of the spectrum for a variety of levels of atmospheric haze.

Absorption is the second process by which the Earth's atmosphere interacts with incoming electromagnetic radiation. Gases such as water vapour, carbon dioxide and ozone absorb radiation in particular regions of the electromagnetic spectrum called *absorption bands*. The processes involved are very complex, and relate to the vibrational and rotational properties of the molecules of water vapour, carbon dioxide, or ozone, and are caused by transitions in the energy levels of the atoms. These transitions occur at characteristic wavelengths for each type of atom and at these wavelengths absorption rather than scattering is dominant. Remote sensing in these absorption bands is thus rendered impossible.

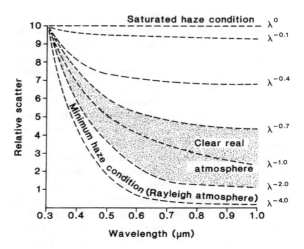

Figure 1.16 Relative scatter as a function of wavelength λ for various levels of atmospheric haze. Based on *Manual of Remote Sensing*, 2nd edition, ed. R.N. Colwell, 1983, figure 6.15; reproduced by permission of the American Society for Photogrammetry and Remote Sensing.

Fortunately, other regions of the spectrum with low absorption (high transmission) can be used. These regions are called 'windows', and they cover the 0.3–1.3 μm (visible/near-infrared), 1.5–1.8, 2.0–2.5 and 3.5–4.1 μm (middle infrared) and 7.0–15.0 μm (thermal infrared) wavebands. The utility of these regions of the electromagnetic spectrum in remote sensing is considered at a later stage.

1.3 INTERACTIONS WITH EARTH-SURFACE MATERIALS

1.3.1 Introduction

Electromagnetic energy reaching the Earth's surface may be reflected, transmitted or absorbed. Reflected energy travels upwards through, and interacts with, the atmosphere; that part of it which enters the field of view of the sensor is detected by the sensor and converted into a numerical value to be transmitted to a ground receiving station on Earth. The amount and spectral distribution of the reflected energy are used in remote sensing to infer the nature of the reflecting surface. A basic assumption made in remote sensing is that specific targets (soils of different types, water with varying degrees of impurities, rocks of differing lithologies, and vegetation of various species) have an individual and characteristic manner of interacting with incident radiation which is described by the spectral response of that target. A spectral reflection curve describes the spectral response of a target for the 0.4–2.5 μm region. In some instances, the nature of the interaction between incident radiation and Earth-surface material will vary from time to time during the year, such as might be expected with vegetation as it develops from the leafing stage, through growth to maturity and, finally, to senescence.

The spectral response of a target will also depend upon such factors as the orientation of the Sun (solar azimuth, Figure 1.17), the height of the Sun in the sky (solar elevation angle), the direction in which the sensor is pointing relative to nadir (the look angle) and the state of health of vegetation, if that is the target. Nevertheless, if the assumption that specific targets are characterised by an individual spectral response were invalid then the Earth observation by remote sensing would be an impossible task. Fortunately, experimental studies in the field and in the laboratory, as well as experience with multispectral satellite imagery, have shown that the assumption is generally a reasonable one. Indeed, the successful development of remote sensing of the environment over the past decade bears witness to its validity. Note that the term *spectral signature* is sometimes

used to describe the spectral response curve for a target. In view of the dependence of spectral response on the factors mentioned above, this term is inappropriate for it gives a misleading impression of constancy and uniqueness.

In this section the spectral reflectance curves of vegetation, soil, rocks and water are examined in order to emphasise their characteristic features. The results summarised in the following paragraphs must not be taken to be characteristic of all varieties of materials or all observational circumstances. One of the problems met in remote sensing is that the spectral reflectance of a given Earth-surface cover type is influenced by a variety of confusing factors. For example, the spectral reflectance curve of a particular agricultural crop such as wheat is not constant over time, nor is it the same for all kinds of wheat. The spectral reflectance curve is affected by factors such as soil nutrient status, the growth stage of the crop, the colour of the soil (which may be affected by recent weather conditions), the azimuth and elevation angles of the Sun, and the look angle of the sensor. The topographic position of the target in terms of slope orientation with respect to solar azimuth and slope

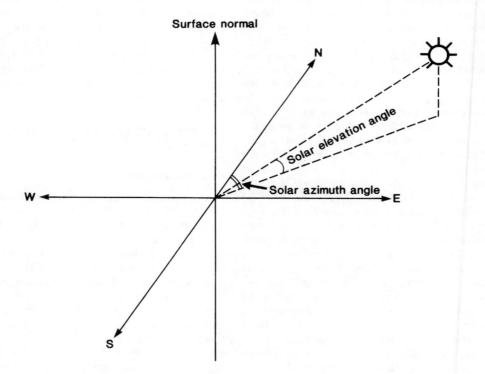

Figure 1.17   Solar elevation and azimuth angles. Elevation is measured upwards from the horizontal plane. Azimuth is measured clockwise from north. The zenith angle is measured from the surface angle, and equals 90° minus elevation angle, in degrees.

angle will also have an effect on the reflectance characteristics of the crop, as will the state of the atmosphere. Methods for dealing with some of these difficulties are described in Chapter 4. Hence, the examples given in this section are idealised models rather than templates.

Before turning to the individual spectral reflectance features of Earth-surface materials, a distinction must be drawn between two kinds of reflectance that occur at a surface. *Specular* reflectance is that kind of reflection in which energy leaves the reflecting surface without being scattered, with the angle of incidence being equal to the angle of reflection (Figure 1.2(b)). Surfaces that reflect specularly are smooth relative to the wavelength of the incident energy. *Diffuse* reflectance occurs when the reflecting surface is rough relative to the wavelength of the incident energy, and the incident energy is scattered in all directions (Figure 1.2(a)). A mirror reflects specularly while a piece of paper reflects diffusely. In the visible part of the spectrum, almost all terrestrial targets are diffuse reflectors, whereas calm water can act as a specular reflector. At microwave wavelengths, however, some terrestrial targets are specular reflectors. A satellite sensor operating in the visible and near-infrared spectral regions does not observe and detect all the reflected energy from a ground target over an entire hemisphere; instead, it records the reflected energy that is returned at a particular angle (see the definition of radiance in section 1.2.1). To make use of such measurements, the distribution of radiance at all possible observation and illumination angles (called the bidirectional reflectance distribution function or BRDF) must be taken into consideration. Details of the BRDF are given by Barnsley (1994), Milton (1986) and Slater (1980) who writes:

> '. . . the reflectance of a surface depends on both the direction of the irradiating flux and the direction along which the reflected flux is detected.'
>
> Slater (1980, p .231)

Hyman and Barnsley (1997) demonstrate that multiple images of the same area taken at different viewing angles provide enough information to allow different land cover types to be identified as a result of their differing BRDF. The MISR (Multi-Angle Imaging SpectroRadiometer) is being developed as part of the US Earth Observing System (EOS). This instrument will collect multidirectional observations of the same ground area over a timescale of a few minutes, at nadir and at fore and aft angles of view of 21.1°, 45.6°, 60.0° and 70.5°, and in four spectral bands in the visible and near-infrared regions of the electromagnetic spectrum.

The instrument will therefore provide data for the analysis and characterisation of reflectance variation of Earth-surface materials over a range of angles (Diner *et al.*, 1991).

It follows from the foregoing that even if the target is a diffuse reflector such that incident radiation is scattered in all directions, the assumption that radiance is constant for any observation angle θ measured from the surface normal does not generally hold. A simplifying assumption is known as Lambert's cosine law, which states that the radiance measured at an observation angle θ is the same as that measured at an observation angle of 0° adjusted for the fact that the projection of the unit surface at a view angle of θ is proportional to $\cos \theta$. Surfaces exhibiting this property are called Lambertian, and a considerable body of work in remote sensing either explicitly or implicitly makes the assumption that Lambert's law applies. However, it is usually the case that the distribution of reflected flux from a surface depends on the geometrical conditions of measurement and illumination. The topic of correction of images for Sun and view angle effects is considered further in Chapter 4.

1.3.2 Spectral reflectance of Earth-surface materials

In this section, typical spectral reflectance curves for characteristic types of Earth-surface materials are discussed. The remarks above (section 1.3.1) should not be overlooked when reading the following paragraphs. The Earth-surface materials that will be considered in this section are vegetation, soil, bare rock and water. The short review by Verstraete and Pinty (1992) is recommended. Hobbs and Mooney (1990) provide a survey of remote sensing of the biosphere, and Belward (1991) summarises the spectral characteristics of soil, water and vegetation in the visible, near-infrared and mid-infrared spectral regions.

1.3.2.1 *Vegetation*

An idealised spectral reflectance curve of vigorous vegetation is shown in Figure 1.18. The curve shows relatively low values in the red and the blue regions of the visible spectrum, with a minor peak in the green spectral band. These peaks and troughs are caused by absorption of blue and red light by chlorophyll and other pigments. Typically, 70–90% of both blue and red light is absorbed to provide energy for the process of photosynthesis. The slight reflectance peak between 0.5 and 0.6 μm is the reason that actively growing vegetation appears green. Reflectivity rises sharply at about

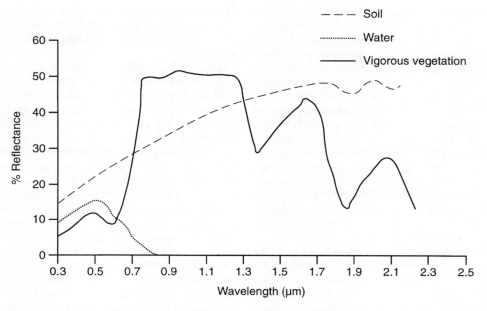

Figure 1.18   Idealised spectral reflectance curves for vigorous vegetation, soil and water.

0.75 μm (the so-called 'red edge'; Filella and Penuelas, 1994), and reflection remains high in the near-infrared region between 0.75 and 1.35 μm as a result of interaction between the internal leaf structure and EMR in this waveband. Between 1.35 and 2.5 μm internal leaf structure has some effect, but the reflectance is largely controlled by leaf-tissue water content, which is the cause of the minima recorded near 1.45 μm and 1.95 μm. As the plant ages, the level of reflectance in the near-infrared region (0.75–1.35 μm) declines first, with reflectance in the visible part of the spectrum not being affected significantly. The reflectance spectra of vegetation can be seen in the laboratory spectra of deciduous leaves, conifer needles, dry grass and green grass (Figure 1.19) and in more detail in the laboratory spectrum of an aspen leaf (Figure 1.20).

As senescence continues, the relative maximum in the green part of the visible spectrum declines as pigments other than chlorophyll begin to dominate, and the leaf begins to lose its greenness and to turn yellow or, occasionally, reddish. Stress caused by environmental factors such as drought or by the presence or absence of particular minerals in the soil can also produce a spectral response that is similar to senescence. The occurence of geobotanical anomalies has been used successfully to determine the location of mineral deposits (Goetz *et al.*, 1983). Hoffer (1978) is a useful source of further information.

The shape of the spectral reflectance curve can be used to distinguish vegetated and non-vegetated areas on remotely-sensed imagery. Differences between species can also be considerable, and may be sufficient to permit their discrimination. Such discrimination may be possible on the basis of relative differences in the spectral reflectance curves of the vegetation or crop types. Absolute reflectance values (section 4.6) may be used to estimate physical properties of the vegetation, such as green leaf area or biomass production. In agriculture, the estimation of crop yields is often a significant economic requirement. It is important to remember, however, the points made in section 1.3.1 – there is no single, ideal spectral reflectance curve for any particular vegetation type, and the recorded radiance from a point on the ground will depend upon the viewing and illumination angles, as well as other variables. The geometry of the crop canopy will strongly influence the bidirectional reflectance distribution function (BRDF) (section 1.3.1), whilst factors such as the transmittance of the leaves, the number of leaf layers, the actual arrangement of leaves on the plant and the nature of the background (which may be soil, or leaf litter, or undergrowth) are also important. In order to distinguish between some types of vegetation, and to assess growth rates from remotely-sensed imagery, it is necessary to use *multitemporal imagery*, that is, imagery of the same area collected at different periods in the growing season.

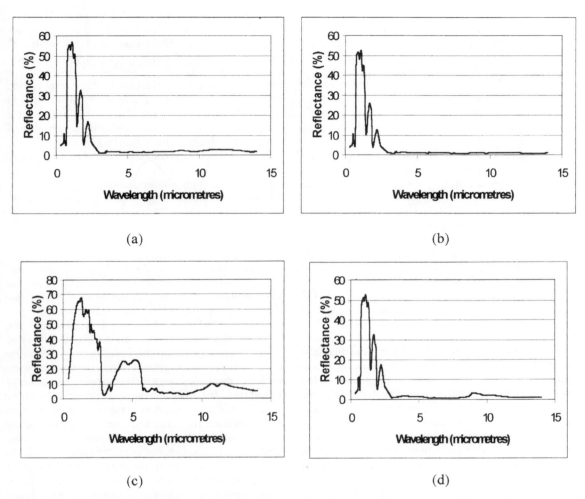

Figure 1.19   Laboratory spectral reflectance curves for vegetation types: (a) deciduous leaves, (b) conifer needles, (c) dry grass and (d) green grass. Data obtained from the NASA Jet Propulsion Laboratory ASTER Spectral Library.

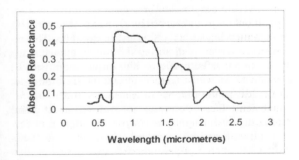

Figure 1.20   Spectral reflectance curve for an aspen leaf. From US Geological Survey Digital Spectral Library.

1.3.2.2   Geology

Geological use of remotely-sensed imagery relies, to some extent, upon knowledge of the spectral reflectance curves of vegetation, for approximately 70% of the Earth's land surface is vegetated and the underlying rocks cannot be observed directly. It was indicated in the preceding section that geobotanical anomalies might be used to infer the location of mineral deposits. Such anomalies include peculiar or unexpected species distribution, stunted growth or reduced ground cover, altered leaf pigmentation or yellowing (chlorosis), and alteration to the phenological cycle, such as early senescence or late leafing in the spring. It would be unwise to suggest that all such changes are due to soil geochemistry; however, the results of a number of studies indicate that the identification of anomalies in the vegetation cover of an area can be a good guide to the presence of mineral deposits. If the relationship between soil formation and underlying lithology has been destroyed, for example by the deposition of glacial material over the local rock, then it becomes difficult to make associations between the phenological characteristics of the vegetation and lithology.

In semi-arid and arid areas the spectral reflectance curves of rocks and minerals may be used directly in order to infer the lithology of the study area, though care should be taken because weathering crusts with spectra that are significantly different from the parent rock may develop. Laboratory studies of reflectance spectra of minerals have been carried out by Hunt and co-workers in the United States (Hunt, 1977, 1979; Hunt and Ashley, 1979; Hunt and Salisbury, 1970, 1971; Hunt *et al.*, 1971). Spectral libraries, accessible over the Internet from the Jet Propulsion Laboratory ASTER Spectral Library and the US Geological Survey Digital Spectral Library, contain downloadable data derived from the studies of Hunt, Salisbury and others. These studies demonstrate that rock-forming minerals have unique spectral reflectance curves. The presence of absorption features in these curves is diagnostic of the presence of certain mineral types. Some minerals, for example quartz and feldspars, do not have strong absorption features in the visible and near-infrared regions, but can be important as dilutants for minerals with strong spectral features such as the clay minerals, sulphates and carbonates. Clay minerals have a decreasing spectral reflectance beyond 1.6 μm, while carbonate and silicate mineralogy can be inferred from the presence of absorption bands in the mid-infrared region, particularly 2.0 to 2.5 μm. Kahle and Rowan (1980) show that multispectral thermal infrared imagery in the 8–12 μm region can be used to distinguish silicate and non-silicate rocks. Some of the difficulties involved in the identification of rocks and minerals from the properties of spectral reflectance curves include the effects of atmospheric scattering and absorption, the solar flux levels in the spectral regions of interest (section 1.2.5) and the effects of weathering. Buckingham and Sommer (1983) indicate that the nature of the spectral reflectance of a rock is determined by the mineralogy of the upper 50 μm, and that weathering, which produces a surface layer that is different in composition from the parent rock, can significantly alter the observed spectral reflectance.

Figure 1.21 was produced from spectral reflectance data downloaded from the NASA Jet Propulsion Laboratory's ASTER Spectral Library. Spectra for three rock types (granite, basalt and black shale) are shown, for the optical and infrared regions of the spectrum. Distinctive reflectance features are apparent; the granite sample shows reflectance values approaching 70% in the near-infrared and similarly high values in the middle infrared (3–5 μm), whereas the basalt sample shows a much reduced reflectance (maximum around 14%). Basalt is also characterised by a broad peak in the thermal infrared waveband spectrum at around 11 μm. Black shale shows reflectance values up to nearly 30% with a number of absorption features.

Good introductions to geological remote sensing are Drury (1993), Goetz (1989) and Gupta (1991).

1.3.2.3 Water bodies

The characteristic spectral reflectance curve for water shows a general reduction in reflectance with increasing wavelength, so that in the near-infrared the reflectance of deep, clear water is virtually zero (Figure 1.18). However, the spectral reflectance of water is affected by the presence and concentration of dissolved and suspended organic and inorganic material, and by the depth of the water body. Thus, the intensity and distribution of the radiance upwelling from a water body are indicative of the nature of the dissolved and suspended matter in the water, and of the water depth. Figure 1.22 shows how the information that oceanographers and hydrologists require is only a part of the total signal received at the sensor. Solar irradiance is partially scattered by the atmosphere, and some of this scattered light (the path radiance) reaches the sensor. Next, part of the surviving irradiance is reflected by the surface of the water body. This reflection might be specular under calm conditions, or its distribution might be strongly influenced by surface waves and the position of the Sun relative to the sensor, giving rise to *sunglint*. Once within the water body, EMR may be absorbed by the water (the degree of absorption being strongly wavelength-dependent) or selectively absorbed by dissolved substances, or backscattered by suspended particles. This latter component is termed the *volume reflectance*. Figure 1.23 shows the spectral distribution curves for visible and near-infrared radiation at the surface and at 20 m depth. At 20 m depth only visible light (mainly blue) is present, as the near-infrared component has been totally absorbed. Particulate matter, or suspended solids, scatter the downwelling radiation, the degree of scatter being proportional to the concentration, although other factors such as the particle size distribution and the colour of

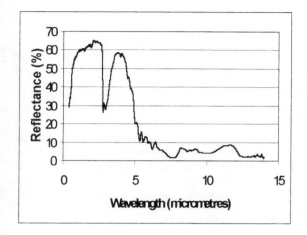

(a)

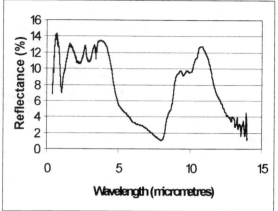

(b)

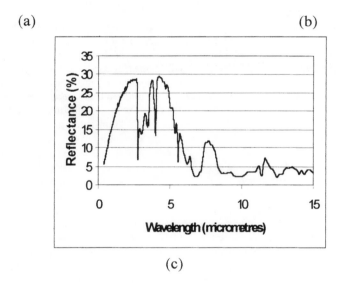

(c)

Figure 1.21   Laboratory spectral reflectance curves for different rock types: (a) granite, (b) basalt and (c) black shale. Data obtained from the NASA Jet Propulsion Laboratory ASTER Spectral Library.

the sediment are significant. Over much of the observed low to medium range of concentrations of suspended matter, a positive, linear relationship between suspended matter concentration and reflectance in the visible and near-infrared bands has been observed, though the relationship becomes non-linear at increasing concentrations. Furthermore, the peak of the reflectance curve moves to progressively longer wavelengths as concentration increases, which may lead to inaccuracy in the estimation of concentration levels of suspended materials in surface waters from remotely-sensed data. Another source of error is the inhomogeneous distribu-

tion of suspended matter through the water body, which is termed *patchiness*.

The presence of chlorophyll is an indication of the trophic status of lakes, and also is of importance in estimating the level of organic matter in coastal and estuarine environments. Whereas suspended matter has a generally broadband reflectance in the visible and near-infrared, chlorophyll exhibits typical absorption bands in the region below $0.5\,\mu m$ and between 0.64 and $0.69\,\mu m$ (Figure 1.24). Detection of the presence of chlorophyll therefore requires an instrument with a higher spectral resolution (section 2.2.2) than would be

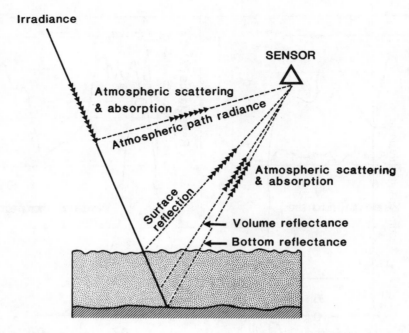

Figure 1.22   Processes acting on solar radiant energy in the visible part of the spectrum over an area of shallow water. From Alföldi, T., 1982, figure 1. Reproduced by permission of the Soil Conservation Society of America.

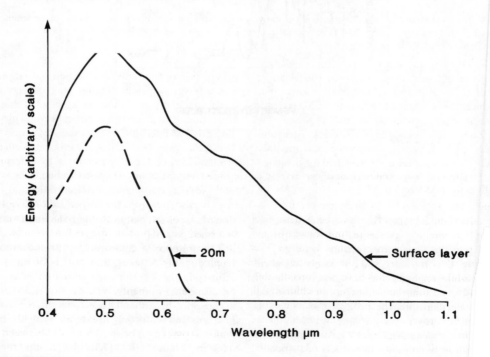

Figure 1.23   Spectral reflectance curve for surface water layer (solid line) and 20 m depth (dotted line). From Alf   ., 1982, figure 3. Reproduced by permission of the Soil Conservation Society of America.

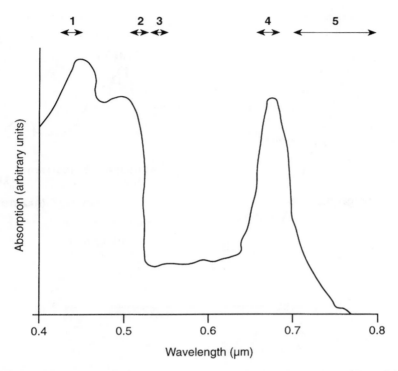

Figure 1.24   Idealised absorption curve for green algae, showing location of the Nimbus-7 Coastal Zone Colour Scanner (CZCS) wavebands 1–5, described in section 2.3.4 Based in part on Alföldi, T., 1982, figure 6. Reproduced by permission of the Soil Conservation Society of America.

required to detect suspended sediment. Furthermore, the level of backscatter from chlorophyll is lower than that produced by suspended sediment – consequently greater radiometric sensitivity is also required. The Coastal Zone Colour Scanner (CZCS) instrument, which was carried on the Nimbus-7 satellite, had wavebands that were specifically designed for the detection of chlorophyll. The position of these bands is shown in Figure 1.24.

Although the spectral reflectance curves for suspended matter and chlorophyll have been described separately, it is not uncommon to find that both are present at one particular geographical locality. The complications in separating out the contribution of each to the total observed reflectance is a considerable one. Furthermore, the suspended matter or chlorophyll may be unevenly distributed in the horizontal plane (the patchiness phenomenon noted above) and in the vertical plane. This may cause problems if the analytical technique used to determine concentration levels from recorded radiances is based on the assumption that the material is uniformly mixed at least to the depth of

penetration of the radiation. In some situations a surface layer of suspended matter may ride on top of a lower, colder, layer with a low suspended matter concentration, giving rise to considerable difficulty if standard analytical techniques are used. Reflection from the bed of the water body can have unwanted effects if the primary aim of the experiment is to determine suspended sediment or chlorophyll concentration levels, for it adds a component of reflection to that resulting from backscatter from the suspended or dissolved substances. In other instances, the reflection from the sea bed might be the primary focus of the exercise, in which case the presence of organic or inorganic material in the water would be a nuisance.

1.3.2.4   Soils

The spectral reflectance curves of soils are generally characterised by a rise in reflectivity as wavelength increases – the opposite, in fact, of the shape of the spectral reflectance curve for water (Figure 1.18). Reflectivity in the visible wavebands is affected by the presence of

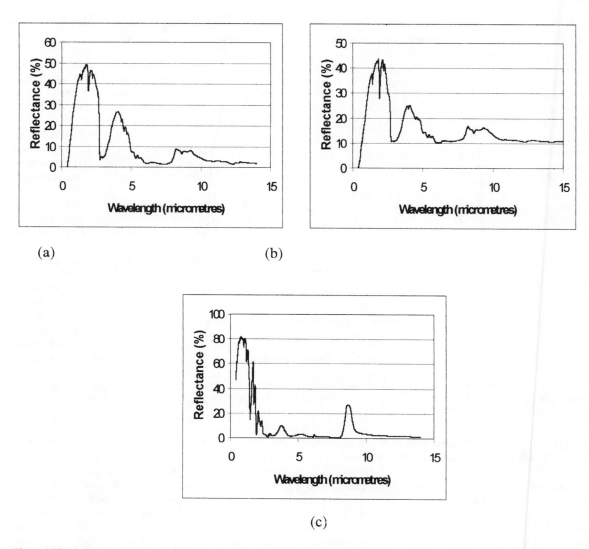

Figure 1.25   Laboratory spectral reflectance curves for different soil types: (a) brown sandy loam, (b) brown/dark brown clay and (c) white gypsum dune sand. Data obtained from the NASA Jet Propulsion Laboratory ASTER Spectral Library.

organic matter in the soil, and by the soil moisture content, while at 0.85–0.93 µm there is a ferric iron absorption band. As ferric iron also absorbs ultraviolet radiation in a broad band, the presence of iron oxide in soils is expressed visually by a reddening of the soil, the redness being due to the absorption of the wavelengths shorter (and longer) than the red. Between 1.3–1.5 µm and 1.75–1.95 µm water absorption bands occur, as mentioned in section 1.3.2.1. Soil reflectance in the optical part of the electromagnetic spectrum is usually greatest in the region between these two water absorption bands, and declines at wavelengths longer than 2 µm, with clay minerals, if present, being identifiable by their typical, narrow-band absorption features in the 2.0–2.5 µm region. Irons *et al.* (1989) provide a comprehensive survey of factors affecting soil reflectance, while Huete (1989) gives a summary of the influence of the soil background on measurements of vegetation spectra. Figure 1.25 shows laboratory soil reflectance spectra for sandy loam and clay soils and for a dune sand.

1.4 SUMMARY

In the opening paragraph of section 1.3.1, a basic principle of applied remote sensing is set out. This states that individual Earth-surface cover types are distinguishable in terms of their spectral reflection characteristics. Changes in the spectral response of objects can also be used as an indication of changes in the properties of the object, for example the health or growth stage of a plant or the turbidity of a water body. In this chapter, the basic principles of electromagnetic radiation are reviewed briefly, and the relevant terminology is defined. An understanding of these basic principles is essential if the methods described in the remainder of this book are to be applied sensibly. Chapter 2 provides details of the characteristics of sensors that are used to measure the level of electromagnetic radiation that is reflected from or emitted by the Earth's surface. It is from these measurements that the spectral reflectance curves of Earth-surface elements can be derived. Further details of the derivation of absolute spectral reflectance values from remotely-sensed measurements are provided in Chapter 4.

1.5 QUESTIONS

1. Describe the differences between analogue and digital images. List and explain the advantages and disadvantages of each for particular applications of your choice.

2. Explain the following terms, using diagrams where appropriate: energy, irradiance, radiance, reflectance, albedo, hertz, wavelength, amplitude, frequency, scattering, absorption, atmospheric window, transmittance, Lambertian reflectance, diffuse and specular reflectance, bidirectional reflectance.

3. Sketch the relationship between wavelength and atmospheric transmittance for the electromagnetic spectrum between 0 and 12.5 μm. List the atmospheric constituents responsible for the major absorption features.

4. Distinguish the characteristics of short-wave or near-infrared and thermal infrared radiation. What are the main remote-sensing applications of these two wavebands?

5. What is the 'red edge'? Why is it important in the remote sensing of vegetation?

6. Plot the spectral reflectance curves for soil, water and vigorous vegetation. Write a short account of the major features of these three curves.

7. Distinguish between *Rayleigh* and *Mie* scattering. What is the effect of scattering on remotely-sensed images acquired in the optical region of the spectrum?

2

Remote-Sensing Platforms and Sensors

2.1 INTRODUCTION

The nature and characteristics of digital images of the Earth's surface produced by satellite-borne sensors operating in the visible, infrared and microwave regions of the electromagnetic spectrum are described in this chapter. Remote sensing of the surface of the Earth has a long history, dating from the use of cameras carried by balloons and pigeons in the eighteenth and nineteenth centuries, but in its modern connotation the term *remote sensing* can be traced back to the aircraft-mounted systems that were developed during the early part of the twentieth century, initially for military purposes. Airborne camera systems are still a very important source of remotely-sensed data (Lillesand and Keifer, 1994, chapter 2), and space-borne camera systems, initially used in low-Earth-orbit satellites for military purposes, have also been used for civilian remote sensing from space, for example the NASA Large Format Camera (LFC) flown on the American Space Shuttle in October 1984.

Although analogue photographic imagery has many uses, this book is concerned with image data collected by scanning systems that ultimately generate digital image products. Film cameras and non-imaging (profiling) instruments are thereby excluded from direct consideration, although hard-copy products from these systems can be converted into digital form by scanning (section 1.1) and the techniques described in later chapters can be applied. Panchromatic and multispectral sensor systems have been, and are presently, carried by aircraft and satellites, by the American Space Shuttle (the Space Transportation System, STS) and by the Russian Space Station, *Mir*. Properties of imaging radar systems are also described in this chapter. Individual instruments, such as the Thematic Mapper instrument, which is carried by the American Landsat satellite, are described briefly and examples of image products are provided. Up-to-date details of current and planned sensor systems carried by individual satellites are provided by national space agencies and by CEOS (the Committee on Earth Observation Satellites) via the World Wide Web, which is described in section 3.2.2.

The general characteristics of imaging remote-sensing instruments, namely, their spatial, spectral and radiometric resolution and the number of spectral bands in which data are collected, are the subject of section 2.2. In section 2.3 the properties of images collected in three regions (optical, infrared and microwave) of the electromagnetic spectrum are described.

Spatial, spectral and radiometric resolution are properties of remote-sensing instruments. An important property of the remote-sensing system, of which the instrument is a part, is the temporal resolution of the system, that is, the time that elapses between successive dates of imagery acquisition. This revisit time may be measured in minutes if the satellite is effectively stationary with respect to a fixed point on the Earth's surface – i.e. in *geostationary* orbit, but note that not all geostationary orbits produce fixed observation points (see Elachi, 1987) – or in days if the orbit is such that the satellite moves relative to the Earth's surface. Meteosat is an example of a geostationary platform, which views an entire hemisphere of the Earth from a fixed position above the equator (Figure 1.11), whereas the NOAA (Figure 1.10), Landsat (Figures 1.8 and 1.9) and SPOT (Figure 2.7) satellites are polar orbiters, each having a specific repeat cycle time (or temporal resolution) of the order of hours (NOAA) or days (Landsat, SPOT). The temporal resolution of a polar orbiting satellite is set by the choice of orbit parameters (such as orbital altitude, shape and inclination) which are related to the objectives of the particular mission. Satellites used in Earth observing missions generally have a near-circular, polar, orbit, though the Space Shuttle flies in an equatorial orbit. Elachi (1987, appendix B) gives details of the mathematics of orbit determination. The relationship between the orbit period T and the orbit radius r is given by his equation B-5:

$$T = 2\pi r \sqrt{\frac{r}{g_s R^2}}$$

where g_s is the acceleration due to gravity $(0.009\,81\,\text{km s}^{-1})$, R is the Earth's radius (approximately 6380 km) and h is the orbit altitude (note that $r = R + h$). If, for example, $h = 800$ km then $T \approx 6052\,\text{s} \approx 100.9$ min. Thus, by varying the altitude of a satellite in a circular orbit the time taken for a complete orbit is also altered; the greater the altitude, the longer the orbital period.

The angle between the orbital plane and the Earth's equatorial plane is termed the *inclination* of the orbit, which is usually denoted by the letter i. Changes in the orbit are due largely to precession, caused mainly by the slightly non-spherical shape of the Earth. If the orbital precession is the same as the Earth's rotation round the Sun then the relationship between the node line and the Sun is always the same, and the satellite will pass over a given point on the Earth's surface at the same Sun time each day. Landsat and SPOT have this kind of orbit, which is said to be *Sun-synchronous*. Figure 2.1(a) shows the principles of a Sun-synchronous orbit, and Figure 2.1(b) illustrates an example of a circular, near-polar, Sun-synchronous orbit. Some Earth observing platforms are not in near-polar, Sun-synchronous orbits. The Space Shuttle has an equatorial orbit, which describes an S-shaped curve between the approximate latitudes 50°N and 50°S. Thus, the orbit selected for a

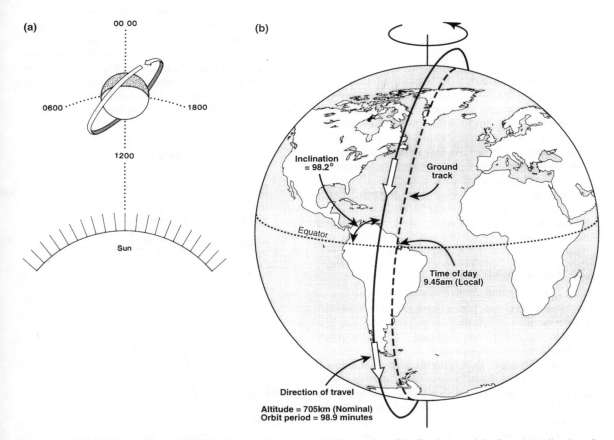

Figure 2.1   (a) Illustrating the principles of a Sun-synchronous orbit. The rotation of the Earth around the Sun, the inclination of the orbit and the orbital period are such that the satellite crosses the Equator at the same local Sun time on each orbit. (b) Example of a Sun-synchronous orbit. This is the Landsat-4 and -5 orbit, which has an equatorial crossing time of 0945 (local Sun time) in the descending node. The satellite travels southwards over the illuminated side of the Earth. The Earth rotates through 24.7° during a full satellite orbit. The satellite completes just over 14.5 orbits in a 24-hour period. The Earth is imaged between 82°N and 82°S latitude over a 16-day period.

particular satellite determines not just the time taken to complete one orbit (which is one of the factors influencing temporal resolution) but also the nature of the relationship between the satellite and the solar illumination direction. The temporal resolution is also influenced by the swath width, which is the length on the ground equivalent to one scan line. Landsat TM has a swath width of 185 km whereas the AVHRR sensor carried by the NOAA satellites has a swath width of 3000 km. The AVHRR can therefore provide much more frequent images of a fixed point on the Earth's surface than can the TM, though the penalty is reduced spatial resolution (1.1 km compared with 30 m). A pointable sensor such as the SPOT HRV can, in theory, provide much more frequent temporal coverage than the orbital pattern of the SPOT satellite and the swath width of the HRV would indicate.

2.2   CHARACTERISTICS OF IMAGING REMOTE-SENSING INSTRUMENTS

The major characteristics of an imaging remote-sensing instrument operating in the visible and infrared spectral region are described in terms of its spatial, spectral and radiometric resolution. Other important features are the manner of operation of the scanning device (electro-mechanical or electronic) and the geometrical properties of the images produced by the system. The interaction between the spatial resolution of the sensor and the orbital period of the platform determines the number of times that a given point on the Earth will be viewed. Allan (1984) gives a useful review of the relationship between spatial and temporal resolution and user requirements.

2.2.1   Spatial resolution

The *spatial resolution* of an imaging system is not an easy concept to define. It can be measured in a number of different ways, depending on the user's purpose. In a comprehensive review of the subject, Townshend (1980) uses four separate criteria on which to base a definition of spatial resolution. These criteria are the geometrical properties of the imaging system, the ability to distinguish between point targets, the ability to measure the periodicity of repetitive targets and the ability to measure the spectral properties of small targets. These will be considered briefly here; a fuller discussion can be found in Billingsley (1983), Forshaw *et al.* (1983), Simonett (1983) and Townshend (1980).

The most commonly used measure, based on the geometric properties of the imaging system, is the *instantaneous field of view* (IFOV) of a sensor. It is defined

as the area on the ground that, in theory, is viewed by the instrument from a given altitude at any given instant in time. The IFOV can be measured in one of two ways, as the angle α in Figure 2.2 or as the equivalent distance XY on the ground (note that Figure 2.2 is a cross-section, and XY is, in fact, the diameter of a circle). The actual, as distinct from the nominal, IFOV depends on a number of factors. No satellite has a perfectly stable orbit; its height above the Earth will vary, often by tens of kilometres. Landsat-1 to -3 had a nominal altitude of 913 km but the actual altitude of these satellites varied between 880 and 940 km. The IFOV becomes smaller at lower altitudes and increases as the altitude increases so, although the spatial resolution of Landsat-1 to -3 Multispectral Scanner (MSS) (section 2.3.6.1) is generally specified as 79 m, the actual resolution (measured by the IFOV) varied between 76 and 81 m.

The IFOV is the most frequently cited measure of resolution, though it is not necessarily the most useful. In order to explain why this is so, we must consider the way in which radiance from a point source is expressed on an image. A highly reflective point source on the ground does not produce a single bright point on an image collected by an aircraft or satellite. The point

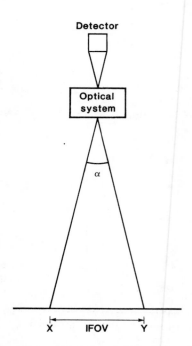

Figure 2.2   Angular instantaneous field of view (IFOV), α, showing the projection XY on the ground. Note XY is the diameter of a circle.

source is seen on an image as a diffused circular region, due to the properties of the optics involved in imaging. A cross-section of the recorded or imaged intensity distribution of a single point source is shown in Figure 2.3, and it can be seen to have a Gaussian-type distribution. The actual shape will depend upon the properties of the optical components of the system and the relative brightness of the point source. The distribution function shown in Figure 2.2 is called the *point spread function* or PSF (Moik, 1980; Slater, 1980; Billingsley, 1983).

An alternative measure of IFOV is, in fact, based on the PSF (Figure 2.3), and the 30 m spatial resolution of the Landsat-4 and -5 Thematic Mapper (TM) (section 2.3.6.2) is based upon the PSF definition of the IFOV. The IFOV of the Landsat MSS using this same measure is 90 m rather than 79 m (Townshend, 1980, p. 9). The presence of relatively bright or dark objects within the IFOV of the sensor will increase or decrease the amplitude of the PSF so as to make the observed radiance either higher or lower than that of the surrounding areas. This is why high-contrast features such as rivers and canals, which are less than 79 m in width, are frequently detectable on Landsat MSS images. Conversely, targets with dimensions larger than the Landsat MSS IFOV of 79 m may not be discernible if they do not contrast with their surroundings. The blurring effects of the PSF can be partially compensated for by image processing involving the use of the Laplacian function (section 7.3.2). Other factors causing loss of contrast on the image include atmospheric scattering and absorption, which are discussed in Chapters 1 and 4.

The definition of spatial resolving power based on the IFOV is therefore not a completely satisfactory one. As it is a geometrical definition it does not take into account the spectral properties of the target. If remote sensing is based upon the detection and recording of the radiance of targets, the radiance being measured at a number of discrete points, then a definition of spatial resolution taking into account the way in which this radiance is generated might be reasonable. This is the basis of the definition of the *effective resolution element* or ERE, which is defined by Colvocoresses (cited by Simonett, 1983) as

'the size of an area for which a single radiance value can be assigned with reasonable assurance that the response is within 5 per cent of the value representing the actual relative radiance'.

Colvocoresses estimated the ERE for the Landsat MSS system as 86 m and 35 m for the TM. These values might be more relevant than the IFOV for a user interested in classification of multispectral images (Chapter 8).

Other methods of measuring the spatial resolving power of an imaging device are based upon the ability of the device to distinguish between specific targets. There are two such measures in use, and both are perhaps more easily defined for photographic sensors. The first method is based on the fact that the PSF of a point source is a bright central disc with bright and dark rings around it. The Rayleigh criterion assumes two equally bright point sources and specifies that the two sources will be distinguishable on the image if the bright central disc of the one falls on the first dark ring of the PSF of the other. The minimum separation between the point sources to achieve this degree of separation on the image is a measure of the spatial resolving power of the imaging system. The second method assumes that the targets are not points but linear and parallel objects with a known separation that is relatable to their spatial frequency. If the objects contrast strongly with their background, then one could consider moving them closer together until the point is reached where they are no longer distinguishable. The spatial frequency of the objects such that they are just distinguishable is a measure of spatial resolving power. This spatial frequency is expressed in terms of line pairs per millimetre

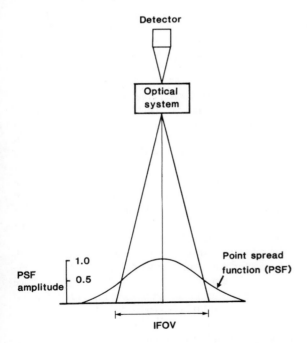

Figure 2.3 Instantaneous field of view based on the amplitude of the point spread function (PSF).

on the image or as cycles per millimetre. In order to calculate the spatial resolving power by this method it is usual to measure the contrast of the targets and their background. The measure most often used is *modulation* (*M*), defined as:

$$M = \frac{E_{max} - E_{min}}{E_{max} + E_{min}}$$

where E_{max} and E_{min} are the maximum and minimum radiance values recorded over the area of the image. For a nearly homogeneous image, *M* would have a value close to zero, whilst the maximum value of *M* is 1.0. Note that a perfectly homogeneous image would have an undefined modulation value since the calculation would involve division by zero. Returning to the idea of the parallel linear targets, we could find the ratio of the modulation measured from the image (M_1) to the modulation of the objects themselves (M_O). This gives the modulation transfer factor. A graph of this factor against spatial frequency shows the *modulation transfer function* or MTF. The spatial frequency at which the MTF falls to a half of its maximum value is termed the *effective instantaneous field of view* or EIFOV. The EIFOV of Landsat MSS imagery has been computed to be 66 m, whilst the 30 m resolution of the Landsat TM is computed from the spatial frequency at the point where the MTF has 0.35 of its maximum value (Townshend, 1980, p. 12), whereas the IFOV measure gives resolutions of 79 and 45 m respectively. The EIFOV measure is based on a theoretical target rather than on real targets, and as such gives a result that is likely to exceed the performance of the instrument in actual applications. Townshend and Harrison (1984) describe the calculation and use of the MTF in estimating the spatial resolving power of the Landsat-4 and -5 Thematic Mapper instrument (section 2.3.6.2).

The IFOV, ERE and EIFOV should not be confused with the pixel size. A digital image is an ordered set of numeric values, each value being related to the radiance from a ground area represented by a single cell or *pixel*. The pixel dimensions need not be related to the IFOV, the ERE or the EIFOV. For instance, the size of the pixel making up a Landsat MSS image is normally specified as 56 m × 79 m. The IFOV at the satellite's nominal altitude of 913 km is variously given as 79 m, 76.2 m and 73.4 m, while the ERE is estimated as 87 m (Simonett, 1983, table 1-2). Furthermore, pixel values can be interpolated over the cells of the digital image to represent any desired ground spacing, using the re-

sampling methods described in Chapter 4. The ground area represented by a single pixel of a Landsat MSS image is thus not necessarily identical to the spatial resolution as measured by any of the methods described above. The discussion in the preceding paragraphs, together with the description given below of the way in which satellite-borne sensors operate, should make it clear that the individual image pixel is not 'sensed' uniquely. This would require a stop/start motion of the satellite, and (in the case of the electro-mechanical scanner) of the scan mirror. The value recorded at the sensor corresponding to a particular pixel position on the ground is therefore not a simple average of the radiance upwelling from that pixel. There are likely to be contributions from areas outside the IFOV (the 'environmental radiance' discussed in section 4.4). Disregarding atmospheric effects (section 4.4), the radiance attributable to a specific pixel is in reality the sum (perhaps a weighted sum) of the contributions of the different land cover components within the pixel, plus the contribution from radiance emanating from adjacent areas of the ground. Fisher (1997) examines the concept of the pixel in some detail.

Spatial resolving power is an important attribute of remote-sensing systems because differing resolutions are relevant to different problems; indeed, there is a hierarchy of spatial problems that can use remotely-sensed data, and there is a spatial resolution appropriate to each problem. To illustrate this point, consider the use of an image with a spatial resolution (however defined) of 1 m. Each element of the image, assuming its pixel size was 1 m × 1 m, might represent the crown of a tree, part of a grass verge by the side of a suburban road, or the roof of a small car. This imagery would be useful in providing the basis for large-scale mapping of urban patterns, analysis of vegetation variations over a small area, or the monitoring of crops in small plots. At this scale it would be difficult to assess the boundaries of, or variation within, a larger spatial unit such as a town; a spatial resolution of 10 m might be more appropriate to this problem. A 10 m resolution would be a distinct embarrassment if the exercise was concerned with the mapping of sea-surface temperature patterns in the North Sea, for which a 500 m resolution could be used, whereas for continental- or global-scale problems a resolution of 1 km and 5 km respectively would produce data that contained the information required (and no more) and, in addition, was present in handleable quantities. The illusion that higher spatial resolution is necessarily better is commonplace; one should always ask 'better for what?' See Atkinson and Curran (1997) for further discussion of this point. The role of geographical scale in remote sensing is considered briefly in the introduction to section 8.5.

2.2.2 Spectral resolution

The second important property of an optical imaging system is its *spectral resolution*. With the exception of microwave images collected by satellites such as Seasat, ERS-1 and -2, JERS and RADARSAT, most of the digital images collected by satellite-borne sensors are multiband or multispectral. That is, individual images are separately recorded in discrete spectral bands, as shown in Figures 1.8 and 1.9. The term *spectral resolution* refers to the width of these spectral bands. An example will illustrate two important points, namely, (i) the position in the spectrum, width and number of spectral bands will determine the degree to which individual targets (vegetation species, crop or rock types) can be discriminated on the multispectral image, and (ii) the use of multispectral imagery can lead to a higher degree of discriminating power than any single band taken on its own.

The reflectance spectra of vegetation, soils, bare rock and water are described in section 1.3. Differences between the reflectance spectra of various rocks, for example, might be very subtle and the rock types might therefore be separable only if the recording device were capable of detecting the spectral reflectance of the target in a narrow waveband. A wide-band instrument would simply average out the differences. Figure 2.4(a) is a plot of the reflection from healthy vegetation against wavelength (dotted line) while the solid line shows the spectral reflectance from an area of diseased or unhealthy vegetation. Most of the difference occurs in the near-infrared region. Figure 2.4(b) shows the reflectance spectra for these two targets as recorded by a sensor such as Landsat's MSS; the difference between the solid and dotted lines is far less marked, and the two vegetation states might well be overlooked. To provide for the more reliable identification of particular targets on a remotely-sensed image the spectral resolution of the sensor must match as closely as possible the spectral reflectance curve of the intended target. This idea is demonstrated by the design of the Coastal Zone Colour Scanner (CZCS) carried by the Nimbus-7 satellite. Recent developments include the use of hyperspectral sensors (imaging spectrometers) such as the AVIRIS (Airborne Visible/Infrared Imaging Spectrometer) (Vane *et al.*, 1993) which acquires data in 224 narrow spectral bands.

Unfortunately, it is not possible to increase the spectral resolution of a sensor simply to suit the user's needs;

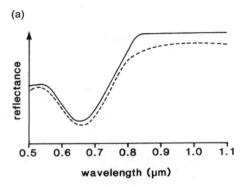

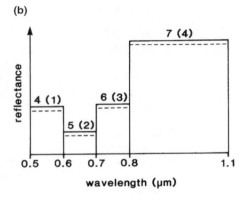

Figure 2.4 (a) Spectral reflectance curve for healthy (dotted line) and senescing (solid line) vegetation. (b) Spectral reflectance curve (as in part (a)) as recorded by an instrument such as Landsat MSS with four broad spectral bands. The Landsat MSS bands were numbered 4–7 for Landsat-1 to -3 and 1–4 for Landsat-4 and -5.

there is a price to pay. Higher spectral resolution reduces the signal-to-noise ratio (SNR) of the sensor output. The signal is the information content of the data received at the sensor, while the noise is the unwanted variation that is added to the signal. Such noise can be either random or systematic in nature, and is caused by the mechanical or electronic components of the instrument being less than perfect in operation. The ratio between the signal level and the noise level is the signal-to-noise ratio. A more technical definition is given by Barker and Gunther (1983, p. 49), who cite a 1978 NASA/Goddard Spaceflight Center document by Weinstein and Bank as follows:

'[The signal-to-noise ratio is defined for . . .] a constant input radiance . . . as the ratio of the output value (in units of radiance) averaged over at least 100 samples to the root mean square (RMS) value of the noise equivalent radiance, which is defined as the RMS of the deviations of the output samples from the average value.'

Smith and Curran (1996) provide details of SNR calculations, and estimate values for the AVHRR, Landsat TM and SPOT HRV as 38:1, 341:1 and 410:1 respectively.

A compromise must be sought between the twin requirements of narrow bandwidth (high spectral resolution) and a low signal-to-noise ratio. The *push-broom* type of sensor is a linear array of individual detectors with one detector per scan line element. The image is formed by the sensor being moved forward with the platform (Figure 2.5(a)). This arrangement provides for a longer 'look' at each scan line element, and it gives a better signal-to-noise ratio (SNR) than does an electro-mechanical scanner employing a single detector that observes each scan line element sequentially (Figure 2.5(b)). The time available to look at each point (the dwell time or integration time) is therefore greater for the push-broom scanner, hence narrower bandwidths and a larger number of quantisation levels are theoretically possible without decreasing the SNR to unacceptable levels. Further discussion of the different kinds of sensor is contained in section 2.3. The use of multivariate rather than univariate measures to help discriminate between groups of objects is well documented in the statistical literature:

> '. . . The prime justification of adopting a multivariate approach [is] that significant differences (or similarities) may well remain hidden if the variables are considered one at a time and not simultaneously.'
>
> Mather (1976, p. 421)

In Figure 2.6(a) the measurements of the spectral reflectance values for individual members of two hypothetical land cover types are plotted separately for two spectral bands, and there is an almost complete overlap between them. If the measurements on the two spectral bands are plotted together, as in Figure 2.6(b), the difference between the types is obvious, and there is no overlap. The use of well-chosen and sufficiently numerous spectral bands is a necessity, therefore, if different targets are to be successfully identified on remotely-sensed images.

2.2.3 Radiometric resolution

Radiometric resolution or radiometric sensitivity refers to the number of digital levels used to express the data collected by the sensor. In general, the greater the number of levels, the greater the detail in the information. At one extreme one could consider a digital image composed of only two levels (Figure 2.7(a)) in which level 0 is shown as black and level 1 as white. As the number of levels increases to 16 (Figure 2.7(b)) so the amount of detail visible on the image increases. With 256 levels of grey (Figure 2.7(c)) there is no discernible difference, though readers should note that this is as much a function of printing technology as of their visual systems. Nevertheless, the eye is not as sensitive to changes in intensity as it is to variations in hue.

The number of grey levels is commonly expressed in terms of the number of binary digits (bits) needed to store the value of the maximum grey level. For two levels the number of bits required per pixel is 1, while for 4, 16, 64 and 256 levels the number of bits required is 2, 4, 6 and 8 respectively. Thus, '6-bit' data have 64 possible levels, represented by the integer values 0 to 63

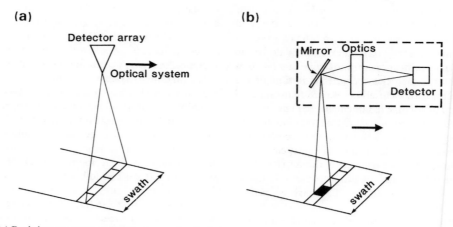

(a)

Detector array

Optical system

swath

(b)

Mirror Optics

Detector

swath

Figure 2.5   (a) Push-broom scanner. (b) Electro-mechanical scanner.

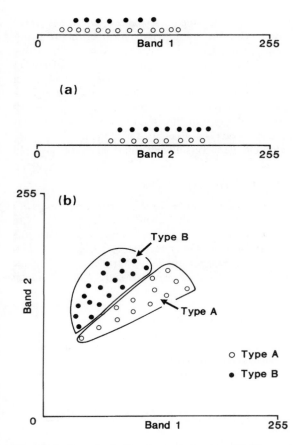

Figure 2.6 Hypothetical land cover types A and B measured in two spectral bands: (a) viewed separately on spectral bands 1 and 2, (b) viewed simultaneously on the same spectral bands. The two land cover types cannot be separated on either band taken separately but are clearly distinguished when the two bands are considered together.

inclusive, with '0' representing black and '63' representing white. The use of binary notation and the storage of digital data in computers are described in Chapter 3. Needless to say, the number of levels used to express the signal received at a sensor cannot be increased simply to suit the user's preferences. The signal-to-noise ratio of the sensor (see above) must be taken into account. The step size from one radiance level to the next cannot be less than the noise level, or else it would be impossible to say whether a change in level was due to a real change in the radiance of the target or to a change in the magnitude of the noise. A low-quality instrument with a high noise level would necessarily, therefore, have a lower radiometric resolution compared with a high-quality,

high signal-to-noise-ratio instrument. Slater (1980) discusses this point in some detail.

Tucker (1979) investigated the relationship between radiometric resolution and the ability to distinguish between vegetation types; he found only a 2–3% improvement for 256 levels over 64-level imagery. Bernstein *et al.* (1984) used a measure known as *entropy* to compare the amount of information (in bits per pixel) for 8-bit and 6-bit data for two images of the Chesapeake Bay area collected by Landsat-4's TM and MSS sensors. The entropy measure, *H*, is given by:

$$H = - \sum_{i=0}^{k} p(i)\log_2 p(i)$$

where *k* is the number of grey levels (for example, 64 or 256) and $p(i)$ is the probability of level *i*, which can be calculated from:

$$p(i) = \frac{F(i)}{nm}$$

In this formula, $F(i)$ is the frequency of occurrence of each grey level from 0 to $k - 1$, and nm is the number of pixels in the image. Moik (1980, p. 296) suggests that the use of this estimate of $p(i)$ will not be accurate because adjacent pixel values will be correlated (the phenomenon of spatial autocorrelation, as described by Cliff and Ord (1973)) and he recommends the use of the frequencies of the first differences in pixel values; these first differences are found by subtracting the value at a given pixel position from the value at the pixel position immediately to the left, with obvious precautions for the leftmost column of the image. Bernstein *et al.* (1984) do not indicate which method of estimating entropy they use but, since the same measure was applied to both the 6-bit and 8-bit images, the results should be comparable. The 8-bit resolution image added, on average, one to one-and-a-half bits to the entropy measure compared with a 6-bit image of the same scene. These results are shown in Table 2.1. Moik (1980, table 9.1) used both measures of entropy given above: for the first measure (based on the levels themselves) he found an average entropy of 4.4 bits for Landsat MSS (6-bit), and for the second measure (based on first differences) he found an average entropy of 4.2 bits. This value is slightly higher than that reported by Bernstein *et al.* (1984) but still less than the entropy values achieved by the 8-bit TM data. It is interesting to note that, using data compression techniques, the 6-bit data could be re-expressed in 4 or so bits without losing any information, whilst the 8-bit data conveyed, on average, approximately 5 bits of

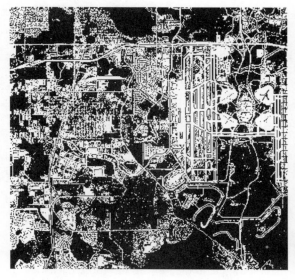

(a)

(b)

(c)

Figure 2.7 SPOT panchromatic image of Orlando, Florida, displayed in (a) 2, (b) 16 and (c) 256 grey levels. © CNES – Spot Image Distribution. Permission to use the data was kindly provided by SPOT Image, 5 rue des Satellites, BP 4359, F-31030 Toulouse, France (http://www.spotimage. fr).

Table 2.1 Entropy by band for Landsat TM and MSS sensors based on Landsat-4 image of Chesapeake Bay area, 2 November 1982 (scene E-40109-15140). See text for explanation. Based on table 1 of Bernstein *et al.* (1984).

Band number	Thematic Mapper		Multi-Spectral Scanner	
	Waveband (μm)	Bits per pixel	Waveband (μm)	Bits per pixel
1	0.45–0.52	4.21	0.5–0.6	2.91
2	0.52–0.60	3.78	0.6–0.7	3.57
3	0.63–0.69	4.55	0.7–0.8	4.29
4	0.76–0.90	5.19	0.8–1.1	3.63
5	1.55–1.75	5.92	–	–
6	10.5–12.5	3.53	–	–
7	2.08–2.35	5.11	–	–
Average	–	4.61	–	3.60

information. Given the enormous amounts of data making up a multiband image (nearly 300 million pixel values for one Landsat TM image) such measures as entropy are useful both for comparative purposes and for indicating the degree of redundancy present in a given data set. Entropy measures can also be used to compare the performance of different sensors. For example, Kaufmann *et al.* (1996) use entropy to compare the German MOMS sensor performance with that of Landsat TM and the SPOT HRV. The idea of reduction of redundancy in an image set underlies the technique of principal components analysis (section 6.4).

2.3 OPTICAL, NEAR-INFRARED AND THERMAL IMAGING SENSORS

The purpose of this section is to provide brief details of a number of sensor systems operating in the visible and infrared regions of the electromagnetic spectrum, together with some sample images, in order to familiarise the reader with the nature of digital image products. Further details of the instruments described in this section, as well as others that are planned for the next decade, are available from the Committee on Earth Observation Satellites (CEOS) via the World Wide Web. A recent comprehensive survey is provided by Joseph (1996). A number of satellite/sensor systems are planned for the next few years, but I have chosen not to include a comprehensive list because much of the information will be out of date within a few years. Attention is focused on sensors that are currently (or have recently been) in orbit, especially the Landsat and SPOT systems, which have generated large and widely used data sets since 1972 and 1986, respectively. Sufficient detail is provided for readers to inform themselves of the main characteristics of each system. More information is provided in the references cited.

2.3.1 Along Track Scanning Radiometer (ATSR)

The ATSR was developed by a consortium of British universities and research institutes led by the Rutherford–Appleton Laboratory (RAL). It has three thermal infrared channels centred at 3.7, 10.8 and 12 μm, plus a near-infrared channel centred at 1.6 μm. These thermal channels correspond to bands 3, 4 and 5 of the NOAA AVHRR. The primary purpose of this instrument is to make accurate measurements of global sea-surface temperatures for climate research purposes. It was flown on the ERS-1 satellite, which was launched in July 1991. ATSR-2 is a development of ATSR, and has additional channels centred at wavelengths of 0.555, 0.659 and 0.865 μm which are intended for land applications. It is carried by the ERS-2 satellite, which was launched in April 1995. ATSR is an interesting instrument (Harries *et al.*, 1994) as it provides two views of the target from different angles – one from nadir and one in the forward direction. Each view includes an internal calibration system, hence the data can be corrected for atmospheric effects (which can be estimated from the two views) and calibrated precisely. The thermal sensor has a nominal accuracy of

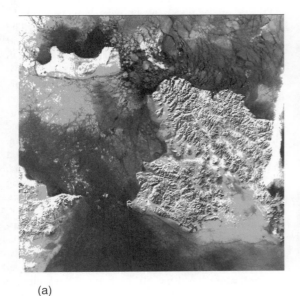

(a)

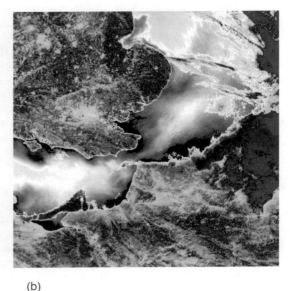

(b)

Figure 2.8 (a) Along Track Scanning Radiometer (ATSR-1) image of the Bering Straits acquired on 5 February 1992. This image shows reflected radiation in the 1.6 μm waveband at a spatial resolution of 1 km. The area imaged is 512 km × 512 km in size. Image courtesy of Rutherford–Appleton Laboratory. (b) Along Track Scanning Radiometer (ATSR-1) thermal (12 μm channel) image of the English Channel and southern North Sea with adjacent coastlands of England, France, Belgium and The Netherlands. Dark areas are coolest. Clouds cover parts of the image and these appear mid-grey. Image courtesy of Rutherford–Appleton Laboratory.

± 0.05 K. The sensor produces images with a spatial resolution of 1 km for a 500 km swath width.

An extended version of ATSR-2, called the Advanced Along Track Scanning Radiometer (AATSR), is to be part of the payload of the European Space Agency's ENVISAT-1 satellite, to be launched in 1999. AATSR has seven channels (0.55, 0.67, 0.86, 1.6, 3.7, 11 and 12 μm). The spatial resolution of the visible and near-infrared channels (bands 1–4) is 500 m (giving a swath width of 500 km). The thermal infrared channels (bands 5–7) have a spatial resolution of 1 km.

Figure 2.8(a) shows an ATSR-1 near-infrared (1.6 μm) image of the Bering Straits, taken on 2 February 1992. An example of an ATSR thermal image is shown in Figure 2.8(b), covering the English and French coasts, the eastern part of the English Channel and the southern North Sea on 9 August 1991.

2.3.2 Advanced Very High Resolution Radiometer (AVHRR)

The AVHRR, which is carried by the American NOAA (National Oceanographic and Atmospheric Administration) series of satellites, was intended to be a meteorological observing system. The imagery acquired by AVHRR has, however, been widely used in land cover monitoring at global and continental scales. Two NOAA satellites are in orbit at any one time, giving morning, afternoon, evening and night-time equatorial crossing times of approximately 0730, 1400, 1930 and 0200 respectively, though it should be noted that the illumination and view geometry is not the same for all of these overpasses. The NOAA satellite orbit repeats exactly after 9 days. The orbital height is 833–870 km with an inclination of 98.7° and a period of 102 minutes. The instrument has five channels and a resolution at nadir of 1.1 km. Data with a 1.1 km nominal resolution are known as *Local Area Coverage* (LAC) data. A lower-resolution sample of the LAC data, called *Global Area Coverage* (GAC) data, is recorded for the entire 102-minute orbit. About 11 minutes of LAC data can be tape-recorded on each orbit, and so LAC data are generally downloaded to ground receiving stations such as that at the Natural Environment Research Council (NERC) ground station at Dundee University in Scotland, which provided the image shown in Figure 1.10. The swath width of the AVHRR is of the order of 3000 km, and so spatial resolution at the edges of the image is considerably greater than the nominal 1.1 km, which refers to pixels viewed at nadir. Correction of the geometry of AVHRR is mentioned further in Chapter 4. The effects of off-nadir viewing are

not only geometrical for, as noted in Chapter 1, the radiance observed for a particular target depends on the angles of illumination and view. Thus, bidirectional reflectance factors should be taken into consideration when using AVHRR imagery.

The five channels of the AVHRR cover the following wavebands: channel 1 (green and red regions of the visible spectrum), 0.58–0.68 μm; channel 2 (near-infrared), 0.725–1.1 μm; channel 3 (thermal), 3.55–3.93 μm; channel 4 (thermal), 10.3–11.3 μm; and channel 5 (thermal), 11.5–12.5 μm. The thermal channels are used for sea-surface temperature determination and cloud mapping, whereas the visible and near-infrared channels are used to monitor land-surface processes, such as snow and ice melt, as well as vegetation status using vegetation indices such as the NDVI (section 6.2.4). The definitive source of information on the AVHRR sensor is Cracknell (1997).

2.3.3 Advanced Visible and Near-Infrared Radiometer (AVNIR)

The Advanced Visible and Near-Infrared Radiometer (AVNIR) was one of the instruments carried on-board the Japanese Advanced Earth Observation Satellite (ADEOS or 'Midori'), which was launched on 17 August 1996 into a Sun-synchronous 800 km orbit. The temporal resolution of the satellite was 41 days, and overpass time was 1015–1045 local time. ADEOS ceased to function in July 1997. AVNIR was a multispectral radiometer operating in the visible and near-infrared spectral bands. Like the SPOT HRV (section 2.3.7), AVNIR used a push-broom design. The instrument had three visible channels operating in the visible blue, green and red wavebands (0.40–0.50 μm, 0.52–0.62 μm and 0.62–0.72 μm) plus a near-infrared channel covering the 0.82–0.92 μm region. In addition, the AVNIR also had a panchromatic channel covering the visible spectrum from 0.52 to 0.72 μm. Spatial resolution is better than either Landsat TM or SPOT HRV, at 16 m in the multispectral bands and 8 m in the panchromatic band. AVNIR was capable of collecting off-nadir images and could tilt up to 40° to either side of the subsatellite track. The instrument had a 5.7° field of view and a swath width of 80 km. Figure 2.9 is an image of the city of Hiroshima, Japan, produced by AVNIR operating in panchromatic mode.

2.3.4 Coastal Zone Colour Scanner (CZCS)

The Coastal Zone Colour Scanner Experiment (CZCS) was carried by the Nimbus-7 satellite, and was primarily

Figure 2.9   The city of Hiroshima imaged by the AVNIR instrument, which was carried by the Japanese ADEOS mission between August 1996 and July 1997. This is a panchromatic image (0.52–0.72 μm) with a spatial resolution of 8 m, produced on 5 September 1996. Image © 1996 National Space Development Agency of Japan, NASDA/MITI.

designed to map the properties of the ocean surface, in particular chlorophyll concentration, suspended sediment distribution, *gelbstoffe* (yellow stuff) concentrations as a measure of salinity, and sea-surface temperatures. Chlorophyll, sediment and *gelbstoffe* were measured by five channels in the optical region, four of which were specifically chosen to target the absorption and reflectance peaks for these materials (Figure 1.24) at 0.433–0.453 μm, 0.510–0.530 μm, 0.540–0.560 μm and 0.660–0.680 μm. The 0.7–0.8 μm band was used to detect terrestrial vegetation, and sea-surface temperatures were derived from the thermal infrared band (10.5–12.5 μm). The spatial resolution of CZCS was 825 m and the swath width was 1566 km. The sensor could be tilted up to 20° off-nadir in order to avoid sunglint, which is specularly reflected sunlight. The Nimbus satellite had a near-polar, Sun-synchronous orbit with an altitude of 955 km. Its overpass time was 1200 local time, and the satellite orbits repeated with a period of 6 days. The CZCS experiment was finally terminated in December 1986.

2.3.5   IRS-1 LISS

The Indian Government has an active and successful remote-sensing programme. The first Indian Remote Sensing (IRS) satellite was launched in 1989, and carried the LISS-1 sensor, a multispectral instrument with a

76 m resolution in four wavebands. A more advanced version of LISS, LISS-2, is on-board the IRS-1B satellite, which was launched in 1991. LISS-2 senses in the same four wavebands as LISS-1 in the optical and near-infrared (0.45–0.52 μm, 0.52–0.59 μm, 0.62–0.68 μm and 0.77–0.86 μm) but has a spatial resolution of 36 m. IRS-1C (launched 1995) carries an improved LISS, numbered 3, plus a 5 m panchromatic sensor. LISS-3 includes a mid-infrared band in place of the blue-green band (channel 1 of LISS-1 and LISS-2). The waveband ranges for LISS-3 are 0.52–0.59 μm, 0.62–0.68 μm, 0.77–0.86 μm and 1.55–1.70 μm. The spatial resolution improves from 36 m to 25 m in comparison with LISS-2. The panchromatic sensor (0.50–0.75 μm) provides imagery with a spatial resolution of 5 m. An example of an IRS-1C panchromatic image of Denver Airport is given in Figure 2.10. IRS-1D carried a similar payload, but did not reach its correct orbit. However, some useful imagery is being obtained from its sensors. The panchromatic sensor is, like the SPOT HRV instrument, pointable so that oblique views can be obtained, and off-track viewing provides the opportunity for image acquisition every 5 days, rather than the 22 days for nadir viewing.

IRS-1C also carries the Wide Field Sensor, WIFS, which produces images with a pixel size of 180 m in two bands (0.62–0.68 μm and 0.77–0.86 μm). The swath width is 774 km. Images of a fixed point are produced every 24 days, though overlap of adjacent images will produce a view of the same point every 5 days.

Figure 2.10   Panchromatic image of Denver Airport obtained from the IRS-1C satellite. The image has a resolution of 5 m. Image courtesy of Space Imaging, Thornton, Colorado, USA.

2.3.6   Landsat instruments

2.3.6.1   *Landsat Multi-Spectral Scanner (MSS)*

Landsat-1, then called ERTS-1 (Earth Resources Technology Satellite), was the first land observation satellite. It was launched into a 919 km Sun-synchronous orbit on 23 July 1972, by the US National Aeronautics and Space Administration (NASA), and operated successfully until January 1978. A second, similar, satellite (Landsat-2) was placed into orbit in January 1975. Landsat-3, -4 and -5 followed in 1978, 1982 and 1984 respectively. A sixth was lost during the launch stage. The Landsat-7 launch is expected in 1999. Landsat-2 and -3 had orbits similar to Landsat-1, but the two later satellites (numbered 4 and 5) used a lower, 705 km, orbit, with a slightly different inclination of 98.2° compared with the 99.09° of Landsat-1 to -3. The Landsat orbit parameters are such that the MSS is capable of imaging the Earth between 82°N and 82°S latitudes. A special issue of the journal *Photogrammetric Engineering and Remote Sensing* (volume 63, number 7, 1997) is devoted to an overview of the Landsat programme.

The Multi-Spectral Scanner (MSS) was carried by all the Landsat satellites presently in orbit. This is a four-band instrument, with two visible channels in the green and red wavebands, respectively, and two near-infrared channels, i.e. (0.5–0.6 μm, 0.6–0.7 μm, 0.7–0.8 μm and 0.8–1.1 μm. These channels were numbered 4–7 in Landsat-1 to -3 because the latter two satellites carried a second instrument, the Return Beam Vidicon (RBV), which was a television-based system. It operated by producing an instantaneous image of the scene, and scanning the image, which was stored on a photosensitive tube. Landsat-4 and -5 did not carry the RBV and so the MSS channels were renumbered 1–4. Since Landsat-4 and -5 operate in a lower orbit than Landsat-1 to -3, the optics of the MSS have been altered slightly to maintain the swath width of 185 km and a pixel size of 79 m (along track) × 57 m (across track). MSS is an electro-mechanical scanner, using an oscillating mirror to direct reflected radiation on to a set of six detectors (one set for each waveband) (Figure 2.11). The six detectors each record the magnitude of the radiance from the ground area being scanned, which represents six adjacent scan lines (Figure 2.11). The analogue signal from the detectors is sampled at a time interval equivalent to a distance of 57 m along-scan, and converted to digital form before being relayed to a ground receiving station. A distance of 57 m represents some oversampling as the instantaneous field of view of the system is equivalent to 79 m on the ground (section 2.2.1). The

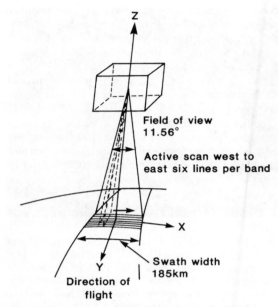

Figure 2.11   Landsat Multi-Spectral Scanner operation.

detectors are active only as the mirror scans in the forward direction. Satellite velocity is such that it moves forward by an amount equal to 6 × 79 m during the reverse scan cycle, so that an image is built up in each band in sets of six scan lines. The detectors deteriorate at different rates, so MSS images may show a phenomenon called sixth-line banding. Methods of removing this kind of 'striping' effect are discussed further in Chapter 4.

Landsat MSS images were collected routinely for a period of over 20 years, from the launch of ERTS-1 in July 1972 to November 1997, when the Australian ground receiving station acquired its last MSS image. Although somewhat old-fashioned by today's standards, the MSS performed well and exceeded its design life. The MSS archive provides a unique historical record, and can be used in studies of temporal change.

2.3.6.2   *Landsat Thematic Mapper (TM)*

The Thematic Mapper (TM) instrument is the primary imaging sensor carried by Landsat-4 and -5. Landsat-4 (launched 16 July 1982) was switched off in August 1993 after failure of the data downlinking system. Landsat-5 continues to operate, though only a direct data downlink facility is available. Like the MSS, TM uses a fixed set of detectors for each band and an oscillating mirror. TM has 16, rather than four, detectors per band (excluding the thermal infrared channel) and scans in both the forward and the reverse directions. It has seven, rather than four, channels covering the visible, near- and mid-infrared, and the thermal infrared, and has a spatial resolution of 30 m. The thermal channel uses four detectors and has a spatial resolution of 120 m. The data are quantised on to a 0–255 range. In terms of spectral and radiometric resolution, therefore, the TM design represents a considerable advance on that of the MSS.

The TM wavebands are as follows: channels 1–3 cover the visible spectrum (0.45–0.52 µm, 0.52–0.60 µm and 0.63–0.70 µm, representing visible blue-green, green and red); channel 4 has a wavelength range of 0.75–0.90 µm in the near-infrared; channels 5 and 7 cover the mid-infrared (1.55–1.75 µm and 2.08–2.35 µm); while channel 6 is the thermal infrared channel (10.4–12.5 µm). The rather disorderly channel numbering is the result of the late addition of the 2.08–2.35 µm band.

Data from the TM instruments carried by the Landsat-4 and -5 satellites are transmitted to a network of ground receiving stations. The European stations are located near Fucino, Italy, and Kiruna, Sweden. Data are also transmitted via the Tracking and Data Relay Satellites (TDRS), which are in geostationary orbits. The two satellites making up the TDRS system are in line of sight of Landsat-4 and -5 except for an area in East Asia which is partly covered by ground stations in India and Thailand. TDRS transmits the image data to a ground station at White Sands, New Mexico, from where it is relayed to the Space Imaging/EOSAT data processing facility at Norman, Oklahoma, using the US domestic communications satellite DOMSAT.

Examples of Landsat Thematic Mapper imagery are provided in Figures 1.8 and 1.9.

2.3.6.3 *Enhanced Thematic Mapper Plus (ETM +)*

The ill-fated Landsat-6, which was lost on launch in October 1993, was carrying a new version of the Landsat TM called the Enhanced Thematic Mapper (ETM). Landsat-7, which is scheduled for a late-1998 launch, will carry an improved version of ETM, called ETM +. The new instrument will extend the capabilities of the Landsat TM by providing a 15 m resolution panchromatic band, and the spatial resolution of the thermal infrared channel will be 60 m rather than 120 m. In addition, an on-board calibration system will allow accurate (±5%) radiometric calibration (section 4.6). Landsat-7 will have substantially the same operational characteristics as Landsat-4 and -5, namely, a Sun-synchronous orbit with an altitude of 705 km and an inclination of 98.2°, a swath width of 185 km, and an equatorial crossing time of 1000.

2.3.7 SPOT sensors

2.3.7.1 *SPOT High Resolution Visible (HRV)*

The SPOT (Satellite Pour l'Observation de la Terre) programme is funded by the governments of France, Belgium and Sweden and operated by the French Space Agency, CNES (Centre National d'Etudes Spatiales), which is located in Toulouse. SPOT-1 was launched on 22 February 1986. It carried an imaging sensor, the High Resolution Visible (HRV) system, which is capable of measuring upwelling radiance in three channels (0.50–0.59 µm, 0.61–0.68 µm and 0.79–0.89 µm) at a spatial resolution of 20 m, or in a single panchromatic channel (0.51–0.73 µm) at a spatial resolution of 10 m. All channels are quantised to a 0–255 scale. HRV does not use a scanning mirror like Landsat MSS and TM. Instead it uses a linear array of charge-coupled devices, or CCDs, so that all the pixels in an entire scan line are imaged at the same time. As it has no moving parts the CCD push-broom system might be expected to last longer than the electro-mechanical scanners carried by Landsats-1 to -5, though all of the Landsat sensors exceeded their design life by substantial amounts, and the TM instrument on Landsat-5 is still producing images after more than 14 years in orbit. A more important consideration is the fact that, since all of the pixel data along a scan line are collected simultaneously rather than sequentially, the individual CCD detector has a longer 'look' at the pixel area on the ground. This increased dwell time means that the signal is estimated more accurately and the image has a higher signal-to-noise ratio.

The SPOT orbit is near-polar and Sun-synchronous at an altitude of 832 km, an inclination of 98.7°, with a period of 101.5 minutes. The field of view of the HRV sensor is 4.13° and the resulting swath width is 60 km. Since two identical HRV instruments are carried, the total swath width is 177 km (there is a 3 km overlap when both sensors operate in nadir-viewing mode). The orbit repeat period is 26 days with an equatorial crossing time of 1030. However, the system can have a higher revisit capability because the HRV sensor is pointable. The sensor can be moved in steps of 0.6° to a maximum of ±27° away from nadir. This feature allows HRV to collect image data within a strip 475 km to either side of nadir. Apart from improving the sensor's revisit capability, the collection of oblique views of a given area (from the left and right sides respectively) provides the opportunity to generate digital elevation models (DEM) of the area, using photogrammetric methods. However, the use of non-nadir views introduces problems when physical values are to be derived from the quantised counts (Chapter 4).

SPOT-1 was retired at the end of 1990, following the successful launch of SPOT-2 in January 1990. SPOT-3 followed in September 1993. All three carry an identical instrument pack. Unfortunately, SPOT-3 was lost following a technical error and SPOT-1 and -2 have been temporarily brought out of retirement. SPOT-4 was successfully launched as this paragraph was being written – the morning of 24 March 1998 – from Kourou, in French Guiana. The HRV instrument on SPOT-4 has been extended to provide an additional 20 m resolution channel in the mid-infrared region (1.58–1.75 μm) and is now called the HRV-IR, which can be used in multi-spectral mode (X), or panchromatic mode (M), or in a combination of X and M modes. SPOT-4 carries an on-board tape-recorder, which provides the capacity to store 20 scenes for downloading to a ground station.

2.3.7.2   *Vegetation (VGT)*

SPOT-4 carries a new sensor, VGT, developed jointly by the European Commission, Belgium, Sweden, Italy and France. VGT operates in the same wavebands as HRV-IR except that the swath width is 2250 km, corresponding to a field of view of 101°, with a pixel size of 1 km. This is called the 'direct' or 'regional' mode. The 'recording' or 'world-wide' mode produces averaged data with a pixel size of around 4 km. It is intended that in this mode VGT will generate data sets for the Earth between 60°N and 40°S (Arnaud, 1994). The 14 daily orbits will ensure that regions of the Earth at latitudes greater than 35° will be imaged daily, whereas equatorial areas will be imaged every other day. The combination of the VGT and HRV-IR sensors on the same platform means that data sets of significantly different spatial resolutions will be obtained simultaneously.

VGT data will complement those produced by the NOAA AVHRR, which is widely used to generate global data sets (Townshend and Skole, 1994). VGT has the advantage – for land applications – of three spectral bands in the optical region plus one in the infrared part of the spectrum.

2.3.8   Other sensors

A number of sensors operating in the visible and infrared regions of the spectrum have been used experimentally over the past two decades. The best known of these is the Modular Optoelectronic Multispectral Stereo scanner (MOMS), developed by the German Space Agency. MOMS-01 was carried by the Space Shuttle in 1983, and a later version (MOMS-02) was carried on another Shuttle mission in 1993. MOMS-2P was also part of the payload of the Russian *Mir* Space Station in 1996. MOMS uses a linear array of CCD elements to form a push-broom scanner, an idea that was later used in the SPOT HRV. The system provides multispectral information in four visible and near-infrared bands, with a spatial resolution of 12.8 m, and a panchromatic band that operates in nadir-, forward- and backward-looking modes to collect along-track stereo information. The SPOT HRV (section 2.3.7.1) also collects stereo information using its pointable sensor, but the two images forming the stereo pair are collected from adjacent orbits, hence the images may be several days or even months apart in time. In technical terms, MOMS-02 is an advanced sensor capable of generating imagery that is suitable for photogrammetric as well as thematic applications (Bodechtel and Zilger, 1996).

Airborne systems provide the opportunity for researchers to become familiar with new instruments, which may be planned for future satellite missions. They also provide the means of collecting multispectral data whenever time constraints do not allow the use of satellite-borne scanners, for example to survey oil spills, floods or forest fires. The UK Natural Environment Research Council's (NERC) airborne remote-sensing facility operates a Daedalus 1268 Airborne Thematic Mapper (ATM). The spatial resolution of this instrument depends upon flying height, but is generally in the range 1–10 m. The instrument has 11 spectral bands, ranging from the visible to the thermal infrared. A CASI (Compact Airborne Spectrographic Imager) is also available. This is a *hyperspectral* instrument, which collects data in 288 spectral channels at 1.9 nm intervals in the spectral range 0.4–1.0 μm (Babey and Anger, 1989). Applications of CASI data in mapping intertidal surfaces are described by Thomson *et al.* (1998).

A significant problem with aircraft remote sensing is the instability of the aircraft platform. Attitude variations around the three axes of the aircraft (known as pitch, roll and yaw – see section 4.3.1.2 and Figure 2.12) as well as variations in the altitude of the aircraft combine to make the geometry of images acquired by such systems rather more complex than is the case for satellite images. The use of inertial navigation systems, giving precise readouts of the aircraft's attitude, and the Global Positioning System provide the means of correcting imagery for the effects of these distortions. A second problem with aircraft imagery results from their wide angular field of view, so that there is a substantial

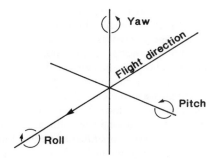

Figure 2.12   Pitch, roll and yaw.

view-angle difference between measurements made at nadir and measurements at the edges of the swath. Reflectance from most natural surfaces is not independent of view angle, thus observations of the same land cover type made at very different view angles will not produce the same radiance levels at the sensor. Additionally, if the field of view of the sensor is divided into N sectors each with the same angular range (i.e. divide the angular field of view by N) then sectors near the nadir point will cover a smaller ground area than those near the extremities of the scan line. These problems must be corrected before aircraft data are used to classify land cover (Chapter 8), or to compute vegetation ratios (section 6.2.4), or to generate any physically meaningful quantity for use in modelling.

The CASI instrument mentioned above is an example of an airborne hyperspectral imaging sensor. Another example is the AVIRIS (Airborne Visible/Infrared Imaging Spectrometer), developed by NASA in the late 1980s. AVIRIS has 224 spectral bands covering the region 0.4 to 2.45 μm. Each image is 614 pixels wide, and pixel size is dependent upon the aircraft's altitude; at a flying height of 20 km the pixel size is 20 m × 20 m. The advantage of imaging spectrometer data is that they are capable of providing very detailed spectral reflectance curves at each pixel position, thus allowing finer discrimination between subtly different Earth-surface cover types including minerals, vegetation, water, ice and snow (Goetz, 1984; Goetz *et al.*, 1985; Vane *et al.*, 1993). The disadvantage is the large amount of data that are generated by such sensors. AVIRIS is described in detail in Vane (1987). Kruse *et al.* (1993) describe a software system designed to handle imaging spectrometer data. An attempt to launch a satellite, LEWIS, carrying an imaging spectrometer with 384 spectral bands (of width 5–6.5 nm) covering the region 0.4 to 2.5 μm, failed in August 1997.

The MODIS (Moderate Resolution Imaging Spectrometer) is part of the US EOS (Earth Observing System) programme, and will be carried by the EOS-AM-01 satellite (morning overpass, to be launched in 1999) and its PM counterpart (afternoon overpass, launch in 2000). MODIS will collect data in 36 spectral bands between 0.4 and 14.5 μm and at a spatial resolution of 250 m, 500 m or 1 km. The 36 bands of MODIS were selected with the aim of providing global data that will be useful in the areas of surface temperature determination, ocean colour, land surface cover, and cloud properties, among others (Running *et al.*, 1994). Multiband spectrometers such as MODIS, AVIRIS and CASI provide the opportunity to target specific diagnostic wavebands that are relevant to a particular problem. Such data will be of increasing importance during the next decade, and new techniques of analysis and visualisation will be needed if reliable and consistent information is to be extracted from such data (Curran, 1994; Toselli and Bodechtel, 1992).

2.4   MICROWAVE IMAGING SENSORS

As noted in section 1.2.3, the microwave region of the electromagnetic spectrum includes radiation with wavelengths longer than 1 mm. Solar irradiance in this region is negligible, although the Earth itself emits some microwave radiation. Imaging microwave instruments do not, however, rely on the detection of solar or terrestrial emissions. Instead, they generate their own source of energy, and are thus examples of *active* sensing devices. In this section, the properties of the operational synthetic aperture radar (SAR) systems carried by the ERS-1/2, JERS-1 and RADARSAT satellites and by the SIR-C system carried by the Space Shuttle in 1994 are presented. General details of imaging radar systems are described in section 1.2.3.

An important property of imaging radar is the instrument's *depression angle*, which is the angle between the direction of observation and the horizontal (Figure 2.13). The angle between the direction of observation and the surface normal (a line at right angles to the Earth's surface) is the *incidence angle*, which is also shown on Figure 2.13. Generally, backscatter increases as the incidence angle decreases. Different depression angles are suited to different tasks, for example ocean and ice monitoring uses lower depression angles than land monitoring. A second important property is the *polarisation* characteristics of the instrument. Polarisation is the orientation of a waveform. If the transmitted waveform is parallel to the Earth's surface (i.e. oscillating from side to side from the point of view of an

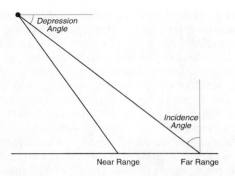

Figure 2.13 Showing radar depression and incidence angles. See text for explanation.

observer on the ground) it is said to be horizontally polarised. If it is oscillating in a plane perpendicular to the Earth's surface ('up and down') then it is vertically polarised. The receiving antenna can be tuned so as to receive the return signal in a particular polarisation, so that imaging radars can be described in terms of the transmit and receive polarisations. Thus HV means 'transmitted polarisation is horizontal, received polarisation is vertical'. Four orthogonal combinations are possible: HH, HV, VV and VH. If sufficient information is provided it is possible to calculate any combination of transmit and receive polarisation from these four pairs, and some characteristics of a target can be more accurately reconstructed from its 'polarisation signature' than from single polarisation (transmit/receive) data. The SIR-C SAR carried by the Space Shuttle in 1994 provided polarimetric radar images in the C and L bands, plus an X-band image (Figure 2.18). Freeman *et al.* (1994) discuss the use of multifrequency and polarimetric radar for the identification and classification of agricultural crops (the topic of classification is treated in Chapter 8). The view direction of a radar sensor is also important in detecting geological features with particular orientations, so that radar images of the same terrain with different 'look angles' will show different features (Blom, 1988; Gauthier *et al.*, 1998; Koch and Mather, 1997; Lowman, 1994).

Radar images show the spatial distribution of the magnitude or amplitude of the radar backscatter. In standard form, a radar image is an average of a number of magnitude images, and the speckle noise magnitude is related to the number of averages (number of looks) used to generate the image. A radar may also receive phase information, representing the difference in travel time from the antenna to the target and back to the antenna for two signals, for example HH and VV

polarised. Phase differences are related to the characteristics of the target. Rao and Rao (1995) show how polarisation phase differences can be used in the discrimination of land cover types. A good survey of synthetic aperture radar principles and applications is contained in the EOS Instrument Panel report published by NASA (1988). See also section 1.2.3 for an introduction to imaging radar principles, and section 4.6 for a brief summary of calibration issues. Other basic sources are Leberl (1990), Oliver and Quegan (1998), Ulaby *et al.* (1981–1986) and Ulaby and Elachi (1990). Kingsley and Quegan (1992) and Schrieir (1993a) provide readable introductory accounts.

2.4.1 ERS SAR

The first space-borne imaging radar was carried by the US Seasat satellite, launched in 1978. It carried an L-band SAR, but operated for only 100 days. Thirteen years later, on 17 July 1991, the first European Remote Sensing Satellite, ERS-1, was launched into an 800 km near-polar orbit. ERS-1 carries the ATSR (see section 2.3.1), which is a low-resolution optical/infrared sensor, and a synthetic aperture radar. A second ERS satellite (ERS-2) was launched in 1994, providing opportunities for 'tandem mode' operation as the two satellites orbited one behind the other to generate SAR images of the same area over a period of a few days. Such data sets are used in interferometric studies (Massonet, 1993; Gens and van Genderen, 1996) to derive digital elevation models and to make measurements of small movements on the Earth's surface.

The ERS SAR operates at a wavelength of around 6 cm in the C band (see Table 1.3) and images an area 100 km wide to the right-hand side of the satellite track (facing the direction of motion). Microwave energy pulses are both transmitted and received in vertical polarisation mode. As noted in section 1.2.3, radar is a side-looking sensor, transmitting pulses of electromagnetic energy in a direction to one side of the spacecraft. The *incidence angle* of this beam in the case of the ERS radar is 20° for near range and 26° at the far range (Figure 2.13). Lillesand and Keifer (1994, p. 670; see also NASA, 1988, p. 113) note that topographic slope effects are greater than radar backscatter effects where the local incidence angle is less than 30°, whereas for incidence angles greater than 70° topographic shadow is dominant. Between these two limits, surface roughness effects are predominant. The ERS SAR depression angle facilitates ocean and ice-sheet observation and is also good for observing some terrestrial features such as agricultural fields.

As microwave energy is transmitted from the SAR antenna it sweeps across the swath from the near to the far range. This is called the range direction. The forward motion of the satellite moves the position of the scanned area in what is called the azimuth direction (the direction of satellite movement). The derivation of SAR images from the data collected by the antenna is very complex and beyond the scope of this book. Suffice it to say that ERS SAR images have a spatial resolution of 25 m, and these are averages of a number of higher-resolution 'looks'. Averaging is performed in order to reduce speckle noise, but results in a lower-resolution image. The variance of the speckle is related to the number of looks used in image generation.

As noted earlier, the advantage of microwave radar is that it is an all-weather sensor operating independently of solar illumination. Figure 2.14 shows an ERS SAR image of the south-east coast of England from Kent to Suffolk. Areas of low radar backscatter are dark. The brighter areas are those with a high surface roughness. Similarly bright areas or points over the land are features that exhibit sharp angles, such as the walls of buildings, walls around fields, or steep slopes. Another ERS SAR image is shown in Figure 2.15. This image was

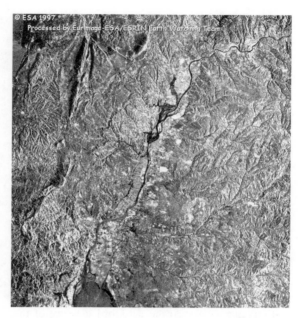

Figure 2.15 ERS-2 SAR image of the Tagus River, Portugal, acquired 7 November 1997. The Tagus valley was subjected to serious flooding as a result of heavy rain during the period 6–8 November 1997. © 1997 European Space Agency.

acquired by ERS-2 in November 1997, and shows an area of Portugal. Relief and geomorphological features are particularly clear. The River Tagus can be seen as it flows down the centre of the imaged area.

2.4.2 JERS SAR

The Japanese (NASDA) ERS satellite, JERS-1, which was operational from 1992 to 1998, also carries a synthetic aperture radar. This SAR has characteristics that are quite different from those of the ERS SAR. First, it operates in the L band (Table 1.3) at a wavelength of 24 cm and its signal is transmitted and received with horizontal (H) polarisation. The orbit of JERS is lower than that of ERS-1 and -2, at 568 km, and its near and far range incidence angles are 32° and 38° respectively. Full-resolution images have a ground resolution of 20 m with four looks. The JERS SAR is therefore more suited to land observations. Figures 2.16 and 2.17 show two images collected by the Japanese JERS SAR during the Global Rain Forest Mapping (GRFM) project, which was begun in 1995. These data sets have been averaged using 250 looks and have a spatial resolution of 100 m, hence speckle noise is reduced considerably. The image in Figure 2.16 shows the upper Amazon region, near the

Figure 2.14 ERS-1 SAR image of the coast of south-east England. Dark areas over the Thames Estuary and the North Sea are regions of calm water. Image acquired on 7 August 1991. © 1991 European Space Agency.

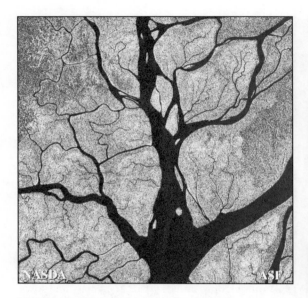

Figure 2.16   JERS-1 SAR image of the upper Amazon, close to the Brazil/Peru border. Collected as part of the Global Rain Forest Mapping Project. © NASDA/MITI. Processing by NASA's Alaska SAR Facility and Jet Propulsion Laboratory.

Brazil/Peru border. The mouth of the Amazon is shown in Figure 2.17. The potential of SAR to provide detailed images of cloudy areas is demonstrated by these images, which, together with many others, are available via the Internet from the Amazon Image Browser, which is maintained by the NASA Jet Propulsion Laboratory (JPL) at

> http://southport.jpl.nasa.gov/amazon/
> imagebrowser/

A useful summary of the relevance of JERS-1 SAR data in geological applications is provided by Moon *et al.* (1994). These authors note that the longer (L-band) wavelength of the JERS-1 SAR makes it useful for geological applications in areas of considerable vegetation, because of the greater penetration at these wavelengths.

2.4.3   RADARSAT

RADARSAT provides a third, different, source of SAR imagery. The RADARSAT programme is funded through the Canadian Space Agency, with NASA and NOAA cooperation. RADARSAT is in a Sun-synchronous orbit at an altitude of 798 km, at an inclination of 98.6° to the equatorial plane. The orbit is such that the satellite should always be in sunlight, and hence

Figure 2.17   JERS-1 SAR image of the mouth of the Amazon. © NASDA/MITI. Processing by NASA's Alaska SAR Facility and Jet Propulsion Laboratory.

power from the solar arrays is continuously available. The SAR sensor carried by RADARSAT can steer its beam so as to collect images at incidence angles in the range 20–50°, with swath widths of between 35 and 500 km, using resolutions ranging from 10 to 100 m. The radar operates in the C band and is HH polarised. RADARSAT is able to collect data in a number of modes, giving fine to coarse resolution and variable inclination angles. Figure 2.18 is an image of south-central Alaska in low-resolution (ScanSAR) mode. The data are recorded on-board the satellite and downlinked to ground receiving stations, of which there are three in North America, three in South America, one in South Africa, two in Australia and seven in Asia.

2.4.4   SIR-C/X-SAR

The Shuttle Imaging Radar (SIR) programme began in 1981 with the SIR-A experiment, which was followed in

Figure 2.18 RADARSAT image of south-central Alaska using low-resolution (ScanSAR) mode, covering an area of 460 km × 512 km, showing Kodiak Island in the south-west (lower left of the image), the Kenai Peninsula and Prince William Sound. This is a reduced-resolution, unqualified and uncalibrated image. A ScanSAR image covers approximately 24 times the area of a standard RADARSAT SAR image. © 1996 Canadian Space Agency/Agence spatiale canadienne. Received by the Alaska SAR Facility, distributed under licence by RADARSAT International.

October 1984 by SIR-B. The third, planned for 1989, was not flown until April 1994. The experiment was repeated in September/October 1994. This was the SIR-C/X-SAR experiment and it is briefly described here because it is the first multiband radar (L, C and X band) flown in space (Evans *et al.*, 1997). The C and L bands operated in polarimetric mode. The L-, C- and X-band images collected by this instrument can be combined together in much the same way as the images produced in the different bands of the Landsat TM or the SPOT HRV. A polarimetric radar transmits in both horizontal and vertical polarisations for each waveband (L and C in the case of SIR-C/X-SAR), and provides images of the magnitude of radar backscatter in HH, HV, VH and VV combinations. These four combinations provide the information required to compute the complete scattering matrix. SIR-C/X-SAR therefore represents a significant development in radar technology.

The orbital altitude of the Space Shuttle during the SIR-C/X-SAR experiment was 225 km. The antenna was adjustable so that swaths of width 15–90 km (C and L band) and 15–40 km (X band) could be derived, at a ground resolution of around 30 m. The antenna depression angle (Figure 2.13) could be selected within the range 73° to 27°. The SIR-C design is expected to be incorporated into operational systems currently being planned for the next decade, such as LightSAR.

2.5 SUMMARY

A variety of information is available from sensors carried on-board satellite and aircraft platforms. Those operating in the visible and near-infrared region of the electromagnetic spectrum provide measurements that correlate with an object's colour, which is often related to the chemical or mineralogical properties of that object. Data from thermal infrared sensors are related to the temperature and thermal properties of a target, while information about surface roughness and (over the land) moisture content can be derived from data collected in the microwave (radar) wavelengths. The spectral range of remotely-sensed data available in the late 1990s from orbiting satellites covers the full range from optical to microwave. Given the number of different remote-sensing programmes around the world, it is rather difficult to keep up to date and surveys such as the one provided in this chapter are only partial, and soon become outdated. Surveys such as Kramer (1994) provide a historical reference, but the only way to keep yourself informed is to join a remote-sensing society and read its newsletter.

Satellite remote-sensing programmes that are presently operational include Landsat, SPOT, NOAA, JERS, ERS, RADARSAT, IRS, Meteosat and other geostationary meteorological satellite systems. By the year 2000, the US Earth Observing System (EOS), the Japanese ADEOS and the European Space Agency's ENVISAT systems should be operational, together with small satellite programmes, which promise panchromatic and multispectral optical data with a resolution of 1–4 m, plus SAR systems with a resolution as high as 3 m. In addition, hyperspectral instruments, capable of acquiring imagery in a large number of channels (typically 256 or 512) have been proposed. Potential technical problems include the development of methods of combining multi-source (multi-sensor) data, handling large volumes of high-resolution data, and selecting the optimum combination of bands to use for a particular application. Clearly, there will be a greater need than ever in the future for research and development in the areas covered by later chapters of this book.

2.6   QUESTIONS

1. Describe the features of the following orbital patterns: geostationary, Sun-synchronous, polar. What is the *inclination angle* of a polar orbit?
2. Give definitions of the following terms: swath width, CCD, push-broom scanner, geometric instantaneous field of view (IFOV), point spread function (PSF), effective resolution element (ERE), polarimetric radar, radar depression angle, signal-to-noise ratio (SNR), speckle.
3. What is meant by 'the spatial resolution of an imaging sensor'? How is this property defined for optical and microwave sensors? Why is it important?
4. Explain the factors that influence (and restrict) the choice of wavebands for the Thematic Mapper sensor. Why does the AVHRR use a different set of wavebands?
5. What is the difference between an 'active' and a 'passive' imaging sensor?
6. List the pros and cons of optical and microwave imaging sensors, giving examples of applications to support your argument.

3

Digital Computers and Image Processing

3.1 INTRODUCTION

During the 10 years or so that have elapsed since the first edition of this book was published there has been a revolution in the world of computers. Desktop machines have become available, offering computing power several orders of magnitude greater than that available in the mid-1980s, while the unit cost of computing has fallen dramatically. Nowadays, an off-the-shelf desktop PC can be used to process and display images at speeds which outstrip anything achievable by a state-of-the-art minicomputer of the mid-1980s, and at a far lower cost. When the first edition of this book was published in 1987 an image processing system (basically video memory plus a monitor) interfaced to a mid-range minicomputer such as a DEC VAX/11-750 (4 Mb RAM, 250 Mb disk) cost around £100 000. Nowadays, the cost of a PC with 64 Mb RAM, a 4 Gb disk and a 17 inch monitor is around £1200. Digital image processing has never been more accessible.

This chapter provides an introduction to the world of computers for those interested in digital image processing. It is not intended to act as a programmer's guide, nor does it explore the more arcane aspects of computer science. A chapter on computers is included because software for digital image processing can only be used sensibly if the principles of operation of the whole computer system are understood. The level of this chapter is such that it should be intelligible to readers with little or no previous experience of computers. Many users of digital image processing systems will not be involved in programming, but all users should be familiar with the way in which digital data are stored and manipulated within an image processing computer system. Programming abilities are essential at research level. The ability to write programs is an important skill that allows researchers to explore ideas rather than have their research agenda dictated by the capabilities of a 'black box' image processing software package. Some of these

packages allow the user to interface his or her own code. This useful compromise reduces the tedium of programming as the input/output and video display elements of the package can be used. Other packages have high-level interfaces that provide the user with access to procedures that perform much of the computation.

Modern computers are conventionally classified into one of two kinds: PCs (personal computers, including IBM-compatible and Apple Macintosh) and workstations. This distinction was originally made based on computing power and cost but, over the past few years, the power of high-end PCs has approached, if not surpassed, that of the low-end workstation. With the introduction of new operating systems for PCs such as Microsoft Windows 95, IBM OS/2 and varieties of UNIX such as Linux, the capabilities of the PC are now similar to those of the workstation. The Microsoft Windows NT operating system runs on both PCs and workstations, and effectively bridges the gap between the two. Hence, we will not be too concerned in this chapter by the distinction between PC and workstation, and will focus instead on the more generic characteristics of computers, beginning in section 3.2 with an introduction to basic concepts, including the hardware components of the computer and proceeding to software aspects. The newly developing field of computer networks is described in section 3.2.2. In section 3.3 more detailed consideration is given to image display functions, which are of central importance to users of remote sensing. Sections 3.4 and 3.5 review data formats for remotely-sensed images, together with the related subject of image data compression. An overview of the MIPS software (provided on the CD accompanying this book) is provided in section 3.6. MIPS is 'educational' software in that it has been written to allow students to learn how digital image processing works from practical experience rather than from a textbook. It is not intended to replace or compete with commercial systems for research purposes or advanced uses, though students moving on to conduct their own project work will find that their experience of using MIPS will make the task of familiarising themselves with commercial software rather less daunting. Learning by doing rather than by reading a textbook is the only way to develop expertise in image processing. The remaining chapters of this book assume a familiarity with MIPS, and readers are encouraged to try out for themselves the different operations and procedures that are described in the text.

3.2   COMPUTER BASICS

3.2.1   Hardware and software

A computer is a device that stores and manipulates data in accordance with a list of rules or instructions called a *computer program* (the American spelling is standard, even in Great Britain). The physical computer and its components are collectively called *hardware* while the logical instructions that form computer programs are known as *software*.

3.2.1.1   Hardware

A typical computer used in image processing consists of a box containing the internal components of the machine, together with a monitor, a keyboard, a mouse and perhaps a printer. The box holds the *central processing unit* or CPU, the *hard disk* (again the American spelling is common), which provides permanent storage for data and programs, the *random-access memory* (RAM), which stores data and programs when the computer is switched on and working, plus *controllers* for the various devices that are attached to the computer, such as the disk drives, the monitor and the printer. The parts of the computer outside the main box are called *peripherals*. Other peripherals may include a tape subsystem (to read and write DAT, Exabyte or industry-compatible $10\frac{1}{2}$-inch (26.67 mm) tapes) and sound systems. The system may also include a network card that allows the computer to link to a local-area or wide-area network. These components are represented diagrammatically in Figure 3.1.

The CPU is a silicon chip that interprets and executes instructions provided by the software (section 3.2.1.2). These instructions may request one of a number of operations, ranging from arithmetic and logical operations to input–output procedures that involve data transfers. These data transfers may be internal (within the random-access memory), or between the random-access memory and the external (disk) memory, or be-

tween the random-access memory and the video memory. The random-access memory, disk memory and video memory are described below. The rate at which these instructions can be executed is related to the 'processor speed', measured in megahertz (MHz). In section 3.2.1.2 we see that software instructions are provided to the CPU in terms of high/low voltage patterns, corresponding to 1/0 (binary) sequences. Each instruction requires a string of at least eight ones or zeros. Each of these ones and zeros is known as a binary digit or *bit*, and a 'processor speed' of 200 MHz means that 200 million bits can be processed per second. Most IBM-compatible PCs use Intel Pentium processors. Apple Macintosh machines use Motorola 'chips', and workstations use proprietary CPUs – for example, Digital Equipment Corporation (DEC) manufactures its own 'Alpha' processors.

Processor clock speed is only one guide to the suitability of a computer for use in image processing, though it is safe to say that the faster the processor the more quickly it will be able to complete image processing operations. There are other considerations, though. Chief among these is the rate at which data can be transferred between the CPU, the random-access memory, and the disk, tape and video subsystems. Early machines transferred information in units of eight bits. Later machines had the capability to transfer information in units of 16 bits, and so data transfer rates were twice as high. A modern PC capable of running Windows 95 or NT uses a 32-bit 'data bus', and some workstations use 64-bit data buses. Clearly, image processing involves moving a considerable amount of data around, from the disk system to the computer's random-access memory, and then to the video system. The data rates achievable by the connections between these basic components of the computer are as important as the processor's clock speed. The apparent rate of data transfer can be increased if the processor is equipped with a small amount of random-access memory that can be accessed very rapidly. This memory is called a *cache* (pronounced like 'cash'). Cache memory can be kept topped up so that there is a high probability that the piece of data that is required by the CPU is in the cache and does not need to be brought from the disk.

The *physical memory* of the computer is analogous to an electronic matrix made up of a large number of cells, each capable of storing a single data item. Each piece of data held in the memory can be accessed independently and equally quickly. Because of this property the memory is called a *random-access memory* or RAM. Its capacity is measured by the number of data items it can hold. The size of the standard data item (a cell of the

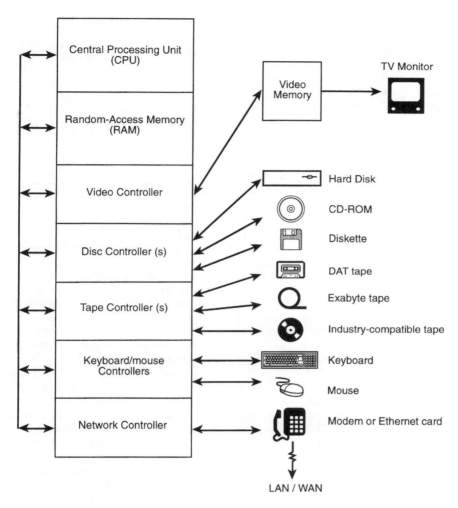

Figure 3.1   Components of a digital computer. The central processing unit (CPU), random-access memory (RAM), and the controllers for the peripheral devices shown on the right-hand side are connected by a high-speed data and address bus, shown by the vertical line with double-headed arrows indicating data flow direction. The different peripheral devices connect the system to the external world. The video controller may have its own memory store (the video memory) which contains one or more bytes per screen pixel. For natural or false-colour display, 24 bits (three bytes) per pixel are required to hold 256 levels of each of the three primary colours (red, green and blue). See Figure 3.2.

matrix) is eight bits. This standard storage unit is called a *byte*. All data items within the RAM (as well as on disk or in the video memory) are stored in units of one or more bytes. It is very easy to use up the available physical memory (usually of the order of 16–64 Mb on a PC) when processing images. Modern operating systems, such as Windows 95, Windows NT, UNIX, VMS and OS/2, use *virtual memory* to augment their physical RAM, but the penalty is reduced speed of access to data. Virtual memory is simply a disk file, and data are swap-ped to and from physical memory as required. That is why it is important to ensure that sufficient physical memory is available in any image processing system. This factor is more important, in many instances, than the speed of the processor, especially as many commercial software systems assume unlimited memory.

A PC used for the analysis and display of remotely-sensed data will be built around a high-speed CPU (at the time of writing the fastest Pentium II processor has a clock speed of 400 MHz), and a RAM capable of storing at least 16 Mb of data. A dedicated remote-sensing workstation should have 128–512 Mb of RAM. The term Mb stands for 'megabyte'; one megabyte is equal to 2^{20} bytes. One megabyte is also equal to 1024 (2^{10}) kilobytes (kb) and one kilobyte is 1024 (2^{10}) bytes. A gigabyte (Gb) is 1024 (2^{10}) megabytes.

Each byte within the memory can store eight *bits*. 'Bit' is an abbreviation for 'binary digit'. The binary digits are 0 and 1. Our everyday number system uses base 10 rather than base 2, and is therefore called the decimal system. The decimal digits are 0, 1, 2, ..., 9. The base 10 number system is positional, and thus we are used to thinking of numbers in terms of 'hundreds, tens and units'. One hundred is equivalent to 10^2, one ten is 10^1 and a unit is 10^0. The units are the rightmost digits in the decimal number, the tens are second from the right and the hundreds are third from the right – that is what is meant by the term 'positional number system'. Hence, the '3' in '369' means 'three hundreds'.

The base 2 number system is also positional, with the digits (counting the rightmost digit as position zero) representing increasing powers of 2. In other words, the base 2 equivalent of hundreds, tens and units is 2^2, 2^1 and 2^0, that is, 4, 2 and 1 respectively in decimal notation. Since the binary digits are 0 and 1, a base 2 number can be made up only of '1's and '0's. The base 2 number 101 means (working from right to left): 1×2^0 plus 0×2^1 plus 1×2^2, giving a value (in base 10 notation) of 5. Table 3.1 gives a worked example of the conversion of an 8-bit base 2 number into its base 10 equivalent.

The importance of a clear understanding of base 2 number representation will become clear when the nature of digital images is considered. The individual elements of a digital image (called *pixels*) are represented within the computer's memory as strings of base 2 digits. For example, a Landsat TM or SPOT pixel is represented by eight bits (one byte), which means that its pixel values range from 00000000 to 11111111 (0 to 255 in base 10 notation). The *video memory* of the computer is also composed of storage units that typically use one byte to represent the dynamic range of each of the three primary colours (red, green and blue). A typical video memory will represent the range of red, for example, in terms of 256 levels from 00000000 (absence of red) to 11111111 (maximum possible intensity of red). Video memory is considered in more detail in section 3.3.

Numerical data are held in the computer as one of two types, integer or real. *Integers* are counting numbers (...−2, −1, 0, 1, 2, ...) and are expressed by direct conversion from base 10 to base 2 format, as shown in Table 3.1. The integer value 255 is converted to base 2 as 11111111 and stored in one byte of memory. Some

Table 3.1 Conversion from base 2 to base 10 number representation. The base 2 number 11010111 contains eight binary digits (bits), shown in the first row of the table. Each bit is multiplied by the power of 2 that corresponds to its position in the chain of bits, with the rightmost bit labelled '0' and the leftmost bit labelled '7'. These positions (powers of 2) are shown in the second row of the table. The third row gives the value of the corresponding power of 2, for example, $2^0 = 1$, $2^1 = 2$, $2^2 = 4$, and so on. Finally, the '1's in row 1 are multiplied by the value of the associated power of 2 (row 3) and the result is shown in row 4. The base 10 equivalent of the base 2 number 11010111 is the sum of the fourth row of the table, i.e. $128 + 64 + 16 + 4 + 2 + 1$, which equals 215.

Base 2 number	1	1	0	1	0	1	1	1
Power of 2	7	6	5	4	3	2	1	0
Value of power of 2	128	64	32	16	8	4	2	1
Row 1 times row 3	128	64	0	16	0	4	2	1

remotely-sensed data are measured on a scale of 0–1023 and the storage of each measurement requires 10 bits. Most computer systems allow integer numbers to be stored as 8-, 16- or 32-bit quantities.

A second kind of number, termed *real* or *floating-point*, is used for measuring and thus has a fractional part. For instance, the x and y coordinates of a point on a map might be written as 25.963 and 82.190 km east and north, respectively, of the origin of the map grid. Real numbers are stored by the computer using scientific notation of the form $a2^b$ in the same way that the base-10 real number 103.4 might be expressed in the form 0.1034×10^3. Real numbers require 32 or 64 bits for their representation, depending on the precision used. Some operations in image processing such as principal components analysis generate real numbers, which are then scaled to a 0–255 scale for display. Modern image processing software can therefore accept image data that are expressed in the form of 8-, 16- or 32-bit integers or as 32-bit (four-byte) real numbers. Some forms of radar imagery may be held in *complex* form, which can be thought of as two real numbers side by side in a storage unit occupying eight bytes. The first four bytes are the real component and the second four bytes are the complex component of the number.

Character data (which are used to represent Western alphabetic characters such as a, b, c together with characters from other alphabets and symbols such as α, β, χ and $\pm$) are also translated into an 8-bit (one-byte) binary code. Most computers use the ASCII code for this purpose. This mnemonic stands for *American Standard Code for Information Interchange*. In ASCII code the character 'A' is represented by the number 65 (01000000 in base 2), while the dollar sign ($) is stored as the number 36 (00100100). The computer does not 'know' whether a particular set of binary digits represents an integer, a real number, an ASCII-coded character or a program instruction, all of which are stored in the computer's memory as base 2 numbers. It is the responsibility of the programmer to ensure that the different kinds of data do not become confused.

Data and programs are organised in *files*. A file is a collection of related bytes, just as a book is a collection of related words. It may be a *text file* which contains data coded as ASCII characters (and can therefore be listed on a screen or printer and read by a human) or a *binary file* in which the information is stored directly in base 2 format. Binary files can contain data, programs, images, or word processor and spreadsheet documents and worksheets. Each file has a name. The MS-DOS and Windows 3.1 operating systems restrict the filename to eight characters or less, but UNIX and Windows 95/NT operating systems allow filenames up to 32 characters long (note that UNIX filenames are case-sensitive, so that `Fred` and `fred` are the names of two different files). Each filename can have an extension (restricted to three letters by MS-DOS and Windows 3.1) and this extension is used to indicate the type of the file. For example, the file extension `.inf` denotes MIPS Image files, and `.txt` indicates a text file.

Files are stored in directories (which also have names). Each directory contains a related set of files, and can be thought of as a section of a library containing books on a given subject. The directory structure used by modern operating systems is hierarchical, with the top directory being the *root directory*. For example, the files on the hard disk of my PC are organised into two subdirectories called *system* and *user*, both of which reside in the main (root) directory at the top of the hierarchy. The *system* directory contains all of the program and other files that make up the operating system as well as the compiler, spreadsheet, word processing, electronic mail and other program and ancillary files that are provided by the vendors of these software packages. The system directory is arranged so that each separate set of files (relating, for example, to my word processor) are kept in separate directories, each of which may have subdirectories. The *user* branch of the directory tree contains my personal files (my own e-mail, my programs, my lectures, letters, accounts and so on). The directory and subdirectory names give the *path* to a file. For instance, the text of this chapter is stored in a file called `chapter3.doc`. Its pathname is `C:\user\book\1987\1997\chap3`. Each subdirectory name is separated from other directory/subdirectory names by a backslash (\) character. UNIX

systems use a similar scheme for identifying files, except that a forward slash (/) is the separator. A UNIX pathname is something like `/usr/user/remote-sensing/mather/programs`.

Files are stored on disk or tape and brought into RAM when required. Disk files can be accessed randomly (that is, the position of the file on the disk does not influence the access time) whereas tape is a sequential device. This means that if the fourth file on a tape is required then the first three files have to be read or skipped before the fourth file can be read. Disk files are preferred for real-time image processing because the speed of access to data on disk is considerably greater than for tape files. Tape is used either for storing large amounts of data, such as remotely-sensed image sets, for delivery to a user via the postal service or for archiving and back-up purposes. Digital Audio Tape (DAT) and Exabyte tapes can hold between four and eight gigabytes of data and are widely used for data archiving and back-up. Most computer users back up the contents of their PC disk system on to tape every few days so that they can recover their work in case of a disk failure. UNIX systems are often shared by a number of users, one of whom is delegated to perform the daily/weekly back-up.

Other hardware that is used in remote-sensing image processing operations is considered separately in specific sections of this chapter. Networks are reviewed in section 3.2.2, video display systems are considered in section 3.3.1, and hard-copy systems used to generate paper or film copies of images are discussed briefly in section 3.3.2.

3.2.1.2 *Software*

Software refers to the instructions that drive a computer. There are two basic kinds of software, called *operating systems software* and *applications software*. Operating systems software provides the basic functionality of the computer, that is, the services that all users expect, such as the ability to list the files in a given directory, move files from one directory (or device) to another, and run applications programs.

Applications programs are those which perform operations of interest to a user (for example, word processing, Internet browsing, or image processing). They are generally written in *programming languages* such as Fortran and C, which have their own syntax and rules. The range of operations that can be executed by a computer is surprisingly limited, being restricted to data transfer, arithmetic and logical comparison. Complex but efficient programs can, nevertheless, be built upon such foundations, provided that the user is aware of the characteristics of the machine.

The operations of data transfer, addition, subtraction and logical comparison are carried out by hardware circuits which provide the basis for the *instruction set* of the computer, which (as far as the programmer is concerned) consists of a number of software procedures, such as 'add two numbers', which can be used in programs. Electronic circuits to carry out these basic operations are built into the CPU of the machine, and these circuits are activated by the software instructions that are brought into the CPU. The instructions form part of a program. Any such program may look readable and almost English-like when printed or displayed on the screen but, before the computer attempts to interpret these instructions, they are translated from the original programming language into a binary code, the workings of which are explained above.

Each element in the instruction set of the computer has a binary code, and a program written in C or Fortran has to be converted ('compiled') into a list of these binary codes before the CPU can interpret, and act upon, the program instructions. A list of binary codes corresponding to a set of program instructions is called a *machine-code* program. Whenever the CPU is presented with a machine-code instruction it performs the corresponding operation, assuming that the instruction is logically possible (for example, the instruction 'divide 10 by 0' is not one that the CPU can execute). The binary machine-code instruction is transmitted to the CPU as a series of positive and negative voltages corresponding to the '1's and '0's of the base-2 representation of the code, and a pulse of high–low voltages triggers off the desired operation, such as addition.

Applications programs draw upon the resources provided by the operating system. Reading and writing data from and to the disk, for example, is carried out by the operating system in order to avoid conflicts between different applications programs. For example, the word processor may be doing an auto-save of your document while you have switched to an image processing application that is writing image data to a disk file. The Windows operating system provides a full range of graphics and imaging operations that the programmer can use to manipulate and display data on the screen. This eliminates the need for every PC manufacturer and compiler writer to include specific software for this purpose. The programs contained on the MIPS CD that accompanies this book were written in Fortran using a compiler developed by Salford Software Ltd. To create two buttons on the screen and determine which one the user has clicked, I simply use the instruction:

$$i = \text{winio}@(`\%bt[THIS\ ONE]\%ta\%bt[THAT\ ONE]')$$

If the user clicks on the button (%bt) THIS ONE then i takes the value 1; if the user clicks on the button labelled THAT ONE then i takes the value 2. The authors of the Salford Fortran compiler and of the Microsoft Windows 95 operating system have included code which will generate the button icons on the screen and return the value 1 or 2, depending on the button that is pressed by the user (with suitable precautions, of course, for those users who will inevitably click the 'x' at the top right corner of the window!).

UNIX systems generally have no built-in 'windowing environment' as UNIX is developed from a text-based system. A third-party program developed at the Massachusetts Institute of Technology (MIT) and called the *X Windows System* (or simply *X*) is widely used and is the *de facto* UNIX windowing system. Most commercial software packages use *X* as their windowing environment. One drawback is the amount of computation required by the *X* Windows System, which makes its use rather slow across a network.

Applications programs are either *stand-alone* or part of a software package. Stand-alone programs (such as those described in Appendix B) are run separately and independently and perform a specific operation, such as principal components analysis or geometric correction. A software package provides an integrated suite of programs that share common functions. The MIPS display program (Appendix A) allows the user to select from various image processing operations and view the result. A recent development is the integration of remote-sensing image processing methods and GIS procedures, so that data of different kinds can be used in a synergistic fashion. In Chapter 4, for example, methods of correcting or rectifying the image coordinate system to fit a given map projection and scale are described. This operation requires access to a map of the area covered by the image and the digitising of selected control points. Chapter 8 reviews techniques of image classification and mentions methods that take account of the characteristics of the image area that can be derived from maps and other data (for example, soil type and land surface elevation). Clearly, if operations such as control point digitising and image/map data combination can be performed within a single system then the user's task is made considerably easier. Some software packages do not offer complete integration of the type mentioned above but instead rely on the use of common file formats, so that an image file generated by an image processing system can be transferred to a GIS software package for further analysis, and vice versa. The subject of file formats is reviewed in section 3.4.

3.2.2 Networks

Both workstations and PCs can be *networked*, that is, linked by cables to other workstations and PCs in a *local-area network* (LAN), which allows the sharing of hard disks and other peripherals such as tape readers. A typical LAN such as the one in the University of Nottingham School of Geography consists of 10–12 workstations and 60–70 PCs. Each workstation has a name – we have given ours the names of famous geographers and explorers such as *Mercator*. One of the workstations is the master machine (paradoxically called the *server*) sharing more than 20 Gb of disk storage organised into one or more filestores, a CD-ROM (compact disk read-only memory), and a tape system for data input, transfer and back-up (i.e. saving the contents of the disks in case of failure), as well as printers, scanners, digitisers and other peripheral devices. The LAN allows any workstation/PC to access any of the disks and other peripherals. The PCs that are linked into the network are less completely integrated into the LAN than are the workstations, and generally are used as terminals that can 'log in' to the network server and behave as a workstation, except that all of the computations are performed on one of the network machines with the PC simply acting as a display terminal and keyboard unit.

The School of Geography LAN is part of the University campus network, so our networked workstations can use facilities on computers in the central computing centre and elsewhere. This is useful when several departments on campus wish to share a software package. The package can be mounted on a central machine and accessed by networked workstations in any of the departments. The access speed depends, of course, on the network capacity and the amount of network traffic.

Developments in networking also allow access to remote computers (that is, computers that are not part of the local-area network, such as the University of Nottingham's campus network) and the transfer of data over long distances using a *wide-area network* (WAN). Where *real-time* image data are required, for example in weather forecasting, the time involved in writing an image to a magnetic tape or CD-ROM and physically transporting it from one place to another is prohibitive and so the data are moved electronically across a *wide-area network*. The best-known wide-area network is the *Internet*, which links computers in many different countries. Individual countries have fast wide-area networks that link into the Internet; in Britain, the Joint Academic Network, JANET, fulfils this role for universities and research establishments. Image data can be sent across the Internet using the *file transfer protocol* (ftp), although transfer rates can be very low when the network is busy. The weather satellite data mentioned above would probably be sent across a dedicated network that was rented from a telecommunications company. In order to use the ftp protocol both the client computer (the one receiving the data) and the server machine (the one providing the data) must have a network connection and a suitable ftp program.

Other facilities besides ftp are available to provide access to remote computers. For example, a *telnet* program allows remote log-ins to a computer anywhere on the Internet providing that the network name or 'IP address' of the host computer is known. For example, to log into the network server at the School of Geography, University of Nottingham, I would type

telnet mercator.geog.nottingham.ac.uk

(the server's network name). Of course, I would need to have a username and password before I could log in and use the system. Computers that are accessible 'across the network' have security features enabled which are designed to prevent 'hackers' from trying to log in illegally.

The most widely used Internet facility is the *World Wide Web* (WWW). To use WWW you need a WWW browser program, such as Netscape or Microsoft Explorer. Each 'web site' has an address, called its URL (Universal Resource Locator), which looks something like this:

http://www.geog.nottingham.ac.uk

When a valid URL is entered into the browser then the appropriate computer is contacted (in this case, the computer called *geog* at the site called *nottingham* which is a university or other academic institution (*ac*) in the United Kingdom (*uk*)) and information is transmitted to the screen of your computer. The WWW is expanding rapidly in an uncontrolled way, and 'pages' of information appear and disappear in an unpredictable way, like fossil species in the geological column. There is, however, a wealth of information to be found on the 'Web', ranging from technical details of remote-sensing programmes provided by space agencies such as the European Community Joint Research Centre's Centre for Earth Observation (CEO), the European Space Agency (ESA), the US Space Agency (NASA), the German Space Agency (DLR), the Japanese Space Agency

(NASDA) and the Canadian Centre for Remote Sensing (CCRS). The best advice to give to the reader is: join a society such as The Remote Sensing Society that has a WWW page containing an up-to-date list of World Wide Web addresses of interest to its members, which is updated regularly, plus pages of remote-sensing information.

One further opportunity opened up by local and global networks is access to *electronic mail* or e-mail. Most computer users can obtain a personal e-mail address of the form

person@organisation.type.country

for example,

Jim.Smyth@some-university.ac.uk

which can be used by an e-mail program to locate them from anywhere in the world, and deliver a message that can consist of ASCII text or images. Sometimes, images are converted from their normal binary form (which is 8-bit base-2 integers) into an ASCII representation using a program called *uuencode*. A 'clever' e-mail program will automatically decode such images and convert them back from ASCII to binary form. If you do not have such a clever e-mail program you will have to run *uudecode* manually.

One advantage of e-mail is that it can be used to run discussion lists that allow correspondence between all of the registered users of the list. One example is The Remote Sensing Society's discussion list

rss-news@geog.nottingham.ac.uk

which provides facilities for members to ask and answer questions, provide information, or make comments on issues of general interest to students and researchers using remote sensing. Other discussion lists of interest to remote-sensing students are IMAGRS-L and GIS-L. As with WWW, details of discussion lists change quickly and so you should join an appropriate society, which will keep you informed.

3.3   IMAGE DISPLAY SUBSYSTEM

3.3.1   Colour display systems

3.3.1.1   Colour monitor and video memory

In section 3.2 the characteristics of a computer system are described. The image display subsystem is a part of

the computer system that is of considerable interest to remote-sensing users, and so it is described here in more detail. In section 2.2.3 the concept of radiometric resolution is explained, and the manner in which radiance values for individual pixels are converted to numbers in the range 0–1023, 0–255 or 0–63 is examined. The process is one of analogue-to-digital (A/D) conversion. The image display subsystem carries out the converse operation, that of taking a digital quantity (an individual pixel value from an image) and converting it to analogue form, in this case a voltage. This is called digital-to-analogue (D/A) conversion. The three separate analogue signal voltages for the red, green and blue components of a colour picture are proportional to the intensities of the red, green and blue digital components of the multispectral image at the point on the screen corresponding to the location of the pixel in the image. These red, green and blue components are termed an RGB triple, and for the sake of illustration it is assumed that each is expressed on a 0–255 scale, with 0 being the minimum intensity level of the colour and 255 being the maximum. If the RGB triple at a given point on the image is {255, 255, 255} then the screen pixel at the corresponding screen position will display maximum red plus maximum green plus maximum blue, which is the brightest white possible for the system to display. Figure 3.2 illustrates the triple {255, 0, 0}. Other colours can be generated as mixtures of RGB values; for example, {255, 255, 0} gives brightest yellow (Table 3.2). Shades of grey can be generated by setting R = G = B so that {127, 127, 127}, for instance, produces a midgrey. Where the image processing software is capable of handling image data in a form other than the 8-bit byte (for example, a 16- or 32-bit integer or a real number) these values are scaled to 8-bit (0–255) form before they are fed to the D/A converter and thence to the monitor screen.

Before the digital image data can be converted to analogue form for display they must first be moved into the memory of the image display subsystem. Each scan line of the image in each of the three bands (representing the red, green and blue components of the displayed image) is transferred from a disk file to the computer's random-access memory, from whence it is moved to the video memory (Figure 3.3). The video memory is, in effect, an electronic map of the remotely-sensed image. It can be considered as a matrix, with m rows and n columns and an origin at the top left corner. Each cell of the matrix is occupied by a pixel value. These values generally lie in the range 0–255 so, in order to be able to hold these values, the number of binary digits associated with each cell of the video memory must be eight or

Table 3.2 Relationship between digital intensity values stored in the red, green and blue channels of a video memory and the perceived colour on the monitor. The figures used in the table assume that eight bits (one byte) per pixel are used in each channel of the video memory, giving 256 brightness levels (0–255) per colour.

Pixel value			
Red	Green	Blue	Perceived colour
255	255	000	brightest yellow
127	127	000	mid-yellow
000	127	127	mid-cyan (turquoise)
255	000	255	brightest magenta (pink)
100	000	000	dull red
100	000	100	dull purple
255	180	120	bright beige
255	255	255	white
127	127	127	mid-grey
000	000	000	black
000	255	000	brightest green
000	000	255	brightest blue

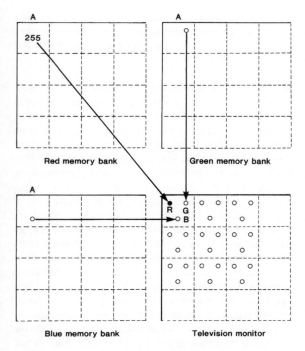

Figure 3.2 Pixel at A has values 255, 0, 0 in the red, green and blue video memory banks. The colour at the corresponding point on the monitor screen is maximum red.

more (as eight bits are needed to represent a number in the range 0–255). The image memory is therefore $m \times n$ bytes in size, as a minimum, and each memory cell is capable of holding an 8-bit integer representing decimal 0 to 255. Each row of the matrix is scanned electronically and the digital quantities (pixel values) are converted to analogue form as voltages. The analogue signal is then sent to the television monitor.

This scanning procedure is carried out repeatedly, for a television picture will fade rapidly if not refreshed. Once the digital image has been loaded into the image memory, the processes of digital-to-analogue conversion and picture refresh are carried out automatically and repeatedly to ensure a flicker-free picture. The time taken to redraw the screen is the *refresh time*.

The monitor supplied with a medium-range PC will display an image of up to 800 pixels wide by 600 scan lines long. This is the Super Video Graphics Adapter (SVGA) standard, which is used by the MIPS software that is provided with this book (section 3.6). More expensive PCs and workstations may have higher screen resolution (1024×768, 1024×1024 or 1280×1024 are quite usual). At these resolutions, and with less expensive monitors, the refresh time may be insufficient for the entire screen to be 'repainted' in one cycle. One solution is to refresh all the odd-numbered scan lines on the screen at cycle i, then the even-numbered scan lines at cycle $i + 1$, and so on. This is called the *interlaced* mode of operation. The picture on the monitor screen is formed by the illumination of a matrix of phosphor dot triplets (sensitive to red, green and blue, respectively) by an electron gun. If the luminance of the phosphor dots decays too rapidly then the interlaced method of refreshing the picture will result in a detectable variation in the luminance of the phosphors from the top to the bottom of the screen. This effect will be seen as a regular flickering of the picture, which can be overcome by the use of long-persistence phosphors. The latter route is the least expensive, but it has the disadvantage that if the monitor is used for graphics (drawings) rather than images then the relatively long delay before the picture leaves the screen may interfere with the perception of animation. A better solution is to buy a monitor that can refresh the whole screen in one cycle in *non-interlaced* mode.

Remotely-sensed images are multispectral – they are not limited to the red, green and blue wavebands of the visible spectrum. For example, SPOT HRV produces images in three spectral bands, the green, red and near-infrared respectively. In order to display these images on a conventional monitor screen, the RGB inputs of the monitor are individually assigned to one of the

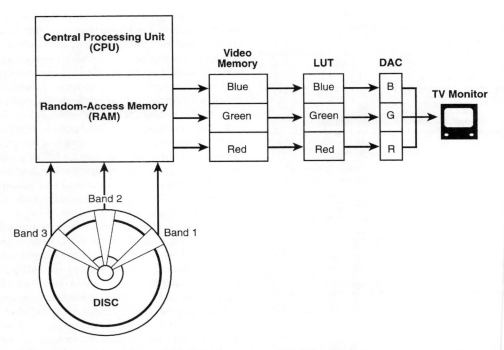

Figure 3.3 Bands 1, 2 and 3 of a multispectral digital image set are read from the disk store and transferred via the random-access memory of the computer to the video memory. The data for the three bands are stored in the blue, green and red video memory banks, respectively. The lookup tables (LUT) modify the values in memory before they are converted from digital to analogue form by the digital-to-analogue converter (DAC) and then transmitted to the monitor screen.

three bands of the multispectral image. Conventionally, the infrared band is shown on the video monitor in red, with the visible red image being displayed on the monitor screen in green, and the visible green band being directed to the monitor's blue input. The result is a *false-colour composite image*. If the value at a given pixel position in the infrared band were 255 and the red and green bands both had zero values at the same point then the colour 'brightest red' would appear on the screen (Figure 3.2). Since there are 256 levels of each primary colour, and because the levels of each colour can be set independently, there are 256^3 or 16 777 216 possible colour combinations, including black and white. A small selection of these colour combinations is shown in Table 3.2. The human eye cannot distinguish 16 million colours; probably no more than 2000 or so are separable. Nevertheless, the availability of 16 million hues means that if a particular hue is required it can be used.

The preceding discussion assumes that the video memory can hold three 8-bit values – representing red, green and blue colour intensities – at each pixel position, giving a '24-bit display'. Cheaper display systems

use fewer bits to represent each of the primary colours. In the case of an 8-bit video memory, the number of colours available is 8^3 or 256. A three-band colour composite image can be displayed on such a system by allocating three of the eight available bits to the red component of the image (giving levels of red), a further three bits to the green component and the remaining two to the blue component. This arrangement gives eight (i.e. 2^3) levels of red, eight (2^3) of green and four (2^2) of blue. Zero intensity indicates absence of colour, or black. The 256 intensity values in the three images forming the colour composite must first be reduced by thresholding to eight levels (in red and green) or four levels (in blue) before the image is placed in the video memory. The choice of threshold values is critical; best results are obtained if a histogram equalisation enhancement is carried out first (section 5.3.2) and the thresholds selected so that an equal proportion of image pixels fall into each of the eight levels (or four, in the case of the image allocated to the blue input of the monitor). Figure 3.4 illustrates the workings of this method. Assume that we are dealing with the image which is to form the red component of the colour

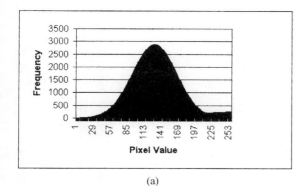

(a)

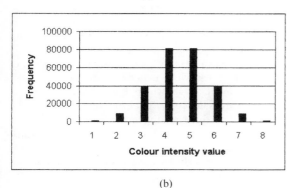

(b)

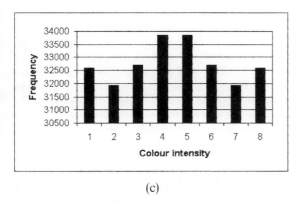

(c)

Figure 3.4 (a) The histogram of the 256 grey levels making up an image of size 512 lines by 512 pixels per line is shown. (b) The 256 levels are reduced to eight by a simple allocation process in which the range 0–255 is subdivided into eight equal parts. The majority of pixels are allocated to levels 3–6 in the reduced histogram. (c) The effect of using unequal class intervals to reduce the 256 grey levels of the original image shown in (a) to eight almost-equal classes. This effect is achieved for the histogram shown in (a) by allocating grey levels 0–85, 86–102, 103–115, 116–127, 128–139, 140–152, 153–168 and 169–255 in the original histogram shown in (a) to classes 0–7 respectively. See text for further explanation.

composite image. There are 256 levels of red in the image. Figure 3.4(a) shows the distribution of pixel values across the 256 levels. The y-axis scale shows the number of pixels that occupy the 256 categories shown on the x-axis. If simple division is used to allocate the 256 levels to eight colour intensities (so that, for instance, a colour level of i in histogram (a) is allocated to level of j in histogram (b), where $j = [i/32]$ and $[\ldots]$ indicates 'the integer part of the result') then the result can be disappointing. If the grey-level values in histogram (a) are thresholded so that around 32 768 pixels are allocated to each of the eight classes, then the result shown in Figure 3.4(c) is obtained. Note that the total number of pixels in a 512×512 image is 262 144 and that 32 768 is equal to 262 144 divided by 8.

Some PC video cards offer 32k or 64k colours, selected from a palette of 256^3 colours. The same method as above applies: the 24-bit image is thresholded and converted from 16 million colour resolution to the appropriate lower resolution. These operations are carried out automatically by the windowing system. For example, the Windows 95 operating system takes a 24-bit bitmap (section 3.5) and displays it on the screen after converting the image to an appropriate number of colours. This allows the software to run on machines that offer 256, 32k, 64k or 16 million colour modes. The MIPS software supplied with this book and described in section 3.6 will perform well if the video card in your PC can support 32k or more colours (16 bits per pixel) at a resolution of 800×600 pixels. Image quality is degraded considerably if only 256 colours (8 bits per pixel) are available at this resolution.

3.3.1.2 *Lookup tables*

If a display subsystem were as simple as that described above then it would be possible to alter the visual characteristics of the displayed image only by altering the values stored in the video memory. For example, the intensity range for each primary colour is 0 to 255 but it might be the case that, in a particular application, the pixel values in the image bands that make up the false-colour composite image occupy only the lower half of the available dynamic range (0–127 from the full range of 0–255). The resulting image would then appear dark; in photographic terms it would be said to be under-exposed. By doubling each pixel value the relative colour intensities would be maintained but the intensity range would increase from 0–127 to 0–254, giving a brighter picture with more contrast. However, if the displayed image had dimensions of 1024 scan lines each of 1024 pixels, then the total number of additions to be performed in order to achieve this rather simple brightness enhancement would be 1 048 576.

A little thought will show that, although there are 1 048 576 pixel values in a 1024 × 1024 false-colour composite image, there are only 256 possible separate intensities for each of the three primary colours, giving a total of 768 values. The default state of the image display subsystem is to map (translate or transform) an input pixel value of 0 to the minimum intensity of the particular colour and an input value of 255 to the maximum colour intensity, with intermediate values being interpolated linearly between these extremes. If this relationship could be quickly and easily modified then the visual appearance of the image could be changed rapidly without altering any of the 1 048 576 pixel values stored in the video memory. Figure 3.5 shows the histograms[1] of the pixels forming the red, green and blue components of a colour composite image, together with the default state of the *lookup tables* or LUT (which are the same as the *palettes* in Windows 95 terminology). There is one LUT per primary colour. The straight lines in Figure 3.5 relate the input pixel value on the x-axis (the value stored in the video memory) to the output colour intensity value shown on the right-hand y-axis (the value sent to the digital-to-analogue converter and thence to the monitor). An input value of 0 is mapped to an output value of 0, and an input value of 255 is mapped to an output value of 255. The input and output values are, in fact, the same for all possible values, and thus the LUTs in their default state have no effect, and the colours seen on the screen are directly proportional to the corresponding image pixel values that are stored in the video memory.

Figure 3.6 shows the same three histograms as are displayed in Figure 3.5. This time, however, the corresponding LUTs are set so that (i) all 256 output values (right-hand y-axis) for each colour are used, and (ii) the shape of the LUT reflects the distribution of the image pixel values as shown by the histogram. This example shows the use of a histogram-controlled contrast enhancement, which is explained further in section 5.3.

An LUT is, as its name implies, a table rather than a graph. Table 3.3 shows the numerical values of the output values, which are sent to the digital-to-analogue converter for the red channel in Figure 3.6. The columns marked 'In' are the input pixel values, as stored in the video memory, while the columns marked 'Out' and shown in italics are the LUT values, which are proportional to the colour intensities on the monitor screen. The table should be read from left to right as four

vertical strips each of four columns. Notice that the LUT is set so that the full output range (the *dynamic range*) of the display screen is used, and that all 256 output values are allocated to pixel values that actually occur in the image. No output values are wasted by being allocated to input pixel values that are not represented in the image.

Figure 3.3 shows the entire image display procedure. Three image files stored on the hard disk are to be allocated to the red, green and blue channels of the image processing system. First, these images are read from the disk and transferred to the appropriate section (R, G or B) of the video memory. The lookup tables are set, as explained in the preceding paragraph. If x_{red}, x_{green} and x_{blue} are the values in the R, G and B sections of the video memory for a given pixel, then the lookup tables output the colour intensity values y^*_{red}, y^*_{green} and y^*_{blue}, the values of y^*_{red} depending on the state of the lookup tables, which may be different for each colour component, as shown in Figure 3.6. The colour intensity values are converted from digital to analogue form, and the three analogue signals are sent to the monitor inputs. The colour corresponding to the RGB triple $\{y^*_{red}, y^*_{green}, y^*_{blue}\}$ appears on the monitor screen. (Table 3.2 shows a selection of colours corresponding to particular $\{y_{red}, y_{green}, y_{blue}\}$ combinations.)

3.3.1.3   *Pointing devices*

One of the operations commonly carried out on a remotely-sensed image involves the establishment of a relationship between the coordinate positions of a set of points which can be located accurately on a map and on a remotely-sensed image. The purpose is to allow the image coordinate system to be transformed so that the map and image can be matched up. The procedure is known as geometric correction, and is explained in Chapter 4. For the purpose of the present discussion we are interested in the mechanism by which coordinate positions of individual pixels on an image can be found. As well as requiring the image coordinates of individual pixels, we might also want to indicate the top right and bottom left corners of a rectangular area in order to define a region to be magnified or enlarged (*zoomed*). A *cursor* is used to perform both tasks. The cursor is generally controlled by a mouse or pointing device, though trackerball and joystick systems are also in use.

Both cursor display and zooming are implemented either by electronic circuits within the image display subsystem (hardware) or by suitable programs (software) depending on the particular equipment used. Hardware operations are much faster and are now al-

---

[1] The histogram of an image is the count of grey levels taking values 0, 1, ..., 255.

Histogram/LUT Plots

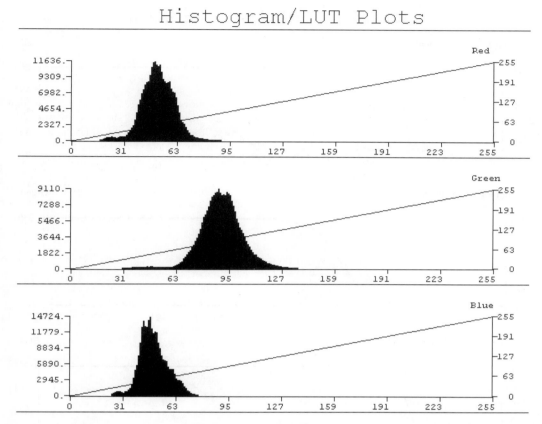

Figure 3.5 Histograms of the red, green and blue components of a colour image (refer to the left-hand scale for frequencies). The entries in the lookup table (LUT) corresponding to the 0–255 input scale (*x*-axis) are graphed. The colour intensities output by the LUT are shown on the right-hand scale. In this case, the LUT is set to its default state, which maps input directly to output with no change. See Figure 3.6 for an example of a non-linear mapping from input pixel value (*x*-axis) to colour intensity (right-hand scale).

most universal. A joystick, mouse or other device is used to move the cursor to the required point. Clicking the left mouse button is the signal for the program controlling the cursor to compute and display the coordinates of the cursor, which may be displayed in the form of an arrow or the intersection of cross-wires. Software-driven cursors use a simple but clever procedure. The pixels forming the cursor shape are combined with the underlying image data using the exclusive-OR (X-OR) operation. When the cursor is moved the cursor pixels are X-ORed again with the image data, a process that makes the cursor disappear and restores the original data.

Zooming a portion of a displayed image involves the use of a pair of memory banks, which appear on the monitor screen as two windows, say window 1 and window 2. Copying pixels from the first memory bank (window 1) in which the displayed image is resident to

the second memory bank (window 2), at the same time replicating each individual pixel in the selected area of window 1 and each line of pixels by a factor called the zoom or magnification factor, will result in the transfer of a magnified version of the window 1 image to window 2. Magnification factors are generally powers of 2. An interpolator could be used rather than a simple copy operation to produce a smoother, less blocky, zoomed image. The cursor can be used on the zoomed image (window 2) as described previously to locate pixels of interest. Naturally, it is easier to get a precise fix on a particular pixel if the image is magnified by a factor of 2 or 4. If a higher magnification factor is used without interpolation the zoomed image may appear blocky, and the eye might be unable to identify patterns or shapes. The MIPS software uses software zooming and restricts the magnification factor to 2.

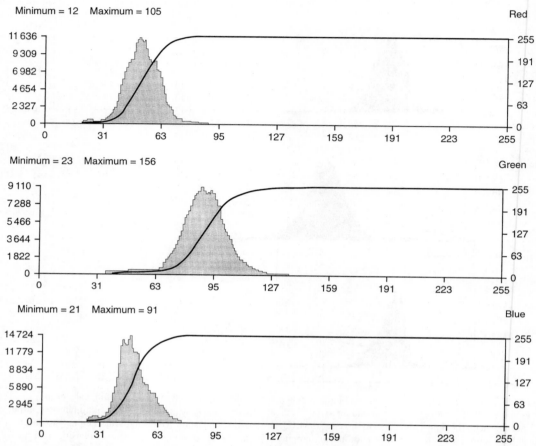

Figure 3.6   Histograms of the colour components of the same image as is used in Figure 3.5. Here, the lookup tables have been altered in a non-linear way to reflect the fact that the input data do not cover the full range (0–255) and thus take account of the distribution of pixel values within each colour component of the image. This is the method of *histogram equalisation*, which is described in Chapter 5. The numerical values held by the modified lookup table are given in Table 3.3.

3.3.2   Hard-copy systems

Users of image display systems may wish to include a paper or photographic (hard) copy of the screen (soft) image in a report or thesis. Several ways of creating hard-copy images are available, ranging in sophistication from the crude to the refined. Crude methods employ a dot-matrix printer, and are rarely used nowadays. Colour *ink-jet printers* use minute nozzles to fire dots of coloured ink onto a sheet of paper. The number of colours that can be derived from four basic ink colours (cyan, magenta, yellow and black, denoted CMYK) is increased by the use of a process known as

dithering,[2] which involves the use of a square pattern of several ink dots to represent one image pixel. If the pattern for each colour of ink is varied then shades of colours can be represented. The quality of images produced by colour ink-jet printers depends on the quality of the printer (more sophisticated printers use better ink

---

[2] Dithering is the process of printing clusters of dots that are close enough together to fool the eye into believing that a continuous grey scale is being viewed. Newspaper pictures, for example, are printed using black dots only. Similarly, a colour can be represented by printing cyan, magenta, yellow and black dots of different intensities close together so that the viewer 'sees' colours such as green.

Table 3.3 Lookup table after manipulation (see Figure 3.6). The table should be read as four vertical strips, each consisting of two columns, labelled 'In' and 'Out'. The columns marked 'In' show the image pixel values. The columns marked 'Out' (with figures in italics) are the corresponding colour values that are displayed on the screen. For example, an image pixel with a value of 62 is displayed as if it had a value of 221. This lookup table refers to the top histogram in Figure 3.6, in which the lookup table is graphed as a solid line.

In	Out	In	Out	In	Out	In	Out	In	Out	In	Out	In	Out	In	Out
0	*0*	1	*0*	64	*232*	65	*236*	128	*255*	129	*255*	192	*255*	193	*255*
2	*0*	3	*0*	66	*239*	67	*242*	130	*255*	131	*255*	194	*255*	195	*255*
4	*0*	5	*0*	68	*244*	69	*246*	132	*255*	133	*255*	196	*255*	197	*255*
6	*0*	7	*0*	70	*248*	71	*249*	134	*255*	135	*255*	198	*255*	199	*255*
8	*0*	9	*0*	72	*250*	73	*250*	136	*255*	137	*255*	200	*255*	201	*255*
10	*0*	11	*0*	74	*251*	75	*252*	138	*255*	139	*255*	202	*255*	203	*255*
12	*0*	13	*0*	76	*252*	77	*253*	140	*255*	141	*255*	204	*255*	205	*255*
14	*0*	15	*0*	78	*253*	79	*253*	142	*255*	143	*255*	206	*255*	207	*255*
16	*0*	17	*0*	80	*254*	81	*254*	144	*255*	145	*255*	208	*255*	209	*255*
18	*0*	19	*1*	82	*254*	83	*254*	146	*255*	147	*255*	210	*255*	211	*255*
20	*1*	21	*2*	84	*254*	85	*255*	148	*255*	149	*255*	212	*255*	213	*255*
22	*2*	23	*3*	86	*255*	87	*255*	150	*255*	151	*255*	214	*255*	215	*255*
24	*3*	25	*4*	88	*255*	89	*255*	152	*255*	153	*255*	216	*255*	217	*255*
26	*4*	27	*5*	90	*255*	91	*255*	154	*255*	155	*255*	218	*255*	219	*255*
28	*5*	29	*6*	92	*255*	93	*255*	156	*255*	157	*255*	220	*255*	221	*255*
30	*7*	31	*7*	94	*255*	95	*255*	158	*255*	159	*255*	222	*255*	223	*255*
32	*8*	33	*9*	96	*255*	97	*255*	160	*255*	161	*255*	224	*255*	225	*255*
34	*10*	35	*11*	98	*255*	99	*255*	162	*255*	163	*255*	226	*255*	227	*255*
36	*13*	37	*15*	100	*255*	101	*255*	164	*255*	165	*255*	228	*255*	229	*255*
38	*18*	39	*23*	102	*255*	103	*255*	166	*255*	167	*255*	230	*255*	231	*255*
40	*28*	41	*34*	104	*255*	105	*255*	168	*255*	169	*255*	232	*255*	233	*255*
42	*40*	43	*48*	106	*255*	107	*255*	170	*255*	171	*255*	234	*255*	235	*255*
44	*56*	45	*64*	108	*255*	109	*255*	172	*255*	173	*255*	236	*255*	237	*255*
46	*73*	47	*82*	110	*255*	111	*255*	174	*255*	175	*255*	238	*255*	239	*255*
48	*92*	49	*104*	112	*255*	113	*255*	176	*255*	177	*255*	240	*255*	241	*255*
50	*115*	51	*126*	114	*255*	115	*255*	178	*255*	179	*255*	242	*255*	243	*255*
52	*136*	53	*147*	116	*255*	117	*255*	180	*255*	181	*255*	244	*255*	245	*255*
54	*157*	55	*166*	118	*255*	119	*255*	182	*255*	183	*255*	246	*255*	247	*255*
56	*174*	57	*183*	120	*255*	121	*255*	184	*255*	185	*255*	248	*255*	249	*255*
58	*191*	59	*199*	122	*255*	123	*255*	186	*255*	187	*255*	250	*255*	251	*255*
60	*207*	61	*214*	124	*255*	125	*255*	188	*255*	189	*255*	252	*255*	253	*255*
62	*221*	63	*227*	126	*255*	127	*255*	190	*255*	191	*255*	254	*255*	255	*255*

nozzle technology and superior dithering procedures), the resolution of the printer, which is usually expressed in dpi (dots per inch or 2.54 cm), and the type of paper used. A colour ink-jet printer now costing less than £200 can produce prints with a resolution of 720 dpi and, if special coated paper is used to stop the ink diffusing, then the output quality is more than acceptable for theses and reports. Care should be taken when using any form of colour printing in order to ensure that the colour balance of the print matches that of the screen image, for a monitor display uses the *luminance* properties of RGB phosphor dots to produce an image whereas any form of colour printing is based on the *absorption* of light by CMYK pigments (Russ, 1995). The *printer driver* that is supplied by the manufacturer of the printer carries out the tasks described in this paragraph; it is no longer necessary for a programmer to be concerned with matching the properties of the image to those of the printer.

More expensive colour *laser printers*, costing several thousands of pounds, will produce a higher-quality image than an ink-jet printer costing maybe one-tenth of the price. Laser printers using the Postscript image description language can generate very high-quality products though the cost per page is significantly higher than the ink-jet.

Most standard image processing software systems provide facilities to output screen images to a range of ink-jet or laser printers. Because ink-jet printers for PCs tend to be cheaper to buy and easier to use than their workstation equivalents (largely due to the availability of excellent image editing and printing programs for PCs), some people might prefer to transfer the image to be printed from a workstation to a PC, using the ftp program mentioned above. In order to do this, the image is first coded into one of several available image formats, such as Windows Bitmap (BMP), JPEG, GIF or TIFF. Caution should be exercised as some of these formats (for example, TIFF) compress the image by reducing the number of colours present, as described in section 3.5. The output from the printer may differ considerably from the image on the screen if a very 'lossy' compression is used.

Photographic slides can be produced from a monitor display using a standard 35 mm SLR (single lens reflex) camera. Provided that reflections on the monitor screen are avoided, reasonable results can be obtained that are adequate for informal presentations. Higher-quality photographic slides are produced using computer-operated systems that accept a digital image (in BMP, GIF, JPEG or TIFF format) as input and generate a photographic slide directly, without the need for a standard camera system, thus eliminating distortions caused by screen curvature and focusing. Even higher quality can be obtained from a film plotter. These devices use a point of light of variable intensity to expose a colour film on a dot-by-dot basis. The cost of the hardware is considerable, although bureau services are available in most large cities.

3.4   DATA FORMATS FOR DIGITAL SATELLITE IMAGERY

Digital data from the various satellite systems described in section 2.3 are supplied to the user in the form of computer-readable magnetic tapes, usually called computer-compatible tapes or CCTs, as well as on miniature tapes such as Exabyte and DAT. CD-ROM is fast becoming an accepted standard because, unlike magnetic tapes, which are sequential devices, CD-ROM allows the user to access any file equally quickly without the need to wind a tape backwards or forwards.

Unfortunately, no world-wide standard for the storage and transfer of remotely-sensed data has been agreed upon, though the CEOS (Committee on Earth Observation Satellites) format is becoming accepted as a standard. In the past, each ground receiving station used its own format and even changed its format specification from time to time. The data on a tape or CD-ROM generally consist of text (header) records and image data. These data are organised into files, each file being made up of a set of records. The records of an image data file each contain the data for one scan line for one or more spectral channels. The records of text or annotation files contain information relating to the date and time of imaging, the Sun azimuth (degrees clockwise from north) and elevation (degrees from the horizontal) together with calibration and processing data.

Image data may be organised either in band-sequential (BSQ) form or in band-interleaved-by-line (BIL) form. BSQ organisation has one complete file (for channel 1) followed by files holding the data for channels 2, 3 and so on in sequential order. BIL format holds image data for all channels for scanline 1 (i.e. scan line 1 for band 1 is followed by scan line 2 for band 1, and so on). Then the data for scanline 2, 3, ..., n, are written to the tape or CD-ROM. Other, more esoteric, formats were used in the 1970s and 1980s and, fortunately, all of them have been abandoned.

The situation has improved considerably over the past 10 years or so, and nowadays most commercial remote-sensing image processing systems provide procedures to read the most commonly used remote-sensing image formats, though users of older data sets such as Landsat-1 to -3 MSS may need to exercise imagination in order to access the data. The CD-ROM accompanying this book contains programs that will read SPOT HRV and Landsat TM images from CDs and write selected parts of the image to separate data files (see Appendix B).

3.5   IMAGE FILE FORMATS AND DATA COMPRESSION

The topic of image file formats is too large for it to be considered here in any detail. Readers should, however, be aware that some file formats involve *compression* of the image data, which may result in some loss of detail. The *JPEG* (Joint Photographers Expert Group) format, for example, first separates image intensity information from colour information. These two components are then compressed separately in order to maintain the intensity (brightness) information. This is because human colour vision can deal more easily with loss of colour information than with loss of intensity information. The two components for each of a number of image sub-regions are compressed by the use of a transformation similar to the two-dimensional Fourier transform that is discussed in section 6.6. This transform expresses the image data in terms of a sequence of components of decreasing importance. The JPEG compression keeps only the first few components and throws the rest away, hence losing some of the detail. Finally, repeating values such as 2, 2, 2 are replaced by counts (such as 3×2). The JPEG compression scheme can achieve compression ratios of around 10:1, which is worth while when dealing with large images. For instance, a 1024×1024 image in red, green and blue and expressed on an 8-bit scale requires 3 Mb of disk storage. A compression ratio of 10:1 means that the image takes up only 314 573 bytes (314 kb) on the disk. The cost is the loss of some information.

Other image formats are PICT (used by the Apple Macintosh), TIFF, GIF and BMP. The BMP type is the Microsoft Windows Bitmap, which is widely used by applications programs running under the Microsoft Windows 95 operating system. The MIPS image display program can write image files in the form of bitmaps, and these can be read by other image processing and display software (for example, to add annotation to and print the image). Users of the BMP format (referred to as a *device-independent bitmap* or DIB) can choose whether or not to use compression. The uncompressed DIB contains a header, giving information about the size of the image and the number of bits per pixel, for example. Next comes a colour table or palette (equivalent to the colour lookup table described in section 3.3.1.2) which maps pixel values on the 0–255 scale on to screen colours, and finally the image pixels in RGB order, with the last scan line being stored first. Appendix B contains a description of a program to convert an image stored as a Windows bitmap into a raw format image that can be read by the MIPS image display software.

Data compression is a useful way of reducing image size before transmission across a network, for example. However, care should be taken when choosing a compression method. It was mentioned above, in the discussion of the JPEG image format, that this method involved loss of information. It is called a 'lossy' compression procedure. Compression methods that preserve all of the data are called 'lossless'. While lossy compressions may be suitable for transmission of digital TV pictures (because the human eye can tolerate some loss of colour information), in general one would choose a lossless compression scheme to encode remotely-sensed data because it is impossible to predict in advance which information can be lost without any cost being incurred. Some image transform methods, including principal components analysis (section 6.4) and the Fourier transform (section 6.6), can be used to compress images. These methods exploit the fact that some redundancy exists in a multispectral image set, and they re-express the data in such a way that large reductions in the volume of transformed data represent only small losses of information. Other methods are *run length encoding* and *Huffman coding*. Run length encoding involves the rewriting of the

image pixels in terms of expressions of the form (l_i, g_i) where l_i is the number of pixels of value g_i that occur sequentially. Thus, a sequence of pixel values along a scanline might be 1112222. This sequence would be encoded as (3,1)(4,2). The end of a scanline is indicated by a special character. Obviously the degree of compression that results from this type of encoding depends on the existence of homogeneous sections of image – in other words, if there are no runs there will be no compression; in fact, if there are no runs there will be expansion rather than compression. Run length encoding is used in fax transmission and may be useful in compressing classified images (Chapter 8) in which individual pixel values are replaced by labels. The quadtree, which is described below, may be a better choice as it is a two-dimensional compression scheme.

The *quadtree* is a form of data structure or organisation that is used in some raster GIS. Its major limitation is that the image to be encoded must be square and the side length must be a power of 2. However, the image can be padded with zeros to ensure that this condition is met. The square image is first subdivided into four component square sub-images of size $2^{n-1} \times 2^{n-1}$. If any of these sub-images is homogeneous (meaning that all of the pixels within the sub-image have the same value) then they are not 'quartered' any further. However, those sub-images that are not homogeneous are again divided into four equal parts and the process repeated (in computing terms, the procedure is a recursive one). When the quadtree operation is completed the individual components, which may be of differing sizes, are given identifying numbers called Morton numbers, and these Morton numbers are stored in ascending order of magnitude to form a *linear quadtree*. For images such as classified images, which generally contain significantly large homogeneous regions, the use of quadtree encoding will result in a substantial saving of storage space. If the image is inhomogeneous then the amount of storage required to store the quadtree may be greater than that required for the raw, uncompressed image. Kess *et al.* (1996) use quadtrees to compress the non-land areas of the Global Land 1-km AVHRR data set. They find that the quadtree gives a reduction in data volume to 6.72% of the original data size, which is better than that achieved by JPEG, GZIP, LZW and UNIX *compress* methods. More details of quadtree-based calculations can be found in Mather (1991), while the definitive reference is Samet (1990).

3.6 MIPS SOFTWARE

MIPS (Mather's Image Processing System) is an educational software package for PCs running the Windows 95 operating system. It has been written specifically for distribution with this book, and may be used freely for teaching and educational purposes. MIPS is not 'supported' in the way that commercial packages are. There is no technical back-up or software support. The MIPS display program uses a fixed image size of 512 scan lines with 512 pixels per line. It is described in Appendix A. Details of other MIPS programs are given in Appendix B.

MIPS runs on a PC with a Pentium processor or better, using the Microsoft Windows 95 operating system. A minimum of 16 Mb memory is required, though MIPS will run at a more acceptable speed if the PC has 24 Mb memory. The software itself takes up only 2 Mb or so of disk space, though 20–30 Mb of free disk space is required to hold the images on which you are currently working. The video system should be capable of supporting at least 256 colours at 800 × 600 pixels resolution, though a system supporting 32k or 64k colours gives a much better result, and your eyes will benefit from a 15 or 17 inch monitor screen. You need a CD reader to extract the programs from the CD-ROM containing the MIPS software. Installation details are given in Appendix C.

MIPS has been developed using the Salford FTN95 Fortran compiler and Clearwin +. It provides an integrated image processing and display system that carries out many of the enhancement operations described in this book. Stand-alone programs are also included; these carry out more lengthy operations including principal components analysis, geometric correction, Fourier transforms and image classification. Some sample images (together with software to read them) are contained on the CD.

MIPS uses a simple system to keep track of image files stored on disk. Each image file that forms part of a set, together with information such as the size of the files, is stored in an 'image dictionary' file, usually with the file extension .inf. The structure of the INF file is described in Appendix A. You must remember to create an INF file (using program make_inf) to reference any image sets that you are going to use in MIPS. Image files are stored without headers as raw binary data, reading sequentially across the rows or scan lines of the image. Thus a 512 × 512 image will be held as a binary file of 262 144 bytes with the pixels on the first image line (reading left to right) stored first, followed by the pixels making up lines 2, 3, 4 and so on. You can create output files (in the form of Windows bitmaps) for use by other programs. Since MIPS does not include any modules to print images, you should consider acquiring a program

to carry out the printing of image files from Windows bitmaps created by MIPS.

3.7 SUMMARY

Computer requirements for digital image processing are summarised in this chapter. It is evident that, at present, these requirements are best met by a dedicated local system rather than by a remote mainframe. The major feature of digital remotely-sensed data is the large size of the data files that are needed. Even for operations that are mathematically simple, the large data volumes imply the need for (i) high data transfer rates from disk to memory and vice versa, and (ii) a high-power CPU to carry out the numerical operations. A basic appreciation of the way in which computers work, and of the design and implementation of algorithms, is needed if the user of a digital image processing system is to gain maximum benefit. Future developments that will undoubtedly influence remote sensing include (i) the use of parallel rather than serial CPUs, (ii) high-powered 'intelligent' workstations that are connected to the central processor by a local-area network (LAN) and (iii) the development of high-speed regional and national networks that will allow the transfer of data and software over a large area. The reducing costs of computer hardware that have been evident over the past few years will, it seems, continue and should lead to the more widespread availability of digital image processing equipment.

3.8 QUESTIONS

1. Give concise definitions of the following: Mb, physical memory, bit, byte, cache, virtual memory, ASCII, CPU, LAN, ftp, refresh time, TIFF, BSQ, quadtree.
2. List the media on which remotely-sensed images can be transferred between computers. Give the advantages and disadvantages of each.
3. Write the following base 2 numbers in base 10 form: 11111111, 00010001, 111110000, 01010101.
4. Graph (a) the default lookup table that is used to display a grey-scale image with a range of pixel values from 0 to 255, and (b) the lookup table that would produce the equivalent of the photographic negative of the image.
5. Explain how a false-colour image with each of the RGB components expressed on a 0–255 scale is displayed on a system with (a) 24 bits and (b) 8 bits of video memory at each pixel position.
6. What is the 'image histogram'? What information can be derived from studying the image histogram?

4

Pre-Processing of Remotely-Sensed Data

4.1 INTRODUCTION

In their raw form, as received from imaging sensors mounted on satellite platforms, remotely-sensed data generally contain flaws or deficiencies. The correction of deficiencies and the removal of flaws present in the data are termed *pre-processing* because, quite logically, such operations are carried out before the data can serve any useful purpose. Despite the fact that some corrections are carried out at the ground receiving station, there is often still a need on the user's part for some further pre-processing. The subject is thus considered here before methods of image display and analysis are examined in later chapters. It is difficult to decide what should be included under the heading of 'pre-processing', since the definition of what is, or is not, a deficiency in the data depends to some extent on the use to which those data are to be put. If, for instance, a detailed map of the distribution of particular vegetation types or a bathymetric chart is required then the geometrical distortion present in an uncorrected remotely-sensed image will be considered to be a significant deficiency. On the other hand, if the purpose of the study is to establish the presence or absence of a particular class of land use (such as irrigated areas in a semi-arid region) then a visual analysis of a suitably processed false-colour image will suffice and, because the study is concerned with determining the presence or absence of a particular land use type rather than its precise location, the geometrical distortions in the image will be seen as being of secondary importance. A second example will show the nature of the problem. An attempt to estimate reflectance from remotely-sensed images will be hindered, if not completely prevented, by the effects of interactions between the incoming and outgoing electromagnetic radiation and the constituents of the atmosphere. Correction of the imagery for atmospheric effects will, in this instance, be considered to be an essential part of data pre-processing whereas, in some other case (for example,

discrimination between land cover types in an area at a particular point in time), the investigator will be interested in relative, rather than absolute, pixel values and thus atmospheric correction would be unnecessary. Measurements of change over time using multitemporal image sets will, in the case of optical imagery, require correction for atmospheric variability, and it will also be necessary to register the images forming the multitemporal sequence to a common geographical coordinate system. In addition, corrections for changes in sensor calibrations will be needed to ensure that like is compared with like.

Because of the difficulty of deciding what should be included under the heading of 'pre-processing methods', an arbitrary choice has been made: correction for geometric, radiometric and atmospheric deficiencies, and the removal of data errors or flaws, will be covered here despite the fact that all of these operations will not necessarily be applied in all cases. This point should be borne in mind by the reader; it should not be assumed that the list of topics covered in this chapter constitutes a menu to be followed in each and every application. The pre-processing techniques discussed in the following sections should, rather, be seen as being applicable in certain circumstances and in particular cases. The investigator should decide which pre-processing techniques are relevant on the basis of the nature of the information to be extracted from the remotely-sensed data; see, for instance, the treatment by Chavez (1986) of digital sonar image data from the GLORIA instrument.

The pre-processing techniques described in section 4.2 are concerned with the removal of data errors and of unwanted or distracting elements of the image. In reality, of course, data errors such as missing scan lines cannot be removed; the data in error are simply replaced with some other data that are felt to be better estimates of the true but unknown values. Similarly, unwanted or distracting elements of the image (such as the sixth-line banding present on Landsat MSS images, as discussed in section 2.3.6.1) can only be eliminated or reduced by modifying all the data values in the image. These errors are caused by detector imbalance, and their correction is termed *radiometric pre-processing*.

Many actual and potential uses of remotely-sensed data require that these data conform to a particular map projection so that information on image and map can be correlated. Two examples will demonstrate the importance of this requirement; in Chapter 8 we see that the classification of a remotely-sensed image is best achieved by establishing the nature of ground cover categories by fieldwork and/or by air-photo and map analysis. In order that the information so derived can be related to the remotely-sensed image, some method of transforming from the scan line/pixel coordinate reference system of the image to the easting/northing coordinate system of the map is required. Secondly, if remotely-sensed data are to be used in association with other data within the context of a geographical information system (GIS), then the remotely-sensed data and products derived from such data (for example, classified images) will need to be expressed with reference to the geographical coordinates that are used for the rest of the data in the information system. In both these cases, there is a need for data pre-processing of a kind known as *geometric correction*. The same arguments can be put forward if the study involves measurements made on images produced on different dates; if information extracted from the two images is to be correlated then they must be registered, that is, expressed in terms of the same geographical coordinate system. Where an image is geometrically corrected so as to have the coordinate and scale properties of a map, it is said to be *georeferenced*. Geometric correction and registration of images are the topic of section 4.3.

Atmospheric effects on electromagnetic radiation (due primarily to scattering and absorption) are described in Chapter 1. These effects add to or diminish the true ground-leaving radiance, and act differentially across the spectrum. If estimates of radiance or reflectance values are to be successfully recovered from remote measurements then it is necessary to estimate the atmospheric effect and correct for it. Such *atmospheric corrections* are particularly important (i) whenever estimates of ground-leaving radiance or reflectance rather than relative values are required, for example in studies of change over time, or (ii) where the part of the signal that is of interest is smaller in magnitude than the atmospheric component. For example, the magnitude of the radiance upwelling from oceanic surfaces is generally very low, being much less than the atmospheric

path radiance (the radiance scattered into the field of view of the sensor by the gaseous and particulate components of the atmosphere). If any useful information about variations in radiance upwelling from the ocean surface is to be obtained from a remotely-sensed image then the component of the signal received at the sensor that emanates from the ocean surface must be separated from the larger atmospheric component (Figure 1.22). It is fair to say that no single method of achieving this aim has yet been established, and it is also true that most of the techniques that are in use today and which produce even approximately realistic results tend to be complex in nature. In my experience, simple techniques cannot solve complex problems. The more complex techniques are well beyond the scope of this book and will thus not be considered in any detail. Section 4.4 contains an introductory review of atmospheric correction techniques.

Sections 4.5 and 4.6 are concerned with the *radiometric correction* of images. If images taken in the optical and infrared bands at different times (multitemporal images) are to be studied then one of the sources of variation that must be taken into account is differences in the angle of the Sun. A low Sun angle image gives long shadows, and for this reason might be preferred by geological users because these shadows may bring out subtle variations in elevation. A high Sun angle will generate a different shadow effect. If the reflecting surface is Lambertian (which is, in most cases, a considerable oversimplification) then the magnitude of the radiant flux reaching the sensor will depend on the Sun and view angles. For comparative purposes, therefore, a correction of image pixel values for Sun elevation angle variations is needed. This correction is considered in section 4.5. The calibration of images to account for degradation of the detectors over time is the topic of section 4.6. Such corrections are essential if multitemporal images are to be compared, for changes in the sensor calibration factors will obscure real changes on the ground. The effects on recorded radiance levels of terrain slope and orientation are reviewed briefly in section 4.7.

The material in this chapter is introductory in scope. Research applications will require more elaborate methods of pre-processing. For example, orbital geometry models of geometric correction (section 4.3.1.1) may use advanced photogrammetric principles, while the use of the more sophisticated atmospheric correction procedures (section 4.4) requires knowledge of higher-level physics. The level of presentation adopted here is general and introductory, and is intended to provide a basic level of appreciation rather than a full understanding. More elaborate treatments are provided by Slater (1980), Elachi (1987) and by various contributors to Asrar (1989).

4.2 COSMETIC OPERATIONS

Two topics are discussed in this section. The first is the correction of digital images that contain either partially or entirely missing scan lines. Such defects can be due to errors in the scanning or sampling equipment, in the transmission or recording of image data, or in the reproduction of the media containing the data, such as magnetic tape or CD-ROM. Whatever their cause, these missing scan lines are normally seen as horizontal black or white lines on the image, represented by sequences of pixel values of 0 or 255. Their presence intrudes upon the visual examination and interpretation of images and also affects statistical calculations based on image pixel values. Methods to replace missing values with estimates of their true (but unknown) values are reviewed in section 4.2.1. This is followed by a brief discussion of methods of 'de-striping' imagery produced by electro-mechanical scanners such as those carried by Landsat (MSS and TM). As noted in Chapter 2, these scanners collect data for several scan lines simultaneously. The Landsat MSS instrument records six scan lines for each spectral band on each sweep of the scanning mirror, while the Landsat TM records 16. The radiance values along each of these scan lines are recorded by separate detectors. In a perfect world, each detector would give the same output if it received the same input. As we know, the world is far from perfect, and so over time the detectors making up the set of six (MSS) or 16 (TM) change at different rates. A systematic pattern is superimposed upon the image, repeating every six (or 16) lines. Techniques to remove this pattern are discussed in section 4.2.2. Note that they cannot be used with images recorded using solid-state (push-broom) scanners such as those carried by the SPOT satellites because each individual pixel across a scan line is recorded by the corresponding detector in the sensor. Hence, each column of pixels in a SPOT HRV image is recorded by the same detector. With more than 6000 columns in an image, the problem of correcting for variations in the detectors is rather more severe than that presented here for electro-mechanical scanners.

4.2.1 Missing scan lines

When missing scan lines occur on an image there is, of course, no means of knowing what values would have been present had the scanner/data recorder/tape drive

been working properly; the missing data have gone for ever. It is, nevertheless, possible to attempt to estimate what those values would be by looking at the image data values in the scan lines above and below the missing values. This approach relies upon a property of spatial data that is called *spatial autocorrelation*. The word 'auto' means 'self', thus autocorrelation is the relationship between one value in a series and a neighbouring value or values in the same series. Temporal autocorrelation is usually present in a series of hourly readings of barometric pressure, for example. The value at 11 o'clock tends to be very similar to the value at 10 o'clock unless the weather conditions are quite abnormal. Spatial autocorrelation is the correlation of values distributed over a two-dimensional or geographical surface. Points that are close in geographical space tend to have similar values of a variable of interest (such as rainfall or height above sea-level). The observation that many natural phenomena exhibit spatial autocorrelation is the basis of the estimation of missing values on a scan line from the adjacent values.

Method 1 is the simplest method to estimate a missing pixel value along a dropped scan line, and involves its replacement by the value of the corresponding pixel on the immediately preceding scan line. If the missing pixel value is denoted by $v_{i,j}$, meaning the value of pixel i on scan line j, then the algorithm is simply

$$v_{i,j} = v_{i,j-1}$$

Method 1 has the virtue of simplicity. It also ensures that the replacement value is one that exists in the near-neighbourhood of pixel (i, j). We will consider an averaging method next, and will see that, where the assumption of positive spatial autocorrelation does not hold, the average of two adjacent pixel values will produce an estimated replacement value that is quite different from either, whereas method 1 produces an estimate that is similar to at least one of its neighbours. Method 1 will need modification whenever the missing line (j) is the first line of an image. In that instance, the value $v_{i,j+1}$ could be used.

Method 2 is slightly more complicated; it requires that the missing value is replaced by the average of the corresponding pixels on the scan lines above and below the defective line, that is

$$v_{i,j} = (v_{i,j-1} + v_{i,j+1})/2$$

(taking the result to the nearest integer). Where the missing line is the first or last line of the image then method 1 can be used. As indicated earlier, if nearby pixel values are not highly correlated then the averaging method can produce hybrid pixels that are unlike their neighbours on the scan lines immediately above or below. This is likely to happen only in those cases where the missing line coincides with the position of a boundary such as that between two distinct land cover types, or between land and water.

Method 3 is the most complex. It relies on the fact that two or more bands of imagery are often available. Thus, Landsat MSS produces four spectral bands, Landsat TM produces seven and SPOT HRV produces three bands of imagery. If the pixels making up two of these bands are correlated on a pair-by-pair basis then high correlations are generally found for bands in the same region of the spectrum. For instance, the Landsat-1 to -3 MSS bands 4 and 5 in the green and red wavebands of the visible spectrum are normally highly correlated. The missing pixels in band k might best be estimated by considering contributions from (i) the equivalent pixels in another, highly correlated band, and (ii) neighbouring pixels in the same band, as in the two algorithms already described. If the neighbouring, highly correlated band is denoted by the subscript r then the algorithm can be represented by

$$v_{i,j,k} = M\{v_{i,j,r} - (v_{i,j+1,r} + v_{i,j-1,r})/2\}$$
$$+ (v_{i,j+1,k} + v_{i,j-1,k})/2$$

The symbol M in this expression is the ratio of the standard deviation of the pixel values in band k and the standard deviation of the pixel values in band r. This algorithm was first described by Bernstein *et al.* (1984) and is examined, together with the two algorithms outlined above, by Fusco and Trevese (1985). The conclusion of the latter authors is that the use of a second correlated band both reduces error and better preserves the geometric structures present in the image. They present some further results and elaborations of the basic algorithm, and readers wishing to go more deeply into the matter are referred to their paper.

The location of missing scan lines might not at first sight seem a topic worthy of serious consideration, for they are usually manifestly obvious when a defective image is examined visually. However, to locate such missing lines interactively using a cursor is a tedious task. The spatial autocorrelation property of images might be used as the basis for formulating a strategy that might allow missing scan lines to be located semi-automatically. If the average of the pixel values along scan line i is computed (with i running from 1 to n, where n is the number of scan lines in the image) then it might be reasonable to expect that the average of scan line i

differs from the average of scan lines $i + 1$ and $i - 1$ by no more than a value e. The parameter e would be determined by looking at the frequency distribution of the scan line averages over a number of images, or for a representative (and non-defective) part of the image under consideration. Step 1 would then involve locating all those scan lines with average values that deviated by more than e from the average of the preceding scan line. The first scan line of the image could either be omitted or compared with the second scan line. At the end of stage 1 we cannot be sure that the unexpectedly deviant behaviour of the scan lines picked out by this comparative method is the result of missing values. Stage 2 thus involves a search along each of the scan lines picked out at stage 1 for unexpected sequences of values. These unexpected sequences are most likely to be strings of extreme values, either 0 or 255. The beginnings and ends of such sequences are marked. At this stage the image can be displayed and a cursor used to mark the start of the suspect sequence. The operator is then able to check that the scan lines or portions of scan lines are indeed defective. Stage 3 consists of the application of one of the three methods described earlier, which allow the defective value to be replaced by an estimate of its true but unknown value. Note that isolated aberrant values such as speckle noise on synthetic aperture radar (SAR) images are removed by the use of filters such as the Lee filter or the median filter. These methods are described in Chapter 7.

4.2.2   De-striping methods

The presence of a systematic horizontal banding pattern is sometimes seen on images produced by electro-mechanical scanners such as Landsat's MSS and TM. This pattern is most apparent when seen against a dark, low-radiance background such as an area of water. The reasons for the presence of this pattern, known as *sixth-line banding* in Landsat MSS images, were given in section 2.3.6.1. It is effectively caused by the imbalance in the six detectors that are used by the Landsat MSS in the scanning process (16 in the case of TM images). The banding can be considered to be a cosmetic defect (like missing scan lines) in that it interferes with the visual appreciation of the patterns and features present on the image. If any statistical analysis of the pixel values is to be undertaken then the problem becomes somewhat more difficult. The pixel values recorded on a magnetic tape or CD-ROM are by no means 'raw' data, for they have been subjected to radiometric and geometric correction procedures at the ground receiving station. Hence, there does not seem to be much force in the argument that 'raw data' should not be interfered with. If we take as our starting point the assumption that the image data should be internally consistent (that is, areas of equal ground-leaving radiance should be represented by equal pixel values in the image, assuming no other complicating factors) then some kind of correction or compensation procedure would appear to be justified. Two reasons can thus be put forward in favour of applying a 'de-striping' correction to Landsat TM and MSS imagery: (i) the visual appearance and interpretability of the image are thereby improved, and (ii) equal pixel values in the image are more likely to represent areas of equal ground-leaving radiance, other things being equal.

Two methods of de-striping Landsat imagery are considered in this section. They are illustrated with reference to MSS images, although the same principles apply to images from the Thematic Mapper (if it is realised that the TM has 16, not six, detectors per band). Both methods are based upon the shapes of the six histograms of pixel values generated by the six detectors; these histograms are calculated from lines 1, 7, 13, 19, ... (histogram 1), lines 2, 8, 14, 20, ... (histogram 2) and so on. The first method characterises the relationship between the scene radiance r_{in} that is received at the detector and the value r_{out} that is output by the sensor system in terms of a linear function. The second method is non-linear in the sense that the relationship between r_{in} and r_{out} is not describable in terms of a single linear function; a piecewise function made up of small linear segments is used instead. Methods based on low-pass filtering (such as those described by Crippen (1989) and Pan and Chang (1992)) are mentioned in Chapter 7.

4.2.2.1 Linear method

The first method uses a linear expression to model the relationship between the input and output values. The underlying idea is quite simple, though it is based upon the assumption that each detector 'sees' a similar distribution of all the land cover categories that are present in the image area. If this assumption is satisfied, and the proportions of pixels representing water, forest, grassland, bare rock and so on are the same for each detector, then the histograms generated for a given band from the pixel values produced by the n detectors should be identical (n is the number of detectors used by the scanning instrument, for example, six for Landsat MSS and 16 for Landsat TM). This implies that the mean and standard deviation of the data from each detector should be the same. Detector imbalance is considered to be the only factor producing differences in means and standard deviations. To get rid of the striping effects of detector

imbalance, the means and standard deviations of the n histograms are equalised, that is, forced to equal a chosen value. Usually the means of the n individual histograms are made to equal the mean of all of the pixels in the image, and the standard deviations of the n individual histograms are similarly forced to be equal to the standard deviation of all of the pixels in the image. The overall mean is simply the mean of the n detector means, while the overall standard deviation is found from:

$$V = \frac{\sum n_i(\bar{x}_i^2 + v_i)}{\sum n_i} - \bar{X}^2$$

The terms in this equation are defined as follows: V is the overall variance (the overall standard deviation, $S = \sqrt{V}$), $\bar{X}$ is the overall mean ($= \Sigma x_i/6$ for Landsat MSS), v_i is the variance of detector i, $\bar{x}_i$ is the mean of detector i and n_i is the number of pixels processed by detector i.

A worked example is shown in Table 4.1. Note that the variance is simply the square of the standard deviation, so that the standard deviation of the digital counts making up the image is the square root of V. These statistics should not be computed for small samples or they will be biased. An image size of at least 512×512 is recommended for efficient estimation.

The individual detector means and standard deviations can be computed in one of two ways. Both methods assume that scan lines 1, 7, 13, ... (in the case of the Landsat MSS, with its six detectors) were produced by detector 1, while scan lines 2, 8, 14, ... were produced by detector 2, and so on. The first method of computing the individual detector means and standard deviations uses the following standard formulae for each of the n sets of scan lines:

$$\bar{x}_i = \frac{\sum r_{ij}}{n_i}$$

$$s_i = \frac{\sum r_{ij}^2}{n_i} - \bar{x}_i^2$$

In these expressions, $\bar{x}_i$ and s_i are the mean and standard deviation respectively of the ith detector where the index i runs from 1 to n_i, r_{ij} is the jth pixel value output by the ith of the n detectors and n_i is the number of pixel values recorded by detector i.

The use of these standard formulae is not recommended. Although they represent the obvious way of calculating the mean and standard deviation of a set of data, they have two defects. First, since the magnitudes of the

Table 4.1   Showing the calculation of individual detector means and standard deviations, and the derivation of the overall mean and standard deviation from the results for individual detectors.

Detector	Pixel values	Mean $(\bar{x}_i)$	Standard deviation (s_i)	Variation (s_i^2)
1	1 3 2 4 6	3.2	1.720	2.96
2	3 6 2 3 8	4.4	2.245	5.04
3	4 3 4 2 9	4.4	2.417	5.84
4	2 4 3 3 7	3.8	1.720	2.96
5	0 2 2 2 6	2.4	1.959	3.84
6	4 3 3 3 9	4.4	2.332	5.44

Overall mean = (sum of the individual detector means)/6 = 3.76
$n_i = 5 \ (i = 1, \ldots, 6)$
$\Sigma n_i(\bar{x}_i + v_i) = 573$
Divided by Σn_i (= 30) = 19.100
Overall mean (3.766) squared = 14.183
Overall variance = 19.100 − 14.183 = 4.912
Overall standard deviation = square root of overall variance = $\sqrt{4.912} = 2.216$*

*Computation of the overall standard deviation using all 30 pixel values directly also gives an answer of 2.216.

summation term $\Sigma r_{ij}^2/n_i$ and the term $\bar{x}_i^2$ are likely to exceed the capacity of an integer store location, the expression for the standard deviation will require the use of floating-point arithmetic, which is inherently slower than integer or fixed-point arithmetic on most small and medium-sized computers, though that is less of a consideration today. Secondly, the expressions were designed as short-cut formulae for use with small data sets on hand calculators. They do not take into account the characteristics of floating-point arithmetic as executed on digital computers. The subtraction of two large floating-point quantities (such as the two terms on the right-hand side of the expression for the standard deviation) can produce results that are in error in both sign and magnitude. Given that the number of pixels in an image is large, then the mean sum of squares ($\Sigma r_{ij}^2/n_i$) and the square of the mean ($\bar{x}_i^2$) are both likely to be large, and so computational errors could occur. These errors are difficult to trap, hence great caution is necessary. Mather (1976, pp. 32–35) discusses this point further and describes a recurrence relationship for the calculation of the mean and variance:

$$\text{mean}(i) = \text{mean}(i-1) + [r(i) - \text{mean}(i-1)]/i$$
$$\text{cssq}(i) = \text{cssq}(i-1) + i(i-1)[r(i) - \text{mean}(i-1)]^2$$

where i runs from 2 to m, the number of elements in the data. The values mean(1) and cssq(1) (the corrected sum of squares) are set to zero before these formulae are applied. The standard deviation is then found by taking the square root of the term cssq(m)/m.

The method can be rejected on the grounds of computational speed and/or accuracy. An alternative method is preferable on both grounds. It involves the calculation of the histograms of the pixel values recorded by the n detectors – giving n frequency histograms – and the use of 'grouped data' formulae to derive the mean and standard deviation. It is both faster and, for large data sets, more accurate than method 1. This observation emphasises the need to consider algorithm choice and design very carefully, for the processing of remotely-sensed images generally involves the handling of large quantities of data. Efficiency and accuracy should be the prime aims. A worked example for method 2 is given in Table 4.2. The expressions for mean and standard deviation derived from grouped data are:

$$\bar{x}_i = \frac{\sum f_j j}{\sum f_j}$$

$$s_i^2 = \frac{\sum f_j (j - \bar{x}_i)}{\sum f_j}$$

where $\bar{x}_i$ and s_{ij}^2 are the mean and variance of detector i. The standard deviation is the square root of the variance. The elements of $\mathbf{f}$ ($= f_j$) are the counts in the histogram classes with j running from 0 to n, the maximum grey level in the image (normally 63, 127 or 255). The second method is inherently faster than the first method because the only processing in which the pixels are involved is the calculation of the n histograms. This processing can be done using integer arithmetic, which

Table 4.2 Computation of mean and standard deviation for grouped data. The mean of the values recorded by detector i is computed first using the formula $\bar{x}_i = \Sigma f_i j / \Sigma f_j$ then the standard deviation is derived from

$$s_i = \sqrt{[\Sigma f_i (j - \bar{x}_i)^2 / \Sigma f_j]}$$

This method is both faster and more accurate than the standard short-cut formulae (see text).

(1)	(2)	(3)	(4)
j	f_j	$f_j j$	$f_j (j - \bar{x}_i)^2$
0	8	0	117.153
1	10	10	79.906
2	14	28	46.719
3	16	48	10.937
4	25	100	0.750
5	32	160	44.047
6	14	84	66.121
7	8	56	80.555
	127	486	446.188

$\bar{x}_i = \Sigma(3)/\Sigma(2) = 486/127 = 3.827$

$s_i = \sqrt{[\Sigma(4)/\Sigma(2)]}$

$\quad = \sqrt{[446.188/127]}$

$\quad = \sqrt{3.513}$

$\quad = 1.874$

is generally faster than floating-point arithmetic. The subsequent calculation of mean and standard deviation is based on the histograms, which have only 256 classes so that the definitional formula for standard deviation can be used with only insignificant overheads.

We now have the statistics needed to carry out the de-striping process. Recall that our aim is to adjust the pixel values r_{ij} so that the n detector means $\bar{x}_i$ become equal to the overall mean $\bar{X}$, and the n detector standard deviations s_i become equal to the overall standard deviation S. The standard deviation and mean can be considered to be equivalent to the gain and offset characteristics of the sensor system:

$$a_i = \frac{S}{s_i}$$

$$b_i = \bar{X} - a_i \bar{x}_i$$

and the corrected r'_{ij} are then calculated from

$$r'_{ij} = a_i r_{ij} + b_i$$

A worked example is given in Table 4.3. Note that the means and standard deviations of the corrected data are not identical, but it is easily verified that if the means

and standard deviations of the corrected data, prior to rounding to the nearest integer, are computed then all detectors have a mean of 7.4 and a standard deviation of 4.484 05. The effects of rounding to the nearest integer are probably exaggerated in a small-scale example, and would not show such variability in a real application.

If method 1 were applied on a pixel-by-pixel basis to an image of any size then it would be inordinately slow, for floating-point arithmetic is involved. Even if a large, fast computer is used, however, one should always endeavour to use efficient algorithms. A little thought will show that, for Landsat MSS images from European Space Agency sources and for Landsat TM images, there are only 256 possible values for each detector (this number may drop to 128 or 64 for Landsat MSS images from other receiving stations). We could build a table giving (i) the input pixel value as the row number in the table and (ii) the n corrected values, one per detector, for that row. The principle is simple. For a given pixel, the row number in the table is the input pixel value while the output (corrected) value is the ith value on that row, where i is the detector number ($1 \le i \le n$). The table is known as a *lookup table*. Lookup tables have already been mentioned in the context of digital image display systems, where they are known as LUTs (section 3.3.1.2), and we will meet them again in later chapters.

4.2.2.2 Histogram matching

The method of de-striping Landsat MSS and TM images described in section 4.2.2.1 was based on the assumption that the output from a detector is a linear function of the input or received value according to the expression

$$r_{\text{out}} = \text{offset} + \text{gain} \times r_{\text{in}}$$

Horn and Woodham (1979) observe that

'... it appears that different gains and offsets are appropriate for different scene radiance [r_{in}] ranges. That is, the sensor transfer curves are somewhat non-linear.'

In other words, the linear relationship between r_{in} and r_{out} used in section 4.2.2.1 is an oversimplification. The method described in this section uses the shape of the cumulative frequency histogram of each detector to find an estimate of the non-linear transfer function. The ideal or target transfer function is taken to be defined by the shape of the cumulative frequency histogram of the whole image, which is easily found by carrying out a class-by-class summation of the n individual detector

Table 4.3   De-striping using equalisation of means and standard deviations.

(a) Observed data: 10 observations, six detectors.

Observation	Detector number					
	1	2	3	4	5	6
1	15	13	15	14	12	13
2	13	12	13	13	9	12
3	8	7	9	9	7	8
4	11	9	12	12	10	10
5	12	10	12	12	11	11
6	3	2	4	3	2	3
7	2	2	2	2	1	3
8	1	0	2	0	0	2
9	5	4	5	4	3	4
10	9	8	10	9	6	9

(b) Means and standard deviations of the six detectors, and overall mean and standard deviations, calculated by method shown in Table 4.1 or 4.2.

Detector number	Mean	Standard deviation
1	7.9	4.6787
2	6.7	4.2673
3	8.4	4.5431
4	7.8	4.8539
5	6.1	4.1581
6	7.5	3.9306
overall	7.4	4.48405

(c) Gains and offsets of the six detectors.

Detector number	Gain	Offset
1	0.9584	−0.1714
2	1.0508	0.3596
3	0.9870	−0.8908
4	0.9238	0.1944
5	1.0784	0.8218
6	1.1408	−1.1560

(d) Corrected data (rounded to the nearest integer).

Observation	Detector number					
	1	2	3	4	5	6
1	14	14	14	13	14	14
2	12	13	12	12	11	13
3	7	8	8	9	8	8
4	10	10	11	11	12	10
5	11	11	11	11	13	11
6	3	2	3	3	3	2
7	2	2	1	2	2	2
8	1	0	1	0	1	0
9	5	5	4	4	4	3
10	8	9	9	9	7	9

(e) Means and standard deviations of the corrected data. Differences are due to rounding error (see text).

Detector number	Mean	Standard deviation
1	7.3	4.2438
2	7.4	4.6519
3	7.4	4.5431
4	7.4	4.4542
5	7.5	4.5880
6	7.3	4.6487

histograms. The histogram for detector 1 is computed from the pixel values on lines 1, 7, 13, ... of the image, while the histogram for detector 2 is derived from the pixel values on lines 2, 8, 14, ... and so on. The histograms are expressed in cumulative form (so that class 0 is the number of pixels with a value of 0, class 1 is the number of pixels with values 0 or 1, and so on). Next, each histogram class frequency is divided by the number of pixels counted in that histogram, thus ensuring that the six individual histograms and the target histogram are all scaled between 0 and 1.

At this stage, we have n individual cumulative histograms and one target cumulative histogram. Our aim is to adjust the individual cumulative histograms so that they match the shape of the target cumulative histogram as closely as possible. This is done by adjusting the class numbers of the individual histograms. Thus, class number x in an individual histogram may be equated with class number y in the target histogram. This means that all pixels scanned by the detector corresponding to that individual histogram, and which have the value x, would be replaced in the de-striped image by the value y, which is derived from the target histogram. In order to determine the class number in the target histogram to be equated to class number x in the individual histogram, we find the first class in the target histogram whose cumulative frequency count equals or exceeds the cumulative frequency value of class x in the individual histogram. The class in the target histogram that is found is class y. An example is given in Table 4.4. The frequency value for cell 3 of an individual histogram is 0.57. This value is compared with the target histogram values until the first class with a value greater than or equal to 0.57 is found. This is class 4 of the target histogram. Class 3 of the detector histogram is thus equated to class 4 of the target histogram, and all pixel values of 3 scanned by that detector are replaced with the value 4. The procedure is applied separately to all 256 values for each of the six (or 16) detectors. The result is generally a reduction in the banding effect, though

Table 4.4 Example of histogram matching for de-striping Landsat MSS and TM images. The target histogram is the cumulative histogram of the entire image or sub-image. The detector histogram is the cumulative histogram using values of pixels scanned by one of Landsat MSS's six detectors. The output pixel value to replace a given input value is found by comparison of the two histograms. For example, the detector histogram for input pixel value 3 is 0.57. The first value in the target histogram to equal or exceed 0.57 is that in row four. Hence, the pixel values in the uncorrected image that are generated by this detector are replaced by the value 4. See Table 3.3 for an example of a lookup table.

Input pixel value	Target histogram value	Detector histogram value	Output pixel value
0	0.09	0.08	0
1	0.18	0.11	1
2	0.33	0.18	2
3	0.56	0.57	4
4	0.60	0.66	4
5	0.76	0.78	6
6	0.95	0.95	6
7	1.00	1.00	7

much depends on the nature of the image. Wegener (1990) gives a critical review of the Horn and Woodham (1979) procedure, and presents a modified form of the algorithm.

The table lookup procedure described at the end of section 4.2.2.1 can be used to make the application of this method more efficient. Thus, for each detector histogram, a table can be constructed so that the output value corresponding to a given input can be easily read. The input value is the pixel value in the image being corrected while the output value is its equivalent in the de-striped image.

A program to implement this method of de-striping (nl_destr) is provided on the CD that accompanies this book (see section B.3 of Appendix B).

4.3   GEOMETRIC CORRECTION AND REGISTRATION

Remotely-sensed images are not maps. Frequently, however, information extracted from remotely-sensed images is to be integrated with map data in a geographical information system or presented to consumers in a map-like form (for example, gridded 'weather pictures' on TV or in a newspaper). The transformation of a remotely-sensed image so that it has the scale and projection properties of a map is called *geometric correction*. A related technique, called *registration*, is the fitting of the coordinate system of one image to that of a second image of the same area. For example, a set of images of a given area might be obtained for different dates and the user may wish to measure changes that have occurred in the period of time that elapsed between the collection of the two images. A map is defined as

'. . . a graphic representation on a plane surface of the Earth's surface or part of it, showing its geographical features. These are positioned according to pre-established geodetic control, grids, projections and scales.'

(Steigler, 1978)

A map projection is a device for the representation of a curved surface (that of the Earth) on a flat sheet of paper (the map sheet). Many different map projections are in common use (see Snyder, 1982; Steers, 1962; Frei *et al.*, 1993). Each projection represents an effort to preserve some property of the mapped area, such as uniform representation of areas or shapes, or preservation of correct bearings. Only one such property can be correctly represented, though several projections attempt to compromise by minimising distortion in two or more map characteristics. The UK Ordnance Survey uses a Transverse Mercator projection. A regular grid, with its origin to the south-west of the British Isles, is superimposed on the map sheet since lines of latitude and longitude plot as complex curves on the Transverse Mercator projection.

Geometric correction of remotely-sensed images is therefore required when the image, or a product derived from the image such as a vegetation index (Chapter 6) or a classified image (Chapter 8), is to be used in one of the following circumstances (Kardoulas *et al.*, 1996):

• to transform an image to match a map projection,
• to locate points of interest on map and image,
• to bring adjacent images into registration,
• to overlay temporal sequences of images of the same area, perhaps acquired by different sensors, and
• to overlay images and maps within a GIS.

The sources of geometric error in digital satellite imagery are (i) instrument error, (ii) panoramic distortion, (iii) Earth rotation, and (iv) platform instability (Bannari *et al.*, 1995a). Instrument errors include distortions in the optical system, non-linearity of the scanning mechanism and non-uniform sampling rates. Panoramic distortion is a function of the angular field of view of the sensor and affects instruments with a wide angular field of view (such as the AVHRR and CZCS) more than those with a narrow field of view (such as the

Landsat MSS and TM and the SPOT HRV). Earth rotation velocity varies with latitude. The effect of Earth rotation is to skew the image. Consider the Landsat satellite as it moves southwards above the Earth's surface. At time t, its sensor scans image lines 1–6. At time $t + 1$, lines 7–12 are scanned. But the Earth has moved eastwards during the period between time t and time $t + 1$, therefore the start of scan lines 7–12 is slightly further west than the start of scan lines 1–6. Similarly, the start of scan lines 13–18 is slightly further west than the start of scan lines 7–12. The effect is shown in Figure 4.1. Platform instabilities include variations in altitude and attitude. All four sources of error contribute unequally to the overall geometric distortion present in an image. In this section, we deal with the geometric correction of high-resolution digital images such as those acquired by the Landsat TM and SPOT HRV instruments. Correction of wide-angle images derived from the NOAA AVHRR is described by Brush (1985), Crawford *et al.* (1996), Moreno and Melia (1993) and Tozawa (1983). Geocoding of SAR images is covered by Dowman (1992), Dowman *et al.* (1993), Johnsen *et al.* (1995) and Schreier (1993a). The use of digital elevation data to correct images for the geometric distortion produced by relief variations is considered by Blaser and Caloz (1991), Itten and Meyer (1993), Kohl and Hill (1988), Palà and Pons (1995), Toutin (1995) and Wong *et al.* (1981). Other useful references are Kropatsch and Strobl (1990), Kwok *et al.* (1987), Novak (1992), Shlien (1979),

van Wie and Stein (1977) and Westin (1990). Williams (1995) provides an excellent treatment of many aspects covered in this section, including geocoding of SAR and AVHRR imagery.

The geometric correction process can be considered to include (i) the determination of a relationship between the coordinate system of map and image (or image and image in the case of registration), (ii) the establishment of a set of points defining pixel centres in the corrected image that, when considered as a rectangular grid, define an image with the desired cartographic properties, and (iii) the estimation of pixel values to be associated with those points. The relationship between the two coordinate systems (map and image) could be defined if the orbital geometry of the satellite platform were known to a sufficient degree of accuracy. Where orbital parameters are known, methods based upon orbital geometry give high accuracy. Otherwise, orbital models are useful only where the desired accuracy is not high, or where suitable maps of the area covered by the image are not available. The method based on nominal orbital parameters is described in section 4.3.1.1, while the map-based method is covered in section 4.3.1.2. The extraction of the locations of the pixel centre points for the corrected image and the estimation of pixel values to be associated with these output points are considered in section 4.3.1.3.

4.3.1 Coordinate transformations

4.3.1.1 *Orbital geometry model*

Orbital geometry methods are based on knowledge of the characteristics of the orbit of the satellite platform. Bannari *et al.* (1995a) describe two procedures based on the photogrammetric equations of collinearity. These equations describe the properties of the satellite orbit and the viewing geometry, and relate the image coordinate system to the geographical coordinate system. They require knowledge of the geographical coordinates of a number of points on the image. Such points are known as *ground control points* or GCP.

A simple method of correcting the coordinate system of remotely-sensed images using approximate orbit parameters is described by Landgrebe *et al.* (1975). It is suitable only for use with images derived from sensors that have a narrow angular field of view, such as the Landsat TM and the SPOT HRV. The description given below refers to Landsat-1 to -3 MSS data, though it is easily extended to deal with imagery from any similar sensor, provided that the nominal orbital parameters of the satellite are known or can be estimated.

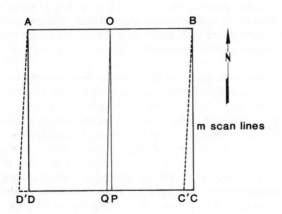

Figure 4.1 Effects of Earth rotation on the geometry of a line-scanned image. At the time the first line AB is scanned the potential mth scan line is in the position CD. By the time the satellite has moved forward to the mth scan line, Earth rotation to the east results in line C'D' being scanned. Hence the line-scanned image covers area ABC'D' and the satellite's ground track is OQ rather than OP.

Since the method is based upon nominal orbital parameters the accuracy of the geometrically corrected image produced by this technique will not be high. Landgrebe *et al.* (1975) suggest that the magnitude of the error is of the order of 1–2%, meaning that if the corrected image is overlaid on a map and both are aligned with reference to a well-defined point, then the error in measured coordinate positions of other points will be 1–2%. The distortions are corrected by the application of the following transformation matrices. This simple transformation is explained in some detail not because of its value as a method of geometric correction but because its derivation illustrates some of the principles involved in image data collection from space.

(*a*) *Aspect ratio.*   Landsat MSS scan lines are nominally 79 m apart, whereas the pixels along each scan line are spaced at a distance of 56 m. Since the instantaneous field of view of the MSS is 79 m there is over-sampling in the across-scan direction. As we generally require square rather than rectangular pixels, we can choose 79 m square pixels or 56 m square pixels to overcome the problem of unequal scale in the *x*- and *y*-directions. Because of the over-sampling in the across-scan direction, it is more reasonable to choose 79 m square pixels. The aspect ratio (the ratio of the *x*: *y* dimensions) is 56:79 or 1:1.41. The first transformation matrix, $\mathbf{M}_1$, which corrects the image to a 1:1 aspect ratio, is therefore

$$\mathbf{M}_1 = \begin{bmatrix} 1.41 & 0.00 \\ 0.00 & 1.00 \end{bmatrix}$$

This correction is not required if the image to be geometrically transformed has square pixels.

(*b*) *Skew correction.*   Landsat MSS images are skewed with respect to the north–south axis of the Earth. Landsat-1 to -3 had an orbital inclination of 99.09° whereas Landsat-4 and -5 have an inclination of 98.2°. The satellite heading (across-scan direction) at the equator is therefore 9.09° and 8.2° respectively, increasing with latitude. The skew angle θ at latitude L is given (in degrees) by

$$\theta = 90 - \cos^{-1}\left(\frac{\sin \theta_E}{\cos L}\right)$$

where θ_E is the satellite heading at the equator and the expression $\cos^{-1}(x)$ means: the angle whose cosine is x, that is, the inverse cosine of x. Given the value of θ the

coordinate system of the image can be rotated through θ degrees anticlockwise so that the scan lines of the corrected image are oriented in an east–west direction using the transformation matrix $\mathbf{M}_2$:

$$\mathbf{M}_2 = \begin{bmatrix} \cos \theta & \sin \theta \\ -\sin \theta & \cos \theta \end{bmatrix}$$

The value of L that should be used in the determination of θ is the centre latitude of the image being corrected. Since the latitude of the satellite is varying continuously, this value will be only an approximation but will serve as a first estimate.

For $L = 51°$, the value of θ is given by:

$$\theta = 90° - \cos^{-1}\left(\frac{\sin(9.09°)}{\cos(51.00°)}\right)$$

$$= 90° - \cos^{-1}\left(\frac{0.1580}{0.6293}\right)$$

$$= 90° - 75.4590° = 14.5410°$$

and the elements of matrix $\mathbf{M}_2$ are:

$$\mathbf{M}_2 = \begin{bmatrix} 0.9680 & 0.2511 \\ -0.2511 & 0.9680 \end{bmatrix}$$

(*c*) *Earth rotation correction.*   As the satellite moves southwards along its orbit, the Earth rotates beneath it in an easterly direction with a surface velocity proportional to the latitude of the nadir point. To compute the displacement of the last line in the image relative to the first line we need to determine (i) the time taken for the satellite sensor to scan the image and (ii) the eastwards velocity of the Earth. The distance travelled by the Earth's surface can then be obtained by multiplying time by velocity. The time taken for the satellite sensor to scan the image can be found if the distance travelled by the satellite and the satellite's velocity are known. Both distance and velocity are expressed in terms of angular measure. If A is a point on the Earth's surface corresponding to the centre of the first scan line in the image, and if B is the corresponding point for the last scan line in the image, then the line AB represents an arc of a circle centred at the Earth's centre. The angle at the Earth's centre O given by angle AOB can be calculated because the Earth's equatorial radius (OA or OB) is 6378 km, and an angle (in radian measure) is equal to arc length divided by radius (Figure 1.4(a)). The arc length AB is the distance between the centre pixels in the first and last scan lines of the image. For Landsat MSS and TM images, AB is 185 km. Hence angle AOB (representing the angular distance moved by the Landsat satellite during the capture of one image) equals 185/6378 or 0.029 radians.

The orbital period (the time required for one full revolution) for Landsats-1 to -3 is 103.267 minutes, so the satellite's angular velocity ω_O is $2\pi/(103.267 \times 60)$ radians per second, or $0.001\,014\,\text{rad s}^{-1}$. The problem is now to find the time required for a satellite travelling at this angular velocity to traverse through an angle of 0.029 radians. The answer is found by dividing the angular distance to be moved by the angular velocity, and the result of this operation is $0.029/0.001\,014 = 28.6$ seconds.

Now the question becomes: how far will point B (the centre of the last scan line) move eastwards during the 28.6 seconds that elapses between the scanning of the first and last scan lines of the MSS or TM image? The answer depends on latitude. For simplicity, we will take the latitude (L) of the centre of the image. The Earth's surface velocity at latitude L is $V_E(L)$ which is defined as:

$$V_E(L) = R \cos(L)\, \omega_E$$

where R is the Earth's radius, defined above, and ω_E is the Earth's angular velocity. Since the Earth rotates once in 23 hours, 56 minutes and 4 seconds (that is, 86 164 seconds), then its angular velocity is simply $2\pi/86\,164$ radians per second, or $0.7292 \times 10^{-4}\,\text{rad s}^{-1}$. If L is 51° then $V_E(L)$ equals $6378 \times 10^3 \times 0.6293 \times 0.72921 \times 10^{-4}\,\text{m s}^{-1}$, which equals $292.7\,\text{m s}^{-1}$. Now that we know (i) that the time taken to scan an entire Landsat TM or MSS image is 28.6 seconds and (ii) the Earth's surface velocity, then the calculation of the eastward displacement of the last scan line in the image can be obtained. At 51°N the surface velocity is $292.7\,\text{m s}^{-1}$ so the distance travelled eastwards is $292.7 \times 28.6 = 8371\,\text{m}$.

These calculations assume that the line AB is oriented along a line of longitude whereas, in fact, the Landsat-1 to -3 satellites had an orbit that is skewed relative to lines of longitude, as noted in the calculation of matrix $\mathbf{M_2}$ above. The skew angle θ for 51° latitude is 14.54° (see above), so the actual eastwards displacement is $8371 \times \cos(14.54°) = 8103\,\text{m}$. These computations are summarised by the term a_{sk}:

$$a_{sk} = \frac{\omega_E \cos L}{\omega_O \cos\theta} = 0.0719\,\frac{\cos L}{\cos\theta}$$

where ω_E is the Earth's angular velocity, ω_O is the

satellite's angular velocity (both given above), and θ and L are defined above. The transformation matrix $\mathbf{M_3}$ is:

$$\mathbf{M_3} = \begin{bmatrix} 1 & a_{sk} \\ 0 & 1 \end{bmatrix}$$

At 51° latitude, $\mathbf{M_3}$ is:

$$\mathbf{M_3} = \begin{bmatrix} 1 & 0.0467 \\ 0 & 1 \end{bmatrix}$$

Note that 'fill pixels' are added to the start of each scan line of a Landsat MSS or TM image by some ground stations to compensate for the Earth rotation effect. If this correction is thought to be sufficient then transformation $\mathbf{M_3}$ can be omitted. Alternatively, since the number of fill pixels is given in the header/trailer data associated with each scan line in the European Space Agency format, these fill pixels can be stripped off and the correction $\mathbf{M_3}$ applied. The effects of Earth rotation are illustrated in Figure 4.1.

The three transformation matrices $\mathbf{M_1}$, $\mathbf{M_2}$ and $\mathbf{M_3}$ given in subsections (a) to (c) above are not applied separately. Instead, a composite transformation matrix, $\mathbf{M}$, is obtained by multiplying the three separate transformation matrices:

$$\mathbf{M} = \mathbf{M_1 M_2 M_3}$$

The corrected image coordinate system is related to the raw image coordinate system by

$$\mathbf{x}' = \mathbf{M x}$$

where $\mathbf{x}'$ ($= x_1'$, x_2') is the vector holding the pixel and line coordinates of the corrected pixel and $\mathbf{x}$ ($= x_1, x_2$) is the original (pixel, line) coordinate. Remember that the origin of the image coordinate system is the top left corner of the image.

4.3.1.2 Transformation based on ground control points

The orbital geometry model discussed in section 4.3.1.1 is based on nominal orbital parameters. It takes into account only selected factors that cause geometric distortion. Variations in the altitude or attitude of the platform are not considered simply because the information needed to correct for these variations is not generally available. Future satellites will carry instruments that will provide precise orbital data, and more complex models than the one described in section 4.3.1.1 will be used to generate geometrically corrected images. An alternative method is to look at the problem from the opposite end and, rather than attempt to define the sources of error and the direction and magnitude of their effects, use a method that compares differences between the positions of points recorded on image and map. From these differences, the nature of the distortions present in the image can be estimated, and an empirical transformation to relate the image and map coordinate systems can be computed. The aim of the procedure described in this section is to produce a method of converting map coordinates to image coordinates, and vice versa. Two pieces of information are required. The first is the location of the image corners on the map. Once the image area is outlined on the map, the map coordinates of the pixel centres (at a suitable scale) can be produced. The map coordinates of the image corners are found by determining an image-to-map coordinate transformation. The map coordinates of the required pixel centres are converted to image coordinates by a map-to-image coordinate transformation. Both transformations are explained in this section. The final stage, that of associating pixel values with the calculated pixel positions, is discussed in the next section under the heading of resampling.

The method relies on the availability of an accurate map of the area at a suitable scale. In some parts of the world suitable maps are not available. It is thus rather paradoxical that accurate 'image maps' can only be produced for areas for which conventional maps are available. Until the position and attitude of the satellite platform with respect to the Earth is known accurately the only alternative to the use of map-derived ground control points is to locate suitable control points on the ground using the Global Positioning System (GPS). Clavet et al. (1993) and Kardoulas et al. (1996) provide details of the use of GPS in locating ground control points for geometric correction. Cook and Pinder (1996) compare the transformation accuracy resulting from the use of control points derived from 1:24 000 US Geological Survey maps and from the use of GPS. They summarise the problems involved in the accurate measurement of control points from maps, and conclude that differential GPS provides substantially better results than map digitising. It is, however, more costly in terms of both equipment and travel time as each control point must be visited and its coordinates measured.

Ground control points are well-defined and easily recognisable features that can be located accurately on a map and on the corresponding image. The symbols (x_i, y_i) will be used to refer to the map coordinates of the ith

ground control point, and the symbols (c_i, r_i) to refer to the image coordinates of the same point. The map coordinates can be expressed in eastings and northings from an arbitrary origin, or in degrees and decimal degrees of longitude and latitude. The image coordinates are expressed as column (pixel number along the scan line) and row. We are seeking a method that will allow us to convert from (x, y) to (c, r) coordinates and vice versa. To achieve this aim, the method of least squares is used. Assume that a sample of s_i and t_i values is available, where s and t are any two variables of interest. Furthermore, assume that values of t_i are easily found (for example the line or column number of a pixel in a digital image) whereas the values s_i are difficult to acquire (for example the map coordinates of the same pixel). The method of least squares allows the estimation of s_i given the corresponding t_i using a function of the form:

$$s_i = a_0 + a_1 t_i + e_i$$

This equation allows the difficult-to-measure s_i to be estimated from the value of the corresponding easy-to-measure t_i. The estimated value of s_i is written as $\hat{s}$. The terms a_0 and a_1 are computed from a sample of values of s and t using the method of least squares, and have the property that the sum of the squared differences between the true (and usually unavailable) values s_i and the values $\hat{s}_i$ estimated from the equation is a minimum for all possible values of s_i and $\hat{s}_i$. These differences are called 'residuals', and they are represented by the term e_i. Thus, we could omit e_i from the equation and express $\hat{s}$ directly as:

$$\hat{s}_i = a_0 + a_1 t_i$$

The linear least-squares function is used to find the least-squares coefficients for the following four expressions:

- map x coordinate as a function of image c and r coordinates,
- map y coordinate as a function of image c and r coordinates,
- image c coordinate as a function of map x and y coordinates, and
- image r coordinate as a function of map x and y coordinates.

If the coefficients of each of these functions are known it is possible to transform from map (x, y) to image (c, r) coordinates or from image (c, r) to map (x, y). The form of each function is similar to the standard least-squares functional form given above. However, two predictor variables are used, so the expression is a little more complicated:

$$\hat{s}_i = a_0 + a_1 t_i + a_2 u_i$$

where one set of three variables as given in the bulleted list above is substituted for $\hat{s}$, t and u in turn. For example, the first expression bulleted in the list above would be written as:

$$\hat{x}_i = a_0 + a_1 c_i + a_2 r_i$$

This equation says that the least-squares estimate of the map x coordinate of the ground control point labelled i can be found from the image column and row coordinates of that ground control point (c_i and r_i) if the least-squares coefficients a_1, a_2 and a_3 are known. The values of these coefficients are determined from a sample of values of x, c and r, as described below, and are then applied to all the image pixels in order to estimate map x coordinates. The same operation is performed to find the map y coordinates of all of the pixels in the image. The steps involved are described in the following paragraphs. First, however, some of the technical details involved in the calculations are described.

The equation used in the preceding paragraphs is a first-order polynomial least-squares function. It is first-order because neither of the predictor variables (on the right-hand side of the equation) is raised to a power greater than one. A first-order function can accomplish scaling, rotation, shearing and reflection but not warping (such as would be necessary to correct for panoramic distortion or for any similar 'bending' effect). A second- or higher-order polynomial can be used to model such distortions, though in practice it is rare for polynomials of order higher than three to be used for satellite imagery. Where a relatively small image area is being corrected (for instance, a 512×512 segment) it should be unnecessary to correct for warping at all. Note that the polynomial method does not correct for relief distortions and is therefore applicable only to relatively flat areas. See Kohl and Hill (1988) and other references listed towards the end of the introduction to section 4.3 for details of a modification to the standard polynomial correction that corrects for relief-induced variations using a digital elevation model (DEM).

A polynomial function in two variables t and u can be written concisely as:

$$\hat{s} = \sum_{j=0}^{m} \sum_{k=0}^{m-j} a_{jk} t^j u^k$$

where m is the order of the polynomial function. A third-order polynomial, written out in full, becomes:

$$\begin{aligned}
\hat{s} = {} & a_{00} t^0 u^0 + a_{01} t^0 u^1 + a_{02} t^0 u^2 + a_{03} t^0 u^3 \\
& + a_{10} t^1 u^0 + a_{11} t^1 u^1 + a_{12} t^1 u^2 + a_{20} t^2 u^0 \\
& + a_{21} t^2 u^1 + a_{30} t^3 u^0
\end{aligned}$$

which can be reduced to a less formidable form as:

$$\begin{aligned}
\hat{s} = {} & a_{00} + a_{10} t + a_{01} u + a_{20} t^2 + a_{11} tu + a_{02} u^2 \\
& + a_{30} t^3 + a_{21} t^2 u + a_{12} tu^2 + a_{03} u^3
\end{aligned}$$

The terms s, t and u are replaced by the appropriate terms in the expressions in the bulleted list above. If, for instance, we wished to estimate y as a function of c and r we would replace s, t and u in the polynomial expansion by y, c and r. There is thus one polynomial function for each of the four coordinate transformations. The first pair gives map coordinates (x, y) in terms of image coordinates (c, r) while the second pair gives image coordinates (c, r) in terms of map coordinates (x, y).

Before we consider methods of evaluating polynomial expressions for given sets of (x, y) and (c, r) coordinates we should consider (i) the size of the sample of control points needed to give reliable estimates of the coefficients a_{ii}, (ii) the spatial distribution of the control points, and (iii) the accuracy with which they are located. In mathematical terms we need to take a sample of n control points where n is the number of coefficients in the polynomial expression. For a first-order polynomial, n is equal to 3. For a second-order polynomial n is 6, and for a third-order equation n is 10. These sizes are necessary purely and simply to ensure that it is *mathematically* possible to evaluate the equations defining the coefficients a_{ij}. It is important to note that the *statistical* requirement, which is concerned not with the feasibility of the calculations but with the interpretability and reliability of the results, sets a much higher standard. Most conventional statistics texts suggest that a sample size of at least 30 is required to achieve reliable estimates, but experience suggests that 10–15 control points will give acceptable results for a first-order fit and a small image area (up to 1024^2) for Landsat MSS or TM and SPOT HRV images. More ground control points will be needed in areas of moderate relief, where a second-order polynomial may be required. Results based on the use of small numbers of ground control points should be treated with caution – they may satisfy the mathematical criteria but they do not satisfy the statistical ones. Mather (1995) gives details of a simulation study which emphasises the importance of adequate sample size.

The second aspect of sampling that should concern users is the spatial distribution of ground control points, a topic that is treated in more detail by Mather (1995). Since the coordinate system being transformed is a two-dimensional one, it seems reasonable to suggest that the control point location should be represented by a two-dimensional pattern. It may, for instance, be possible to take a large number of control points along a transect. The information contained in these control points refers only to one dimension. Results derived from such information could say nothing about variations in the direction perpendicular to the transect line. Another possibility would be to locate control points in clusters. Again, large areas of the image would be unrepresented and results would thus be biased. In extreme cases, a condition known as 'singularity' would be signalled during the least-squares computations. This is equivalent in scalar terms to trying to divide by zero. Mather (1976, pp. 124–129) considers this point in detail, and also provides the technical background for two-dimensional least-squares problems. We can conclude from this brief examination not only that control points should be sufficient in number but also that they should be evenly spread, as far as possible, over the image area. This presents problems where substantial parts of the image cover sea or water areas where control points are absent. The same could be said of images covering any relatively featureless region. Control points should be well distributed over the image. Unwin and Wrigley (1987) introduce the concept of the *leverage* of a control point, which measures the influence of the point on the overall fit of the polynomial function and allows the user to determine which points require most care and attention.

The accuracy with which control points are measured is also considered by Mather (1995). In his simulation experiment, increasing amounts of 'noise' were added to the known locations of the ground control points, and (particularly when the distribution of control points was linear or clustered) the effects of the noise were severe. Some users are encouraged by package software to remove or weed out control points that are identified as erroneous by the fact that their associated residual error is considered to be unacceptably high. The residual error can be calculated for the line/pixel coordinates of the image or the easting/northing coordinates of the map. Using the appropriate polynomial function se-

lected from the four listed above, the map easting coordinate of a control point can be estimated from the image row (line) and column values of the same control point. The difference between the measured and estimated values of the map easting of the control point is the residual value for the map easting. This residual value is expressed in the same units as the map eastings, for example kilometres. A similar operation can be carried out for the map northings and the image line and pixel coordinates. It may seem sensible to eliminate control points that have high residual values (sometimes 'high' is defined as a value more than one standard deviation distant from the mean). As Morad *et al.* (1996) show, however, this is not necessarily the case. Instead of eliminating control points one should seek reasons for the error, which may be the result of erroneous or inaccurate digitisation or measurement. Bolstad *et al.* (1990) discuss positional uncertainty in digitising map data. Since remotely-sensed images are likely to be co-registered with maps in a geographical information system (GIS), it is even more important that the geometric quality of products derived from remotely-sensed images matches the requirements of the project. Overlay operations within a GIS are prone to error if misregistration is present. Users should, therefore, take considerable care in the digitising and measurement of control point locations because locational errors are magnified by any departures from randomness of the spatial pattern of the control points, and the net result may well be unacceptable.

We should also pay attention to the factors to be taken into consideration when control points are selected. If there is a substantial difference in time between the date of imaging and the last full map revision then it would be unwise to select as control points any feature subject to change, for example, a point on a river meander, at the edge or corner of a forest plantation or on a coastline. The land/sea boundary goes back and forth with the tide – sometimes for several hundred metres – and it is thus difficult to correlate a point shown on a map as being on the coastline and the equivalent point on the image. The best control points are intersections of roads or airport runways, edges of dams, or isolated buildings and other permanent features. All are unlikely to change in position, and all are capable of being located accurately on both map and image. At the 1:50 000 scale or better, control points should be located to within 20 m, while a cursor can be used to fix the location on the image to within one resolution element. If a zoom function is used to enlarge the image, the exact position of a pair of intersecting linear features can be estimated to within half a pixel.

It would be sensible to test the accuracy of the image–map and map–image coordinate transformations before using them to convert from one system to the other. It seems illogical to generate the least-squares coefficients a_i using a set of (x, y, c, r) control point coordinates and then use the same data to test the adequacy of the calculation. An independent assessment is necessary. A second set of control points should be used to assess the goodness of fit of the transformation by taking each of the test points in turn and converting its c and r image coordinates to map x and y coordinates, then calculating the residual values. The same should be done for the reverse transformation. The standard of the map coordinate residuals and of the image coordinate residuals will give a measure of the goodness of fit. If the residual values are normally distributed, which can be checked by producing a histogram of these values, then 68% of all calculated values should lie within one standard deviation of the mean. The mean residual is zero. If a root-mean-square error or standard deviation of 0.5 pixel or 10 m, for example, is cited then one should be careful not to interpret this statement to mean that *all* the coordinate transformation results fall within the quoted range.

In some instances, control points may already be available for the required map area and for an earlier image of the region of interest. Rather than go through the tedious procedure of collecting control point information again, the map coordinates of the control points could be reused without difficulty. However, owing to fluctuations in the orbit of the satellite, it is unlikely that the control point positions on the second image would be the same as those on the first. Benny (1981) points out that the variation in altitude and heading for Landsat-1 to -3 was quite small (and the same is true of Landsat-4 and -5 and SPOT), so (i) the pixels in two separate Landsat MSS images of the same area should have nearly equal sizes, and (ii) the linear features whose intersection defined the control point should have approximately the same orientation from image to image of the same area. Benny (1981) therefore used 'chips' or extracts of images of size 19 × 19 pixels, each chip containing a control point. The map coordinates of the centre of the chip are known. The problem is to find the image coordinates of the centre of the 19 × 19 chip on the new image. This problem is solved by a correlation procedure. Given that it is possible to estimate the position of the control point on the new image, and maximum deviations from this position in the column and row directions, then a rectangular search area can be defined. The control point is thought to be somewhere inside this search area. The 'chip' is placed over the top left 19 × 19 pixels of this search area and the correlation between the chip pixels and the pixels of the image that underlie the chip pixels is calculated and recorded, together with the row and column coordinates of the image pixel that lies below the centre pixel of the chip (Figure 4.2). The chip is then moved one column to the right and the correlation coefficient is recalculated. Once the chip has reached its furthest right position it is moved down one row and back to its leftmost position. The process continues until the chip

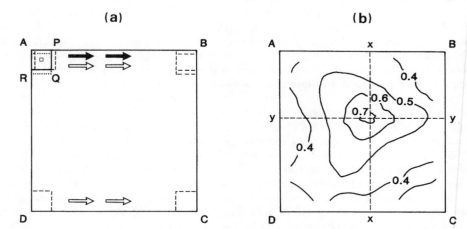

Figure 4.2   (a) Chip APQR overlying top left position of search area ABCD. Chip is moved successively rightwards until it is in the rightmost position on the top line. It is then moved down one row and back to the left of the search area. This procedure continues until the whole search area has been covered. (b) Correlation surface resulting from contouring the correlation coefficients located at the centres of the chip positions shown in (a). Most probable position of ground control point is *x, y*.

has moved to the bottom right-hand corner of the search area. At this stage, we have (i) the row coordinate of the chip centre, (ii) the corresponding column coordinate, and (iii) the correlation value for that position. Values (i) and (ii) could be used as conventional x and y coordinates and contours of correlation could be drawn on this 'map'. The most likely match would be given by the point that had the highest correlation. If the maximum correlation is low (less than about $+0.4$) then it might be concluded that no match has been found. Benny (1981) describes a 'spiral search' algorithm that differs from the regular search algorithm outlined above. If the regular search algorithm were coded efficiently then it would be as fast as the spiral technique.

An efficient method of conducting the regular search would be to compute the terms x_i, y_i, x_i^2, y_i^2 and $x_i y_i$ for the first (top left) chip position. The x_i are the chip pixels and the y_i are the pixels in the search area lying below the chip pixels. The label i is counted sequentially across the rows, so that for a 19×19 chip the label 1 refers to the first pixel on row 1, label 20 is the first pixel on row 2 and so on. The correlation coefficient r_{xy} is then computed from:

$$r_{xy} = \frac{N \sum x_i y_i - \sum x_i \sum y_i}{\left\{ \left[N \sum x_i^2 - \left(\sum x_i \right)^2 \right] \left[N \sum y_i^2 - \left(\sum y_i \right)^2 \right] \right\}^{1/2}}$$

The term $N\Sigma x_i^2 - (\Sigma x_i)^2$ needs to be computed only once for each chip as the chip pixel values are fixed – they do not change as the chip is moved over the search area. Reference to Figure 4.2 will show that as the chip is moved from position 1 to position 2 then a column of 19 image pixels enters the region covered by the chip and a column of 19 image pixels leaves that region. If we compute all the terms in the expression for r_{xy} that involve x for position 1 then we can update them by subtracting the values relating to the pixels leaving the region and by adding the values relating to the pixels entering the region. For example, we would update the calculations for the term x_i, y_i as follows:

$$x_i y_i = x_i y_i - x_{out} y_{out} + x_{in} y_{in}$$

The number of summations is thus reduced from $19 \times 19 = 361$ to only $19 + 19 = 38$ for each position of the chip. If the values of the terms involving x for the start position of each horizontal sweep across the search area are remembered then the corresponding values for the start of the next horizontal sweep can be computed by a similar method, this time involving the top row

going out and a new bottom row coming in. Since all the calculations in the derivation of r_{xy} except the final division can be performed exactly in integer arithmetic the method is both fast and accurate. This 'moving window' technique is used again in Chapter 7 when filtering techniques are considered.

All the control points could be located by this correlation-based method, which requires the user to provide an estimate of the position of the control point and of the dimensions of the search area for each control point. Benny (1981) goes on to consider a method for the automatic relocation of all control points without user intervention, provided one control point can be located accurately. He estimates that, for a typical image containing 100 control points, the visual location of the control points would take about 80 hours of effort, plus 1 minute of computer time to carry out the coordinate transformation method described below. For the semi-automatic method, with the user supplying information about the approximate location of each control point, the user man-hours drop from 80 to 8, but computer time goes up to 20 minutes. Finally, in the automatic procedure, in which the user supplies information relating to only one control point, the number of man-hours required falls to 0.1 (6 minutes) while the computer time requirement remains at 20 minutes. Orti *et al.* (1979) also provide a discussion of automatic control point location. The automatic identification of ground control points is also the subject of a paper by Motrena and Rebordão (1998).

Procedures for estimating the coefficients a_{ij} in the least-squares functions in the bulleted list above relating map and image coordinate systems can now be considered. The following description assumes that we wish to estimate the map easting e from the image column and row coordinates r and c for a set of n control points. The set of control point map eastings is denoted by the vector **e**, while the powers and cross-products of the c and r values are considered to form the matrix **P**. The coefficients a_{ij} are the elements of the vector **a**. For a second-order fit we would have the system shown in Table 4.5. The method of least squares is used to find the vector of estimates **e** according to the following model:

$$\mathbf{e} = \mathbf{Pa}$$

P and **e** are explained above, while **a** is a vector of unknown coefficients, which are to be estimated from the data. The standard formula for the evaluation of **a** is:

$$\mathbf{a} = (\mathbf{P'P})^{-1} \mathbf{P'e}$$

The elements of vector **a** can be found by the application

Table 4.5   Matrix **P** and vectors **e** and **a** required in solution of second-order least-squares estimation procedure. In this example the map eastings vector **e** is to be estimated from the powers and cross-products of the image column (**c**) and row (**r**) vectors that form the matrix **P**. The measurements of **c**, **r** and **e** are performed at n ground control points.

$$\mathbf{e} = \begin{bmatrix} e_1 \\ e_2 \\ e_3 \\ e_4 \\ \cdot \\ \cdot \\ \cdot \\ e_n \end{bmatrix} \qquad \mathbf{a} = \begin{bmatrix} a_{00} \\ a_{10} \\ a_{01} \\ a_{20} \\ a_{11} \\ a_{02} \end{bmatrix}$$

$$\mathbf{P} = \begin{bmatrix} 1 & c_1 & r_1 & c_1^2 & c_1 r_1 & r_1^2 \\ 1 & c_2 & r_2 & c_2^2 & c_2 r_2 & r_2^2 \\ 1 & c_3 & r_3 & c_3^2 & c_3 r_3 & r_3^2 \\ 1 & c_4 & r_4 & c_4^2 & c_4 r_4 & r_4^2 \\ \cdot & \cdot & \cdot & \cdot & \cdot & \cdot \\ \cdot & \cdot & \cdot & \cdot & \cdot & \cdot \\ 1 & c_n & r_n & c_n^2 & c_n r_n & r_n^2 \end{bmatrix}$$

of a standard subroutine for solving linear simultaneous equations. Such routines work efficiently with well-conditioned equations but can produce significant errors if the equations are ill-conditioned. The condition of a matrix such as $\mathbf{P'P}$ in the expression above can be considered to be a measure of the degree of independence of its columns. Ill-conditioning is indicated by a large determinant, or by a high value of the ratio of the largest and smallest eigenvalues of $\mathbf{P'P}$. The degree of independence of a pair of column vectors can be visualised by analogy with the crossing point of two lines. The intersection can be measured accurately if the lines are perpendicular, whereas if the lines cross at a very acute angle the exact point of intersection is difficult to specify. In a similar way, the solutions of a set of well-conditioned equations (with the columns of the matrix $\mathbf{P'P}$ being independent, or nearly so) can be found accurately but the solution vector could be substantially in error if the matrix is badly conditioned. Various studies have indicated that the accuracy of standard procedures (using matrix inversion techniques) for the solution of

the least-squares equations depends critically on the condition of the matrix. The Gram–Schmidt procedure described by Mather (1976) appears to be a more reliable algorithm, and this conclusion is confirmed by later work (Mather, 1995).

It is likely that the matrix $\mathbf{P'P}$ will be ill-conditioned in our particular case because its columns are the powers and cross-products of two variables, c and r. Hence, standard methods may well produce poor or even misleading results and these will be worsened if the distribution of control points is linear or clustered, as noted above. The Gram–Schmidt method is thus to be preferred. The application of this method gives estimates of the coefficients of vector **a**, which can then be used to find the map easting coordinate **e**. The other unknowns (n estimated from c and r, c estimated from e and n, and r estimated from e and n) can be found in a similar fashion.

Note that the procedures described in this section are applicable only to images obtained from relatively stable platforms such as satellites and for areas with a low relative relief. Scanned images from aircraft contain distortions caused by rapid variations in the aircraft's attitude as measured by pitch, roll and yaw. Such imagery will contain defects such as non-parallel scan lines which cannot easily be corrected by polynomial least-squares methods. The influence of terrain on the results obtained by the use of the methods described above are summarised earlier in this section.

A program (`geomcorr`) to implement the polynomial-based geometric correction procedure is provided on the CD accompanying this book. Appendix B contains details of the use of this program.

4.3.1.3   *Resampling procedures*

Once the four transformation equations relating image and map coordinate systems are known, and the results tested, the next step is to find the location on the map of the corners of the image area to be corrected, and to work out the number of and spacing (in metres) between the pixel centres necessary to achieve the correct map scale. We can now work systematically across the map area, starting at the top left, and locate (in map coordinates e and n) the centre of each pixel in turn. Given the (e, n) location coordinates of a pixel centre on the map we can apply the transformations described in the preceding section to generate (c, r) image coordinates corresponding to the position of the pixel's centre. These (c, r) coordinates are the column and row position in the uncorrected image of the new (geometrically corrected) pixel centres (Figures 4.3 and 4.4). It is unlikely that c

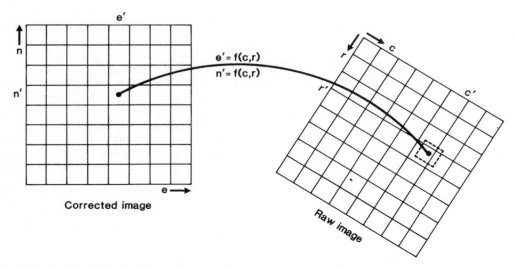

Figure 4.3   Resampling of digital image. Coordinate transformation gives e', n' in the corrected image as the centre of a pixel whose brightness value is associated with a point that does not correspond to any single pixel centre in the uncorrected image. Resampling is the process of assigning to point e', n' in the corrected image a brightness value interpolated from the uncorrected (raw) image.

and r are integers; if they were, it would be possible to take the pixel value at (c, r) and transfer it to the corrected image. Non-integral values of c and r imply that the corrected pixel centre lies between the columns and rows of the uncorrected image, so that a method of interpolation is needed to estimate the pixel value at (c, r). Three methods are in common use. The first is simple – take the value of the pixel in the input image that is closest to the computed (c, r) coordinates. This is called the *nearest-neighbour* method. It has two advantages; it is fast and its use ensures that the pixel values in the output image are 'real' in that they are copied directly from the input image. They are not 'fabricated' by an interpolation algorithm such as the two described next. On the other hand, the nearest-neighbour method of interpolation tends to produce a rather blocky effect as some pixel values are repeated.

The second method of resampling is *bilinear interpolation*. This method assumes that a surface fitted to the pixel values in the immediate neighbourhood of (c, r) will be planar, like a roof tile. The four pixel centres nearest to (c, r) lie at the corners of this tile; call their values v_{ij} (Figure 4.5). The interpolated value V at (c, r) is obtained from:

$$V = (1 - a)(1 - b)v_{i,j} + a(1 - b)v_{i,j+1}$$

$$+ \, b(1 - a)v_{i+1,j} + abv_{i+1,j+1}$$

where $a = c - j, j = |c|, b = r - i, i = |r|$ and $|x|$ is the absolute value of x, as shown in Figure 4.5. Note that the method breaks down if the point (c, r) is coincident with any of the four grid points v_{ij} and so a test for this should be included in any computer program. Bilinear interpolation results in a smoother output image because it is essentially an averaging process. Thus, sharp boundaries in the input image may be blurred in the output image. The computational time requirements are greater than those of the nearest-neighbour method.

The third spatial interpolation technique that is in common use is called *bicubic* because it is based on the fitting of two third-degree polynomials to the region surrounding the point (c, r). The 16 nearest pixel values in the input image are used to estimate the value at (c, r) on the output image. This technique is more complicated than either the nearest-neighbour or the bilinear methods discussed above, but it tends to give a more natural-looking image without the blockiness of the nearest-neighbour or the over-smoothing of the bilinear method, though – as is the case with all interpolation methods – some loss of high-frequency information is involved. The interpolator is essentially a low-pass filter (Chapter 7). The penalty to be paid for these improvements is considerably increased computing requirements. Table 4.6 shows a sample grid. The point to be interpolated is somewhere within the central cell of this grid. Step 1 is interpolation in the x or column direction;

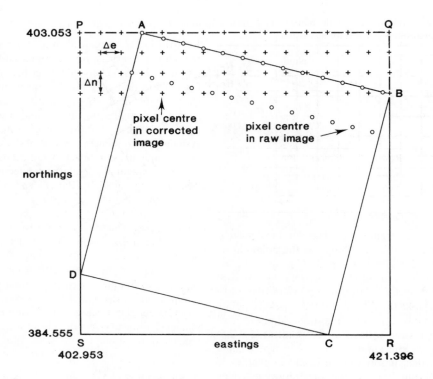

Figure 4.4 Schematic representation of resampling process. The extreme east, west, north and south map coordinates of the uncorrected image ABCD define a bounding rectangle PQRS, which is north-orientated with rows running east–west. The elements of the rows are the pixels of the geometrically corrected image. Their centres are shown by the + symbol and the spacing between successive rows and columns is indicated by Δe and Δn. The centres of the pixels of the corrected image are marked by the symbol o. See text for discussion.

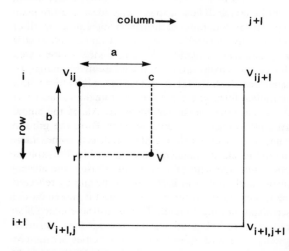

Figure 4.5 Bilinear interpolation. Value V at (c, r) is obtained from the values of the four closest points, as described in the text.

four intermediate values (denoted by V_{cm}) are obtained; these correspond to the values along the line $y = r$ at the points $c = i - 1$, i, $i + 1$ and $i + 2$. The quantity i is simply the integer part of the value r. These four intermediate values are obtained from:

$$V_{c,m} = - a(1 - a)^2 v_{j-1,m} + (1 - 2a^2 + a^3)v_{j,m}$$
$$+ a(1 + a - a^2)v_{j+1,m} - a^2(1 - a)v_{j+2,m}$$

where $m = i - 1$, i, $i + 1$, $i + 2$; and i and a are defined above. Next, these intermediate values are used to estimate the value at (c, r):

$$V_{c,r} = - b(1 - b)^2 V_{c,i-1} + (1 - 2b^2 + b^3)V_{c,i}$$
$$+ b(1 + b - b^2)V_{c,i+1} + b^2(b - 1)V_{c,i+2}$$

Although the expression looks formidable, it is easily programmed, although its incorporation into an efficient computer program is more problematical especially for a small computer. To illustrate the differences

Table 4.6 Test grid for use with bilinear and bicubic interpolation algorithms. See text for discussion.

Row	Column			
	(1)	(2)	(3)	(4)
(1)	4.8	4.0	4.0	4.8
(2)	4.0	3.0	3.0	4.0
(3)	4.0	3.0	3.0	4.0
(4)	4.8	4.0	4.0	4.8

Column	Row	Bilinear	Bicubic
2.50	2.50	3.0000	2.4875
2.99	2.99	3.0000	2.9801
2.30	2.70	3.0000	2.5712

between the results from the bilinear and bicubic techniques, consider Table 4.6. Values at the points at $(c, r) = (2.5, 2.5)$, $(2.99, 2.99)$ and $(2.3, 2.7)$ are to be estimated. The bilinear method assigns the value 3.0 to all of these points, since all of the four grid values closest to the point (c, r) take the value 3.0. If the closest 16 points are taken then it is apparent that a surface in the form of a circular depression with its lowest point at $(2.5, 2.5)$ would fit the 16 grid points quite closely. The bicubic method uses this information to determine values of 2.4875, 2.9801 and 2.5712 at the three test points. This example ignores the fact that the output values would be rounded to the nearest integer; it is intended to make a simple point, not to demonstrate the operational use of the techniques. The choice between the three methods depends upon two factors – the use to which the corrected image is to be put, and the computer facilities available. If the image is to be subjected to classification (Chapter 8) then the replacement of raw data values with artificial, interpolated values might well have some effect on the subsequent classification (although, if remote-sensing data are to be used alone in the classification procedure, and not in conjunction with map-based data, then it would be more economical to perform the geometric correction after, rather than before, the classification). If the image is to be used solely for visual interpretation, for example in the updating of a topographic map, then the resolution requirements would dictate that the bicubic method be used. The value of the end-product will, ultimately, decide whether the additional computational cost of the bicubic method is justified. Khan *et al.* (1995) consider the effects of resampling on the quality of the resulting image.

A simple program to illustrate the workings of the bicubic interpolation method is provided on the accompanying CD (program `cubconv`). Refer to Appendix B for details of its use.

4.3.1.4 *Image registration*

Registration of images taken at different dates (multi-temporal images) can be accomplished by image correlation methods such as that described in section 4.3.1.2 for the location of ground control points. Although least-squares methods such as those used to translate from map to image coordinates, and vice versa, could be used to define a relationship between the coordinate systems of two images, correlation-based methods are more frequently employed. A full account of what are termed sequential similarity detection algorithms (SSDA) is provided by Barnea and Silverman (1972) while Anuta (1970) describes the use of the fast Fourier transform (Chapter 6) in the rapid calculation of the inter-image correlations. Eghbali (1979) and Kaneko (1976) illustrate the use of image registration techniques applied to Landsat MSS images. Townshend *et al.* (1992) point out the problems that may arise if change detection techniques are based on multitemporal image sets that are not properly registered. As noted earlier, any GIS overlay operation involving remotely-sensed images is prone to error because all such images must be registered to some common geographical reference frame. Misregistration provides the opportunity for error to enter the system.

4.4 ATMOSPHERIC CORRECTION METHODS

An introductory description of the interactions between radiant energy and the constituents of the Earth's atmosphere was given in Chapter 1. From this discussion, one can conclude that a value recorded at a given pixel location on a remotely-sensed image is not a record of the true ground-leaving radiance at that point, for the magnitude of the ground-leaving signal is attenuated due to atmospheric absorption and its directional properties are altered by scattering. Figure 4.6 shows, in a simplified form, the components of the signal received by a sensor above the atmosphere. All of the signal appears to originate from the point P on the ground whereas, in fact, scattering at S_2 redirects some of the incoming electromagnetic energy within the atmosphere into the field of view of the sensor (the atmospheric path radiance) and some of the energy reflected from point Q is scattered at S_1 so that it is seen by the sensor as coming from P. To add to these effects, the radiance from P (and Q) is attenuated as it passes through the atmosphere. Other difficulties are caused by variations in the illumination geometry (the geometrical relationship between the Sun's elevation and azimuth angles, the slope of the ground and the disposition

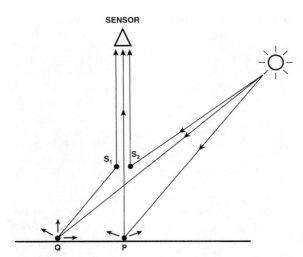

Figure 4.6   Components of the signal received by a satellite-mounted sensor. See text for discussion.

of topographic features). If the sensor has a wide field of view, such as the NOAA AVHRR, or is capable of off-nadir viewing (for example, the SPOT HRV), then further problems result from the fact that the reflectance of a surface will vary with the view angle as well as with the solar illumination angle (this is the bidirectional reflectance property, noted in Chapter 1). Given this catalogue of problems one might be tempted to conclude that quantitative remote sensing is the art of the impossible. However, there are a variety of techniques that can be used to estimate the atmospheric and viewing geometry effects, although a great deal of research work remains to be done in this highly complex area. A good review of terminology and principles is contained in Deschamps *et al.* (1983), while Woodham (1989) gives a lucid review of problems. The problem of the 'adjacency effect', that is, EMR that is scattered into the field of view of the sensor from pixels other than the target, is considered by Milovich *et al.* (1995).

Atmospheric correction might be a necessary preprocessing technique in three cases. In the first, we may want to compute a ratio of the values in two bands of a multispectral image (section 6.2.4). As noted in Chapter 1, the effects of scattering increase inversely with wavelength, so the two channels are affected differentially by scattering. The computed ratio will thus be a biased estimate of the true ratio. In the second situation, a research worker may wish to relate upwelling radiance from a surface to some property of that surface in terms of a physically based model. To do this, the atmospheric

component present in the signal recorded by the sensor must be estimated and removed. The third case is that in which results or ground measurements made at one time (time 1) are to be compared with results achieved at a later date (time 2). Since the state of the atmosphere will undoubtedly vary from time 1 to time 2 it is necessary to correct the radiance values recorded by the sensor for the effects of the atmosphere. In addition to these three cases, it may well be necessary to correct multispectral data for atmospheric effects even if it is intended for visual analysis rather than any physical interpretation.

The atmosphere is an extremely complex and dynamic system. A full account of the physics of the interactions between the atmosphere and electromagnetic radiation is well beyond the scope of this book, and no attempt will be made to provide a comprehensive survey of the progress that has been made in the field of atmospheric physics. Instead, techniques developed by remote-sensing researchers for dealing with the problem of estimating atmospheric effects on multispectral images in the 0.4–2.4 µm reflective solar region of the spectrum will be reviewed. We will begin by summarising the relationship between radiance received at a sensor above the atmosphere and the radiance leaving the ground surface:

$$L_s = H_{tot}\rho T + L_p$$

Here H_{tot} is the total downwelling radiance in a specified spectral band, ρ is the reflectance of the target (the ratio of the downwelling to the upwelling radiance), T is the atmospheric transmittance and L_p is the atmospheric path radiance. The downwelling radiance is attenuated by the atmosphere as it passes from the top of the atmosphere to the target. Further attenuation occurs as the signal returns through the atmosphere from the target to the sensor. Some of the radiance incident upon the target is absorbed by the ground-surface material, a proportion ρ being reflected by the target. Finally, the radiance reaching the sensor includes a contribution made up of energy scattered within the atmosphere; this is the path radiance term (L_p) in the equation. In reality, the situation is more complicated, as Figure 4.6 shows.

The path radiance term L_p varies in magnitude inversely with wavelength, for scattering increases as wavelength decreases. Hence, L_p will contribute differing amounts to measurements in individual wavebands. In terms of a Landsat TM image the blue-green band (TM band 1) will generally have a higher L_p component than the green band (TM band 2). The first method of atmospheric correction that is considered is the estimation of the path radiance term, L_p. Two relatively simple techniques are described in the literature. The first is the histogram minimum method, and the second is the regression method.

Histograms of all bands are computed for the full image, which generally contains some areas of low reflectance (clear water, deep shadows or exposures of dark-coloured rocks). These pixels will have values very close to zero in near-infrared bands, for example Landsat TM band 4 or SPOT HRV band 3. If the histograms of the other bands in the visible and near-infrared regions of the electromagnetic spectrum are plotted they will generally be seen to be offset progressively towards higher levels. The lowest pixel values in the histograms of visible and near-infrared bands are a first approximation to the atmospheric path radiance in those bands, and these minimum values are subtracted from the respective images. Path radiance is very much reduced in mid-infrared bands such as Landsat TM bands 5 and 7, and these bands are not normally corrected.

The regression method is applicable to areas of the image that have dark pixels as described above. In terms of the Landsat TM sensor, pixel values in the near-infrared band (numbered 4) are plotted against the values in the other bands in turn, and a best-fit (least-squares) straight line is computed using standard regression methods. The offset *a* on the *x*-axis represents an estimate of the atmospheric path radiance term for that particular band (Figure 4.7).

Chavez (1988) describes an elaboration of the haze correction procedure based on the 'histogram minimum' methods described above. Essentially Chavez's method is based on the fact that Rayleigh scattering

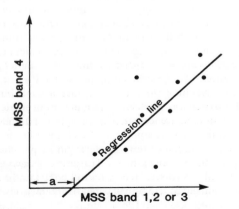

Figure 4.7 Regression method for computation of atmospheric path radiance, which is proportional to the offset term *a*. Dots represent sample points taken from dark areas of image.

is inversely proportional to the *n*th power of the wavelength, the value of *n* varying with atmospheric turbidity. He defines a number of 'models' ranging from 'very clear' to 'hazy' and each of these is associated with a value of *n*. A 'starting haze' value is provided for one of the short-wavelength bands, and the haze factors in all other bands are calculated analytically using the Rayleigh scattering relationship. The method requires the conversion from pixel values to radiances, and in the program chavez on the accompanying CD the gains and offsets that are used to perform this conversion are the pre-flight calibrations for the Landsat TM. Program chavez works for Landsat TM data only. Further details are given in Appendix B. Chavez (1996) gives details of modifications to the method which are claimed to enhance its performance.

The first two of the methods described above rely on information from the image itself in order to determine the path radiance for each spectral band. However, a considerable body of theoretical knowledge concerning the complexities of atmospheric radiative transfer is in existence, and has found expression in numerical models such as LOWTRAN (Kneizys *et al.*, 1983) and 5S/6S (Tanré *et al.*, 1986). Operational (as opposed to research) use of these models is limited by the need to supply data relating to the condition of the atmosphere at the time of imaging. The cost of such data-collection activities is considerable, hence reliance is placed upon the use of 'standard atmospheres' such as 'mid-latitude summer'. The use of these estimates results in a loss of accuracy, and the extent of the inaccuracy is not assessable. See Popp (1995) for further elaborations. Richter (1996) shows how spatial variability in atmospheric properties can be taken into account by partitioning the image.

The following example of the use of the 5S model of Tanré *et al.* (1986) is intended to demonstrate the magnitudes of the quantities involved in atmospheric scattering. Two hypothetical test cases are specified, both using Landsat TM bands 1–5 and 7 for an early summer (1 June) date at a latitude of 51°. A standard mid-latitude summer was chosen, and the only parameter to be supplied by the user was an estimate of the horizontal visibility in kilometres (which may be obtainable from a local meteorological station). Values of 5 km and 20 km were input to the model, and the results are shown graphically in Figure 4.8. The apparent reflectance is that which would be detected by a scanner operating above the atmosphere, while the pixel reflectance is an estimate of the true target reflectance which, in this example, is green vegetation. The intrinsic atmospheric reflectance is analogous to the path radiance. At 5 km

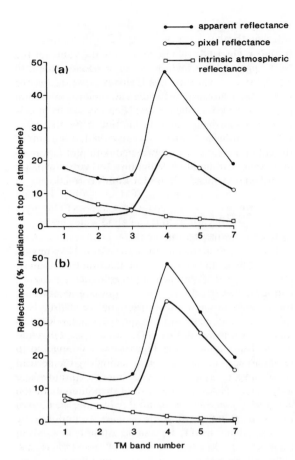

Figure 4.8 Examples of output from atmospheric model (Tanré *et al.*, 1986). Reflectance, expressed as per cent irradiance at the top of the atmosphere in the spectral band, is shown for the atmosphere (intrinsic atmospheric reflectance), the target pixel (pixel reflectance) and the received signal (apparent reflectance). The difference between the sum of pixel reflectance plus intrinsic atmospheric reflectance and the apparent reflectance is the background reflectance from neighbouring pixels. Examples show results for (a) 5 km visibility and (b) 20 km visibility.

visibility the intrinsic atmospheric reflectance is greater than pixel reflectance in TM bands 1 and 2. Even at the 20 km visibility level, intrinsic atmospheric reflectance is greater than pixel reflectance in TM band 1. In both cases the intrinsic atmospheric reflectance declines with wavelength, as one might expect. Note that the difference between the apparent reflectance and the sum of the pixel reflectance and the intrinsic atmospheric reflectance is equal to reflectance from pixels neighbouring the target pixels being scattered into the field of view

of the sensor (see Figure 4.7). The *y*-axis in Figure 4.8 shows per cent reflectance, that is, the proportion of solar irradiance measured at the top of the atmosphere that is reflected.

4.5   ILLUMINATION AND VIEW ANGLE EFFECTS

The magnitude of the signal received at a satellite sensor is dependent on several factors, particularly:

• reflectance of the target,
• nature and magnitude of atmospheric interactions,
• slope and aspect of the ground target area relative to the solar azimuth,
• angle of view of the sensor, and
• solar elevation angle.

Variations in the spectral reflectance of particular types of Earth-surface cover materials are discussed in Chapter 1, where a general review of atmospheric interactions is also to be found, while section 4.4 contains an introduction to the correction of image data for atmospheric effects. In this section we consider the effects of (i) the solar elevation angle, (ii) the view angle of the sensor, and (iii) the slope and aspect angles of the target.

In the absence of an atmosphere, and assuming no complicating factors, the magnitude of the radiance reflected from or emitted by a fixed target and recorded by a remote sensor will vary with the illumination angle and the angle of view of the sensor. The reflectance of the target will vary as these two angles alter, and one could therefore consider a function that described the magnitude of the upwelling radiance of the target in terms of these two angles. This function is termed the *bidirectional reflectance distribution function* (described in Chapter 1). When an atmosphere is present an additional complication is introduced, for the irradiance at the target will be reduced as the atmospheric path length increases. The path length (which is the distance that the incoming energy travels through the atmosphere) will increase as the solar elevation angle decreases, and so the degree of atmospheric interference will increase.

The radiance upwelling from the target also has to pass through the atmosphere. The angle of view of the sensor will control the upward path length. A nadir view will be influenced less by atmospheric interactions than would an off-nadir view; for example, the extremities of the scan lines of a NOAA AVHRR image are viewed at angles of up to 56° from nadir, while the SPOT HRV instrument is capable of viewing angles of ±27° from nadir. Amounts of shadowing will also be dependent upon the solar elevation angle. For instance, shadow effects in row crops will be greater at low Sun elevation angles than at high angles. Ranson *et al.* (1985) report the results of a study of variations in the spectral response of soybeans with respect to illumination, view and canopy geometry. They conclude that the spectral response depended greatly on solar elevation and azimuth angles and on the angle of view of the sensor, the effect being greater in the visible red region of the spectrum than in the near-infrared. In another study, Pinter *et al.* (1983) show that the spectral reflectance of wheat canopies in MSS and TM wavebands is strongly dependent on the direction of incident radiation and its interaction with vegetation canopy properties, such as leaf inclination and size. Clearly, reflectance from vegetation surfaces is a complex phenomenon which is, as yet, not fully understood and methods for the estimation of the effects of the various factors that influence reflectance are still being evaluated at a research level. It would obviously be desirable to remove such effects by pre-processing before applying methods of pattern recognition (Chapter 8) particularly if results from analyses of images from different dates are to be compared, or if the methods are applied to images produced by off-nadir viewing or wide angle-of-view sensors.

The effects of variation in the solar elevation angle from one image to another of a given area can be accomplished simply if the reflecting surface is Lambertian (section 1.3.1). This is rarely the case with natural surfaces, but the correction may be approximate to the first order. If the solar zenith angle (measured from the vertical) is θ and the desired angle is x then the correction is simply:

$$L' = \frac{\cos x}{\cos \theta}$$

This formula may be used to standardise a set of multi-temporal images to a standard solar illumination angle. See Frulla *et al.* (1995) for a discussion of illumination and view angle effects on the NOAA AVHRR imagery.

Barnsley and Kay (1990) consider the relationship between sensor geometry, vegetation canopy geometry and image variance with reference to wide field-of-view imagery obtained from aircraft scanners. The effect of off-nadir viewing is seen as a symmetric increase in reflectance away from the nadir point, so that a plot of the mean pixel value in each column (y-axis) against column number (x-axis) shows a parabolic shape. A first-order correction for this effect uses a least-squares procedure to fit a second-order polynomial of the form:

$$\hat{y}_x = a_0 + a_1 x$$

to the column means. The term $\hat{y}_x$ is the value on the least-squares curve corresponding to image column x. Let $\hat{y}_{nad}$ be the value on the least-squares curve at the nadir point (the centre of the scan line and the minimum point on the parabolic curve). Then the values in column x of the image can be multiplied by the value $\hat{y}_{nad}/\hat{y}_x$ in order to 'flatten' the curve so that the plot of the column means of the corrected data plot (approximately) as a straight line. Program `nadircor`, which is described in Appendix B, performs this correction.

4.6 SENSOR CALIBRATION

Sensor calibration has one of three possible aims. First, the user may wish to combine information from images obtained from different sensor systems such as Landsat TM and SPOT HRV. Secondly, in studies of change detection it may be necessary to compare pixel values from images obtained at different times. Thirdly, remotely-sensed estimates of surface parameters such as reflectance are now being used in physical models. Generally speaking, sensor calibration is combined with atmospheric and view angle correction (e.g. Teillet and Fedosejevs, 1995) in order to obtain estimates of target reflectance. These authors also provide details of the correction required for variations in solar irradiance over the year. A useful starting point is the review by Duggin (1985). In this section, attention is focused on the Landsat TM optical bands and the SPOT HRV. Calibration of the Landsat TM thermal band (conventionally numbered 6) is discussed by Schott and Volchok (1985) and Itten and Meyer (1993). Consideration of calibration issues relating to the NOAA AVHRR is provided by Che and Price (1992) and by Vermote and Kaufman (1995). Gutman and Ignatov (1995) compare the pre- and post-launch calibrations for the NOAA AVHRR, and show how the differences in these calibrations affect the results of vegetation index calculations (Chapter 6). A special issue of the *Canadian Journal of Remote Sensing* (volume 23, number 4, December 1997) is devoted to calibration and validation issues.

Atmospheric correction methods are considered in section 4.4, while corrections for illumination and view angle effects are covered in section 4.5. In the present section, the topic of sensor calibration is reviewed. Recall from Chapter 1 that the values recorded for a particular band of a multispectral image are simply counts: they are the representation, usually on a 0–255 scale, of equal-length steps in a linear interpolation between the minimum and maximum levels of radiance recorded by the detector. If more than one detector is used then differences in calibration will cause 'striping' effects on the image (section 4.2). The factors used to calibrate these image data vary over time, and consequently the relationship between the pixel value recorded at a particular location and the reflectance of the material making up the surface of the pixel area will not be constant. The determination of these factors is not an easy task. Pre-launch calibration factors are obtainable from ground receiving stations and from vendors of image data, and some image processing software packages incorporate these values into their sensor calibration routines. However, studies such as Thome *et al.* (1993) indicate that these calibration factors are time-varying for the Landsat-5 TM sensor, and that substantial differences in the outcome of calibration calculations depend on the values of the calibration factors that are used. The fact that a number of different processing procedures have been applied to Landsat TM data by ground stations further complicates the issue (Moran *et al.*, 1992, 1995). Thome *et al.* (1993, see table 2 of Teillet and Fedosejevs, 1995) provide the following expressions for Landsat-5 TM sensor gains in bands 1–5 and 7 (the offset is assumed to be constant over time):

$$G_1 = (-7.84 \times 10^{-5})D + 1.409$$
$$G_2 = (-2.75 \times 10^{-5})D + 0.7414$$
$$G_3 = (-1.96 \times 10^{-5})D + 0.9377$$
$$G_4 = (-1.10 \times 10^{-5})D + 1.080$$
$$G_5 = (7.88 \times 10^{-5})D + 7.235$$
$$G_7 = (7.15 \times 10^{-5})D + 15.63$$

D is the number of days since 1 March 1984, the date of the launch of the Landsat-5 satellite. It can be calculated using the program `julday`, which is described in Appendix B. These gain coefficients are used in the equation:

$$L_n^* = (\text{PV} - \text{offset})/G_n$$

This equation, which differs from the relationship presented in the next paragraph, uses the symbol L_n^* to denote apparent radiance at the sensor, while PV is the pixel value, G_n is the sensor gain and the offsets are as follows (values in brackets): TM 1 (2.523), TM 2 (2.417), TM 3 (1.452), TM 4 (1.854), TM 5 (3.423) and TM 7 (2.633). Note that the values of gain and offset given in Table 4.7 refer to the procedure that is described in the following paragraphs.

Hill (1991) and Hill and Aifadopoulou (1990) present a methodology for the calibration of the Landsat TM optical bands and SPOT HRV data. They note that the

Table 4.7   Landsat-5 TM offset (a_0) and gain (a_1) coefficients. The results of pre-flight calibration are given in the two leftmost columns. The values in the two rightmost columns are derived from observations at White Sands, New Mexico. Note that the gain and offset coefficients for the TM bands 5 and 7 are unchanged.

Band No.	Pre-launch		Post-launch	
	a_0	a_1	a_0	a_1
1	−0.1009	0.0636	−0.1331	0.0727
2	−0.1919	0.1262	−0.2346	0.1385
3	−0.1682	0.0970	−0.1897	0.1102
4	−0.1819	0.0914	−0.1942	0.0885
5	−0.0398	0.0126	−0.0398	0.0126
7	−0.0203	0.0067	−0.0203	0.0067

relationship between radiance and pixel value (PV, sometimes called a digital signal level (DSL) or a digital count (DC)) can be defined for spectral band *n* of the Landsat TM as follows:

$$L_n^* = a_0 + a_1\text{PV}$$

where a_0 and a_1 are the offset and gain coefficients and L_n^* is apparent radiance at the sensor measured in units of mW m^{-1} sr^{-1} μm^{-1}. The work of Thome *et al.* (1993) on the determination of the values of the gain and offset coefficients is noted above; their gains and offsets use the relationship between L_n^* and PV described earlier. Other relevant references are Hall *et al.* (1991), Holm *et al.* (1989), Markham *et al.* (1992), Moran *et al.* (1990, 1992), Muller (1993), Olsson (1995), Price (1987, 1988), Rao and Chen (1996), Sawter *et al.* (1991), Slater *et al.* (1987), Teillet and Fedosejevs (1995) and Woodham (1989). Table 4.7 shows the Landsat-5 TM calibration coefficients (i) as determined before launch and (ii) as estimated from ground-based measurements at White Sands, New Mexico.

An alternative to the absolute calibration procedures for the determination of reflectance factors is to use one of the images in a multitemporal image set as a reference and adjust other images according to some statistical relationship with the reference image. Thus, Elvidge *et al.* (1995) use a regression procedure to perform relative radiometric normalisation of a set of Landsat MSS images. Their procedure assumes that land cover has not changed and that the vegetation is at the same phenological stage in all images. A similar approach is used by Olsson (1993).

The corresponding expression for SPOT HRV uses

calibration coefficients derived from the header data provided with the image. Table 4.8 shows a part of the output generated by the program `read_spot_cd`, which is included on the CD provided with this book. Further details of this program are given in Appendix B. The header data refer to a SPOT-1 multispectral (XS) image of la Camargue, France, acquired on 12 January 1987, and gain values a_i for the three multispectral channels are provided. Note that all of the offset values are zero. The apparent radiance of a given pixel is calculated from:

$$L = PV/a_1$$

where a_1 is the gain coefficient, and PV and L are as defined earlier. These coefficients are updated regularly (Begni, 1986; Begni *et al.*, 1988; Moran *et al.*, 1990). These apparent radiance values must be further corrected if imagery acquired from SPOT in off-nadir viewing mode is used (Moran *et al.*, 1990; Muller, 1993).

Given the value of radiance at the sensor (L) it is usual to convert to apparent reflectance. This is accomplished for each individual band using the expression:

$$\rho = \frac{\pi L d^2}{E_s \cos \theta_s}$$

Note that ρ is not corrected for atmospheric effects by this process. L is the radiance computed as described earlier, E_s is the exoatmospheric solar irradiance (Markham and Barker, 1987; Price, 1988; Table 4.9), d is the relative Earth–Sun distance in astronomical units (mean distance is 1.0 AU) for the day of image acquisi-tion, and θ_s is the solar zenith angle. The values of ρ derived from this procedure should be corrected for atmospheric effects. The Earth–Sun distance correction factor is required because there is a variation of around 3.5% in solar irradiance over the year. The value of d is provided by the formula:

$$d = \frac{1}{1 - 0.016\,74 \cos[0.9856(JD - 4)]}$$

JD is the 'Julian day' of the year, that is, the day number counting 1 January = 1. A program, `julday`, is provided on the accompanying CD to calculate the Julian day given the calendar date. The program actually calculates the number of days that have elapsed since a reference calendar date since the start of the first millennium, though most readers will wish to compute elapsed days since the start of a given year. The program will also be useful in computing the time-dependent offset and gain values from the formulae given in Thome *et al.* (1993), described above, which require the number of days that have elapsed since the launch of Landsat-5 (1 March 1984).

Calibration of synthetic aperture radar (SAR) imagery requires the recovery of the normalised radar cross-section (termed sigma0 and measured in terms of decibels, dB). Sigma0 values range from $+5$ dB (very bright target) to -40 dB (very dark target). Meadows (1995) notes that the purpose of calibration is to determine absolute radar cross-section measurements at each pixel position, and to estimate drift or variation

Table 4.8 Extract from SPOT header file showing radiometric gains and offsets. The output was generated using the program `read_spot_cd` *described in Appendix B.*

Scene ID:	S1H1870112102714		
Scene centre latitude:	N0434026		
Scene centre longitude:	E0043615		
Spectral Mode (XS or PAN):	XS		
Preprocessing level identification:	1B		
Radiometric calibration designator:	1		
Deconvolution designator:	1		
Resampling designator:	CC		
Pixel size along line:	20		
Pixel size along column:	20		
Image size in Map projection along Y axis:	059792		
Image size in Map projection along X axis:	075055		
Sun calibration operation date:	19861115		
This is a multispectral image			
Absolute calibration gains:	00.86262	00.79872	00.89310
Absolute calibration offsets:	00.00000	00.00000	00.00000

Table 4.9 Exoatmospheric solar irradiance for (a) Landsat TM and (b) SPOT HRV (XS) bands (Markham and Barker, 1987; Price, 1988; Teillet and Fedosejevs, 1995). The centre wavelength is expressed in micrometres (μm) and the exoatmospheric solar irradiance in $mW\,cm^{-2}\,sr^{-1}\,\mu m^{-1}$. See also Guyot and Gu (1994, table 2).

(a) Landsat Thematic Mapper.

Landsat TM band number	Centre wavelength	Centre wavelength (Teillet and Fedosejevs, 1995)	Exoatmospheric irradiance	Exoatmospheric irradiance (Teillet and Fedosejevs, 1995)
1	0.486	0.4863	195.70	195.92
2	0.570	0.5706	192.90	182.74
3	0.660	0.6607	155.70	155.50
4	0.840	0.8832	104.70	104.08
5	1.676	1.677	21.93	22.075
7	2.223	2.223	7.45	7.496

(b) SPOT High Resolution Visible (HRV).

SPOT HRV band number	Centre wavelength	Exoatmospheric irradiance
1	0.544	187.48
2	0.638	164.89
3	0.816	110.14

over time in the radiometric performance of the SAR. Calibration can be performed in three ways: by imaging external calibration targets on the ground, by the use of internal calibration data, or by examining raw data quality.

4.7 TERRAIN EFFECTS

The corrections required to convert TM and SPOT data described in section 4.6 assume that the area covered by the image is a flat surface, imaged by a narrow field-of-view sensor. It was noted that apparent reflectance depends also on illumination and view angles, as target reflectance is generally non-Lambertian. This discussion did not refer to the commonly observed fact that the Earth's surface is not generally flat. Variations in reflectance from similar targets will occur if these targets have a different topographic position, even if they are directly illuminated by the Sun. Therefore, the spectral reflectance curves derived from multispectral imagery for what is apparently the same type of land cover (for example, wheat or coniferous forest) will contain a component that is attributable to topographic position, and the results of classification analyses (Chapter 8) will be influenced by this variation. Various corrections have been proposed for the removal of the 'terrain illumination effect'. See Li *et al.* (1996), Proy *et al.* (1989), Teillet *et al.* (1982), Woodham (1989) and Young and Kaufman (1986) for reviews of the problem.

Correction for terrain illumination effects requires a digital elevation model (DEM) that is expressed in the same coordinate system as the image to be corrected. Generally, the image is registered to the DEM, as the DEM is likely to be map-based. The DEM should also be of a scale which is close to that of the image, so that accurate estimates of slope angle and slope direction can be derived for each pixel position in the image. A number of formulae are in common use for the calculation of slope and aspect 'images' from a DEM, and different results may be determined by different formulae (Bolstad and Stowe, 1994; Carara *et al.*, 1997; Hunter and Goodchild, 1997). The accuracy of slope and aspect 'images' depends also on the accuracy of the DEM, a question that is considered by Li (1993). A simple method to correct for terrain slope in areas that receive direct solar illumination is simply to use the Lambertian assumption (that the surface reflects radiation in a diffuse fashion, so that it appears equally bright from all feasible observation angles). This cosine correction is mentioned above. It involves the multiplication of the apparent reflection for a given pixel by the ratio of the cosine of the solar zenith angle (measured from the vertical) by the cosine of the incidence angle (measured from the surface normal, which is a line perpendicular to the sloping ground). Teillet *et al.* (1982, p. 88) note that this correction is not particularly useful in areas of steep terrain where incidence angles may approach 90°.

Non-Lambertian models include the *Minnaert correction*, which is probably the most popular method of computing a first-order correction for terrain illumination effects (though the method does not include any correction for diffuse radiation incident on a slope). The Lambertian model can be written as:

$$L = L_N \cos(i)$$

where L is the measured radiance, L_N is the equivalent radiance on a flat surface with incidence angle of zero, and i is the exitance angle. The Minnaert constant, k, enters into the non-Lambertian model as follows:

$$L = L_N \cos^k(i) \cos^{k-1}(e)$$

where i, L and L_N are defined as before and e is the angle of exitance. If both sides of the equation are multiplied by $\cos(e)$ and the terms rearranged then we get:

$$L_N = \frac{L \cos(e)}{\cos^k(i) \cos^k(e)}$$

The value of the Minnaert constant k is the slope of the least-squares line relating $\log[L \cos(e)]$ and $\log[\cos^k(i) \cos^k(e)]$. Most surfaces have k values between 0 and 1; $k = 1$ implies Lambertian reflectance while $k > 1$ implies a dominance of the specular reflectance component. Once k has been determined then the equivalent radiance from a flat surface can be calculated. However, the value of k depends on the nature of the ground cover, and so will vary over the image. Since a sample of pixels is required in order to estimate the slope of the least-squares line relating $\log[L \cos(e)]$ and $\log[\cos^k(i) \cos^k(e)]$ it might be necessary to segment the image into regions of similar land cover type and calculate a value of k for each type. Often, however, the purpose of performing a terrain illumination correction is to improve the identification of land cover types, hence the problem takes on circular proportions. An iterative approach might be possible, in which classification accuracy assessment (Chapter 8) is used as a criterion. However, a simpler approach would be to calculate an average k value for the whole image.

Parlow (1996b) describes a method for correcting terrain-controlled illumination effects using a simulation model of solar irradiance on an inclined surface. The SWIM (short wave irradiance model) computes both direct and diffuse components of irradiance for given atmospheric conditions and allows the conversion of satellite-observed radiances to equivalent radiances for a flat surface. Note that the Minnaert method, described in the preceding paragraphs, does not consider diffuse illumination. Parlow (1996b) shows that correction of the image data for terrain illumination effects produces superior classification performance (see Chapter 8 for a survey of image classification methods). Further references are Conese *et al.* (1993), Egbert and Ulaby (1972), Hay and Mackay (1985), Hill *et al.* (1995), Jones *et al.* (1988), Katawa *et al.* (1988) and Smith *et al.* (1980).

4.8   SUMMARY

Methods of pre-processing remotely-sensed imagery are designed to compensate for one or more of (i) cosmetic defects, (ii) geometric distortions, (iii) atmospheric interference, and (iv) variations in illumination geometry, to calibrate images for sensor degradation, and to correct image pixel values for the effects of topography. The level of pre-processing required will depend on the problem to which the processed images are to be applied. There is therefore no fixed schedule of pre-processing operations that are carried out automatically prior to the use of remotely sensed data. The user must be aware of the geometrical properties of the image data and of the effects of external factors (such as the level of, and variations in, atmospheric haze) and be capable of selecting an appropriate technique to correct the defect or estimate the external effect, should that be necessary.

The material covered in this chapter represents the basic transformations that must be applied in order to recover estimates of ground-leaving radiance. The development of models requiring such estimates as input has expanded in recent years. Equally importantly, the use of remote sensing to measure change over time is becoming more significant in the context of global environmental change studies. Multitemporal analysis, the comparison of measurements derived by different sensors at different points in time, and the determination of relationships between target radiance and growth and health characteristics of agricultural crops are examples of applications that require the application of corrections described in this chapter. Procedures to accomplish these corrections are not well formulated at present, and the whole area requires more research and investigation.

4.9   QUESTIONS

1. Define and explain the following: geo-referenced, registration, map, RMSE, resampling, panoramic distortion, least squares, DEM, singular matrix, 5S, nadir, Julian day, AU.

2. When is de-striping necessary? What are the alternative methods for de-striping an image?

3. Why is geometric correction of remotely-sensed images needed? What precautions should you take to ensure that the correction is adequate for a particular purpose? How would you measure the adequacy of the correction?

4. How can GPS technology be used to assist the process of geometric correction?

5. Give a worked example to show that you understand the principles of (a) bilinear and (b) bicubic interpolation for resampling.

6. What is atmospheric path radiance? How does it affect remote sensing of the Earth's surface?

7. What factors affect the magnitude of the signal received by a sensor carried by a satellite or aircraft?

8. How is the Minneart equation derived? What is its application in remote sensing? Are there any alternatives?

9. Why is sensor calibration important? What are the problems that you would face in correcting (a) a Landsat TM band 1 image and (b) a SPOT HRV band 1 image? Give details of the methods you would use in each case.

5

Image Enhancement Techniques

5.1 INTRODUCTION

The way in which environmental remote-sensing satellite and aircraft systems collect multi-spectral digital images of the Earth's surface is described in Chapter 2; it will be recalled that a digital image is a numerical record of the radiance leaving each of a large number of rectangular areas on the ground (called pixels) in each of a number of spectral bands. The range of radiance values is represented (quantised) in terms of a scale that is normally 6, 7, 8 or 10 bits in magnitude, depending on the type of scanner that is used and on the nature of any processing carried out at the ground station. For example, the AVHRR instrument carried by the NOAA satellites uses a 10-bit scale, while the Landsat MSS is basically a 6-bit instrument, although processing at the ground station may alter the apparent dynamic range to 7 or 8 bits. The SPOT HRV and Landsat TM each have a radiometric resolution of 8 bits. Associated with each pixel of a digital image, therefore, is a set of numbers, with one number per pixel per spectral band. For example, Landsat MSS acquires data in four bands, while Landsat TM provides seven bands of multispectral data and multispectral SPOT imagery has three spectral bands.

A digital image can thus be considered as a three-dimensional rectangular array or matrix of numbers, the x- and y-axes representing spatial dimensions and the z-axis the quantised spectral radiance (pixel value). A fourth dimension, time, could be added since satellite data are collected on a routine and regular basis (every 16 days for Landsats-4 and -5, for example). The elements of this matrix are numbers in the range 0–63, 0–127, 0–255 or 0–1024 depending on the instrument used. In the remainder of this chapter and in the software provided on the CD-ROM it is assumed that the image data lie on a 0–255 scale. All the techniques discussed are applicable to any scale; all that is necessary is to introduce a scaling factor of 0.5 or 0.25 if the

image data are measured on the 0–127 or 0–63 scales, respectively. Data recorded on a 0–1024 scale will need slightly different programming techniques, as the 6-, 7- and 8-bit data are normally stored in a single byte (BYTE, CHARACTER*1 or INTEGER*1 data type in Fortran; or unsigned char in C) whereas the data for a 10-bit pixel are stored in a 16-bit word (INTEGER*2 data type in Fortran; short int in C) unless sophisticated programming techniques are used to pack the 10-bit pixel values. For example, four bytes (32 bits) might be used to store three 10-bit pixels, leaving two bits unused per 32-bit storage unit. However, RAM is now cheap enough for the mental wrestling associated with such operations to be almost forgotten.

Visual analysis and interpretation are often sufficient for many purposes to extract information from remotely-sensed images in the form of standard photographic prints. If the image is digital in nature, such as the satellite- and aircraft-acquired images considered in this book, a computer can be used to manipulate and display the image data and to generate pictures that satisfy the particular needs of the interpreter. The way in which the computer is used to display a digital image is described in Chapter 3. In this chapter, methods of enhancing digital images are considered. The term *enhancement* is used to mean the alteration of the appearance of an image in such a way that the information contained in that image is more readily interpreted visually in terms of a particular need. Consequently, no single standard method of enhancement can be said to be 'best', for the needs of each user may differ. Also, the characteristics of each image in terms of the distribution of pixel values over the 0–255 range will change from one area to another; thus, enhancement techniques suited to one image (for example, covering an area of forest) will differ from the techniques applicable to another kind of area (for example, the Antarctic ice-cap).

There are a number of general categories of

enhancement technique and these are described in the following sections. As in many other areas of knowledge, the distinction between one type of analysis and another is a matter of personal taste; some kinds of image transformations (Chapter 6) or filtering methods (Chapter 7) can, for instance, reasonably be described as enhancement techniques. In this chapter we will concentrate on ways of improving the visual interpretability of an image by one of two methods:

- altering the contrast of an image, and
- converting from black-and-white to colour representation.

The first group of techniques consists of those methods which can be used to compensate for inadequacies of what, in photographic terminology, would be called 'exposure'; some images are intuitively felt to be 'too dark', while others are over-bright. In either case, information is not so easily comprehended as it might be if the contrast of the image were greater. In this context, contrast is simply the range and distribution of the pixel values over the 0–255 (0–63, 0–127, 0–1024) grey scale. The second category includes those methods which allow a black-and-white image to be re-expressed in colour. This is sometimes desirable, for the eye is more sensitive to variations in hue than to changes in brightness.

The chapter begins with a brief description of the human visual system, since the techniques covered in the following sections are fundamentally concerned with the visual comprehension of information displayed in image form.

5.2   HUMAN VISUAL SYSTEM

There are a number of theories that seek to explain the manner in which the human visual system operates. The facts on which these theories are based are both physical (to do with the external, objective world) and psychological (to do with our internal, conscious world). Concepts like 'red' and 'blue' are an individual's internal or sensory response to external stimuli. Light reaching the eye passes through the pupil and is focused on to the retina by the lens (Figure 5.1). The retina contains large numbers of light-sensitive photoreceptors, termed *rods* and *cones*. These photoreceptors are connected via a network of nerve fibres to the optic nerve, along which travel the signals that are interpreted by the brain as images of our environment. There are around 100 million rod-shaped cells on the retina, and five million cone-shaped cells. Each of these cells is connected to a nerve, the

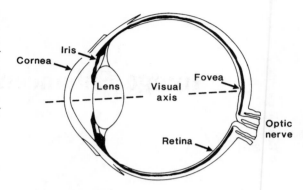

Figure 5.1   Simplified diagram of the human eye.

junction being called a synapse. The way in which these cells respond to light is through alteration of a molecule known as a chromophore. Changes in the amount of light reaching a chromophore produce signals, which pass through the nerve fibre to the optic nerve. Signals from the right eye are transmitted through the optic nerve to the left side of the brain, and vice versa.

It is generally accepted that the photoreceptor cells, comprising the rods and cones, differ in terms of their inherent characteristics. The rod-shaped cells respond to light at low illumination levels, and provide a means of seeing in such conditions. This type of vision is called *scotopic*. It does not provide any colour information, though different levels of intensity can be distinguished. Cone or *photopic* vision allows the distinction of colours or hues and the perception of the degree of saturation (purity) of each hue as well as the intensity level. However, photopic vision requires a higher illumination level than does scotopic vision. Colour is thought to be associated with cone vision because there are three kinds of cones, each kind being responsive to one of the three primary colours of light (red, green and blue). This is called the *tri-stimulus theory* of colour vision. Experiments have shown that the number of blue-sensitive cones is much less than the number of red- or green-sensitive cones, and that the areas of the visible spectrum in which the three kinds of cones respond do, in fact, overlap (Figure 5.2). There are other theories of colour (Land, 1977; Overheim and Wagner, 1982; Wasserman, 1978) but the tri-stimulus theory is an attractive one not merely because it is simple but because it provides the idea that colours can be formed by adding red, green and blue light in various combinations.

A model of 'colour space' can be derived from the idea that colours are formed by adding together differing amounts of red, green and blue light. Figure 5.3 shows a

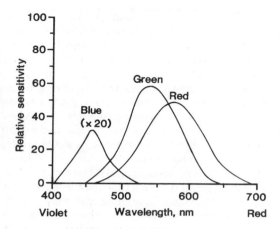

Figure 5.2 Sensitivity of the eye to red, green and blue light.

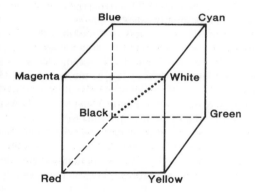

Figure 5.3 Red–green–blue colour cube.

geometrical representation of the RGB (red, green, blue) colour cube. The origin is at the vertex of the cube marked 'black' and the axes are black–red, black–green and black–blue. A colour is specified by giving its coordinates along these three axes. These coordinates are termed (R, G, B) triples. Notice that white light is formed by the addition of maximum red, maximum green and maximum blue light. The line joining the black and white vertices of the cube represents colours formed by the addition of equal amounts of red, green and blue light; these are shades of grey. Colour television makes use of the RGB model of colour vision. A television screen is composed of an array of dots, each of which contains red-, green- and blue-sensitive phosphors. Colours on the screen are formed by exciting the red, green and blue phosphors in differing proportions. If the proportions of red, green and blue were equal at each point

(but varying over the area of the screen) a grey-scale picture would result. A colour picture is obtained when the amounts of red, green and blue at each point are unequal.

The RGB colour cube model links intuitively with the tri-stimulus theory of colour vision and also with the way in which a colour television monitor works. Other colour models are available which provide differing views of the nature of our perception of colour. The HSI model uses the concepts of hue (H), saturation (S) and intensity (I) to explain the idea of colour. Hue is the dominant wavelength of the colour we see; hues are given names such as red, green, orange and magenta. The degree of purity of a colour is its saturation. A pure colour is 100% saturated. Intensity is a measure of the brightness of a colour. Figure 5.4 shows a geometrical representation of the HSI model. Hue is represented by the top edge of a six-sided cone (hexcone) with red at 0°, green at 120° and blue at 240°. Pure colours such as red, green and blue lie around the top edge of the hexcone. Addition of white light produces less saturated, paler, colours and so saturation can be represented by the distance from the vertical axis of the hexcone. Intensity (sometimes called value) is shown as a distance above the apex of the hexcone, increasing upwards as shown by the widening of the hexcone. The point marked black has no hue, nor do any of the shades of grey lying on the vertical axis between the black and white points. All these shades of grey, including white and black, have zero saturation.

The RGB model of colour is that which is normally used in the study and interpretation of remotely-sensed images, and in the rest of this chapter we will deal exclusively with this model. The use of the HSI model is considered in Chapter 6 in common with other image

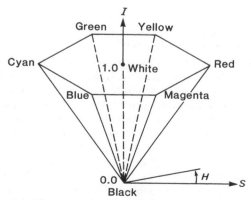

Figure 5.4 Hue–saturation–intensity hexcone.

transforms, for the representation of colours in terms of the HSI model can be accomplished by a straightforward transformation of the RGB colour coordinates. The HSI transform can be used to enhance multispectral images in terms of their contrast (section 5.3), and this operation is implemented in the MIPS software contained in the accompanying CD-ROM.

Further details of the material covered in this section can be found in Foley *et al.* (1990, 1994), Smith (1978), Williams and Becklund (1972) and Stimson (1974). Drury (1993) discusses the properties of the human visual system in relation to the choice of techniques for processing remotely-sensed images for geological applications.

5.3   CONTRAST ENHANCEMENT

The sensors mounted on board aircraft and satellites have to be capable of detecting upwelling radiance levels ranging from low (for example, over oceans) to very high (for example, over snow or ice). For any particular area that is being imaged it is unlikely that the full dynamic range of the sensor will be used and the corresponding image is dull and lacking in contrast or over-bright. In terms of the RGB colour cube model of section 5.2 the pixel values are clustered in a narrow section of the black–white axis (Figure 5.3). Not much detail can be seen on such images, which are either under-exposed or over-exposed in photographic terms. If the range of grey levels could be altered so as to fit the full range of the black–white axis of Figure 5.3, then the contrast between the dark and light areas of the image would be improved while maintaining the relative distribution of the grey levels. It is indeed possible to do this by manipulating the colour lookup tables (LUTs) described in Chapter 3 using the technique of linear contrast stretching.

5.3.1   Linear contrast stretch

In its basic form the linear contrast stretching technique involves the translation of the image pixel values from the observed range V_{min} to V_{max} to the full range of the display device (generally 0–255, which is the range of values representable in an eight-bit display device). V is a pixel value observed in the image under study, with V_{min} being the smallest pixel value in the image and V_{max} the largest. The pixel values are scaled so that V_{min} maps to a value of 0 and V_{max} maps to a value of 255. Intermediate values retain their relative positions, so that the observed pixel value in the middle of the range from

V_{min} to V_{max} maps to 127. Notice that we cannot map the middle of the range of the observed pixel values to 127.5 (which is exactly half-way between 0 and 255) because the display system can store only the discrete levels 0, 1, 2, ..., 255.

An uncritical approach to the mapping problem would involve the application of the mapping function to every pixel in the image by computing a 'stretched' pixel value and replacing the original value in the video memory with the 'stretched' value. This would involve considerable computational effort and, in addition, would mean that the original pixel values would be lost. Since enhancement is often a trial-and-error procedure the pixel values would have to be read back into the display system, possibly from a relatively slow disk system, if the linear contrast stretch proved to be unsuitable for the particular need.

A more reasonable approach would be one which realised that the number of separate pixel values in the image could be simply calculated as ($V_{max} - V_{min} + 1$), which must be 256 or less. All lookup table (LUT) output values corresponding to input values of V_{min} or less are set to zero, while LUT output values corresponding to input values of V_{max} or more are set to 255. The range V_{min} to V_{max} is then linearly mapped on to the range 0–255, as shown in Figure 5.5. Table 5.1 shows an extract from the LUT corresponding to Figure 5.5. Using the LUT shown in this figure, any pixel in the image having the value 16 is transformed to an output value of 0 before being sent to the digital-to-analogue converter and thence to the monitor display. All input values of

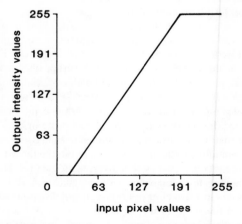

Figure 5.5   Graphical representation of lookup table to map input pixel values 16–191 on to the full intensity range 0–255. Input values less than 16 are set to 0 on output. Input values of 191 or greater are set to 255 on output. Input values between 16 and 191 inclusive are interpolated to output values 0–255.

Table 5.1 Extracts from lookup table values corresponding to the graph shown in Figure 5.5. Column marked 'Input' shows the input pixel value. Column marked 'Output' (figures in italic) shows the corresponding output intensity value.

Input	Output	Input	Output	Input	Output	Input	Output	Input	Output
0	*0*	1	*0*	2	*0*	3	*0*	4	*0*
5	*0*	6	*0*	7	*0*	8	*0*	9	*0*
10	*0*	11	*0*	12	*0*	13	*0*	14	*0*
15	*0*	16	*1*	17	*2*	18	*4*	19	*5*
20	*7*	21	*8*	22	*10*	23	*11*	24	*13*
25	*14*	26	*16*	27	*17*	28	*18*	29	*20*
30	*21*	31	*23*	32	*24*	33	*26*	34	*27*
35	*29*	36	*30*	37	*32*	38	*33*	39	*34*
...	...	...	...	...	...	...	...	...	...
135	*174*	136	*175*	137	*177*	138	*178*	139	*180*
140	*181*	141	*183*	142	*184*	143	*186*	144	*187*
145	*189*	146	*190*	147	*191*	148	*193*	149	*194*
150	*196*	151	*197*	152	*199*	153	*200*	154	*202*
155	*203*	156	*205*	157	*206*	158	*207*	159	*209*
160	*210*	161	*212*	162	*213*	163	*215*	164	*216*
165	*218*	166	*219*	167	*221*	168	*222*	169	*223*
170	*225*	171	*226*	172	*228*	173	*229*	174	*231*
175	*232*	176	*234*	177	*235*	178	*237*	179	*238*
180	*239*	181	*241*	182	*242*	183	*244*	184	*245*
...	...	...	...	...	...	...	...	...	...
221	*255*	222	*255*	223	*255*	224	*255*	225	*255*
226	*255*	227	*255*	228	*255*	229	*255*	230	*255*
231	*255*	232	*255*	233	*255*	234	*255*	235	*255*
236	*255*	237	*255*	238	*255*	239	*255*	240	*255*
241	*255*	242	*255*	243	*255*	244	*255*	245	*255*
246	*255*	247	*255*	248	*255*	249	*255*	250	*255*
251	*255*	252	*255*	253	*255*	254	*255*	255	*255*

191 and more are transformed to output values of 255. The range of input values between 16 and 191 is linearly mapped on to the full dynamic range of the display device, assumed in this example to be 0–255. The operation of lookup tables is explained in more detail in section 3.3.1.

The technique as described can be applied to a single-band, grey-scale image, where the image data are mapped to the display via all three colour LUTs. These three LUTs are set to identical values. When a false-colour composite image is to be displayed, a separate linear contrast stretch is normally applied to each colour component of the false-colour image, and so the three LUTs have different settings. The overall effect of this simple technique is to brighten up an image that is under-exposed (too dark) or to darken an otherwise over-bright image.

A slight modification can be used to provide a little more user interaction. The basic technique, as described above, does not take into consideration any characteristic of the image data other than the maximum and minimum pixel values. These values may be outliers,

located well away from the rest of the image data values. If this were so, it could be observed if the image histogram were computed and displayed. The *image histogram* for a single-band grey-scale image or for one band (red, green or blue) of a false-colour image is a plot of the number of pixel values having the values 0, 1, 2, ..., 255 against those class values. The histogram is an important tool in digital image processing, and we will return to its uses in later sections. For the present, it is sufficient to note that it is a relatively straightforward matter to find the 5th and 95th percentiles of the distribution of pixel values from inspection of the image histogram. The 5th percentile point is that value exceeded by 95% of the image pixel values, while the 95th percentile is the pixel value exceeded by 5% of all pixel values in the image. If, instead of V_{max} and V_{min} we use V_{95} and V_5 then we can set the LUT so that all pixel values equal to or less than V_5 are output as zero while all pixel values greater than V_{95} are output as 255. Those values lying between V_5 and V_{95} are linearly mapped, as before, to the full brightness scale of 0 to 255. Again, this technique is applied separately to each

(a)

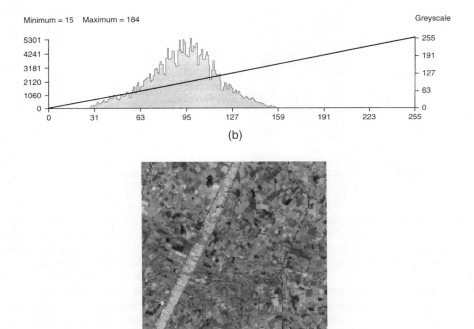

Minimum = 15   Maximum = 184   Greyscale

(b)

(c)

Minimum = 15   Maximum = 184   Greyscale

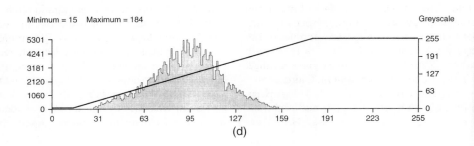

(d)

(a)

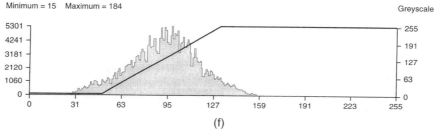

(f)

Figure 5.6 Linear contrast stretch. (a) Unenhanced Landsat TM band 4 (near-infrared) image of the Littleport area of eastern England (see Figure 1.9). (b) Plot of image histogram and default lookup table corresponding to (a). (c) Image shown in (a) after a 0–100% (default) linear contrast stretch. (d) Plot of image histogram showing lookup table after linear contrast stretch shown in (c). (e) Image shown in (a) after a 5–95% linear contrast stretch. (f) Plot of image histogram showing lookup table after linear contrast stretch shown in (e). Original data © ESA 1994, distributed by Eurimage.

component (red, green or blue) of a false-colour image. Of course, values other than the 5th and 95th percentiles could be used; for example, one might elect to choose the 10th and 90th percentiles, or any other (usually symmetric) pair.

Figure 5.6 shows the effects of these linear contrast stretches, using an extract from a Landsat Thematic Mapper band 4 image showing the Littleport area of Cambridgeshire in eastern England. The first image/histogram pair shows the raw data, with the default LUT superimposed on the histogram. The second image/histogram pair shows the effects of the automatic (0–100%) linear stretch. The minimum and maximum pixel values in the image are 15 and 184, so the LUT sets all values of 15 and below to zero and all values of 184 and above to 255. The intermediate values 15–184 are interpolated on to the 0–255 output brightness range. The last of the three image/histogram pairs uses a 5–95% linear contrast stretch, which maps the image pixel values 51–133 on to the 0–255 output brightness range.

In terms of the RGB colour model described in section 5.2 the original black-and-white image data lay along, but did not completely use up, the colour space between the corners of the cube labelled 'black' and 'white'. After stretching, the distribution of the brightness values along this diagonal of the RGB cube is much more even. The linear contrast stretching technique is a relatively simple one, and can be executed rapidly by changing the 256 LUT values for each of the RGB colour components. If the modified method is used then the image histogram must be computed; this is not an instantaneous procedure if the image is at all large. Consequently, the image histogram is often computed routinely when the image is read from tape or CD, and stored in a file for future reference. When image histograms are calculated it must be remembered that the largest positive integer value that can be stored in a 16-bit word (Fortran INTEGER*2 data type, C short int) is 32 767. Frequency counts larger than this are inevitable given the capabilities of most remotely-sensed images. Errors, which are not picked up by some operating systems, will occur if a number greater than 32 767 is placed in a Fortran `INTEGER*2` store location; for example, 32 768 may be stored as $-32\,768$ without any indication of error (though more sophisticated operating systems will flag an 'integer overflow' error). Hence, in computing image histograms a 32-bit (`INTEGER*4`) array must be used to store the class frequencies.

The linear contrast stretch operation is implemented in the MIPS software provided with this book. An automatic stretch (using the minimum and maximum histogram values) can be selected, or the user can select specific upper and lower levels for the stretch, or the 5–95% limit can be used. The image histograms are calculated during the creation of the image dictionary file, using the program `make_inf`, and so the operation is virtually instantaneous. Appendix A provides further details.

5.3.2   Histogram equalisation

The whole image histogram, rather than its extreme points, is used in the more sophisticated methods of contrast enhancement. Hence, the shape as well as the extent of the histogram is taken into consideration. The first of the two methods described here is called *histogram equalisation*. Its underlying principle is straightforward. It is assumed that each level in the displayed image should contain an approximately equal number of pixel values, so that the histogram of these displayed values is almost uniform (though not all 256 classes are necessarily occupied). If this is done, the entropy of the image, a measure of the information content of the image, will be increased (section 2.2.3). Because of the nature of remotely-sensed digital images, whose pixels can take on only the discrete values 0, 1, 2, ..., 255, it may be that there are 'too many' pixel values in one class, even after equalisation. However, it is not possible to take some of the values from that class and redistribute them to another class, for there is no way of distinguishing between one pixel value of 'x' and another of the same value. It is rare, therefore, for a histogram of the pixel values to be exactly uniformly distributed after the histogram equalisation procedure has been applied.

The method itself involves, first, the calculation of the target number of pixel values in each class of the equalised histogram. This value (call it n_t) is easily found by dividing N, the total number of pixels in the image, by 256 (the number of histogram classes, which is the number of intensity levels in the image). Next, the histogram of the input image is converted to cumulative form

with the number of pixels in classes 0 to j represented by C_j. This is achieved by summing the number of pixels falling in classes 0 to j of the histogram (the histogram classes are labelled 0–255 so as to correspond to the pixel values on an 8-bit scale):

$$C_j = n_0 + n_1 + \ldots + n_j$$

where n_j is the number of pixels taking the grey-scale value j. The output level for class j is simply C_j/n_t.

The method is not as complicated as it seems, as the worked example in Table 5.2 demonstrates. By setting the LUT rather than altering the individual pixel values in the image memory, considerable computation time is saved, and the effects of the histogram equalisation can be undone, if necessary, simply by resetting the LUT to its default state. The original pixel values (16 levels in the example) are shown in column 1 of Table 5.2(a), and the number of pixels at each level is given in column 2. The cumulative number of pixels is listed in column 3. The values in column 4 are obtained by determining the target number of pixels ($=$ total number of pixels divided by the number of classes, that is, $262\,144/16 = 16\,384$) and then finding the integer part of C_j divided by n_t, the target number. Thus, input levels 0 to 2 are all allocated to output level 0, input levels 3 and 4 are allocated to output level 1, and so on. Notice that the classes with relatively low frequency have been amalgamated while the classes with higher frequency have been spaced out more widely than they were originally. The effect is to increase the contrast in the centre of the range while reducing contrast at the margins. Table 5.2(b) gives the numbers of pixels assigned to each output level. In this example, which uses only 16 levels for ease of understanding, the output histogram is not uniform. This is not surprising, for the number of pixels at five of the input levels considerably exceeds the target number of 16 384. The LUT is set originally to the values shown in column 1 of Table 5.2(a); if the values shown in column 4 are sent to the LUT the effect on the displayed image will be as if all the pixel values stored in the image memory bank had been altered from the values in column 1 to the values in column 4.

The example shows that the effect of the histogram equalisation procedure is to spread the range of pixel values present in the input image over the full range of the display device; in the case of a colour monitor this is normally 256 levels for each primary colour (red, green, blue). The relative brightness of the pixels in the original image is not maintained. Also, in order to achieve the uniform histogram the number of levels used is reduced (see example, Table 5.2). This is because those histogram

Table 5.2   The histogram equalisation procedure.

(a) Illustrating calculations involved in histogram equalisation procedure. $N = 262\,144$, $n_t = 16\,384$. See text for explanation.

Old LUT value	Number in class	Cumulative number	New LUT value
0	1 311	1 311	0
1	2 622	3 933	0
2	5 243	9 176	0
3	9 176	18 352	1
4	13 108	31 460	1
5	24 904	56 364	3
6	30 146	86 510	5
7	45 875	132 385	8
8	58 982	191 367	11
9	48 496	239 863	14
10	11 796	251 659	15
11	3 932	255 591	15
12	3 932	259 523	15
13	2 621	262 144	15
14	0	262 144	15
15	0	262 144	15

(b) Number of pixels allocated to each class after the application of the equalisation procedure shown in (a). Note that the smaller classes in the input have been amalgamated, reducing the contrast in those areas, while larger classes are more widely spaced, giving greater contrast. The number of pixels allocated to each non-empty class varies considerably, because discrete input classes cannot logically be split into subclasses.

Intensity	Number	Intensity	Number
0	9 176	8	45 875
1	22 284	9	0
2	0	10	0
3	24 904	11	58 982
4	0	12	0
5	30 146	13	0
6	0	14	48 496
7	0	15	22 281

classes with relatively few members are amalgamated to make up the target number, n_t. In the areas of the histogram that have the greatest class frequencies the individual classes are stretched out over a wider range. The effect is to increase the contrast in the densely populated parts of the histogram and to reduce it in other, more sparsely populated, areas. If there are relatively few occupied LUT entries after the equalisation process then the result may be unsatisfactory compared to the simple linear contrast stretch.

Sometimes it is desirable to equalise only a specified part of the histogram. For example, if a mask is used to eliminate part of the image (for example, water areas may be set to zero) then a considerable number of pixel values of zero will be present in the histogram. If there are N zero pixels then the output LUT value corresponding to an input pixel value of zero after the application of the procedure described above will be N/n_t, which may be large; for instance, if N is equal to 86 134 and n_t is equal to 3192 then all the zero (masked) values in the original image will be set to a value of 27 if the result is rounded to the nearest integer. A black mask will thus be transformed into a dark grey one, which may be undesirable. The calculations described above can be modified so that the input histogram cells between, say, 0 and a lower limit L are not used in the calculations. It is equally simple to eliminate input histogram cells between an upper limit H and 255; indeed, any of the input histogram cells can be excluded from the calculations.

Figure 5.7 shows the same image as is used in Figure 5.6 after a histogram equalisation stretch. Note the non-linear shape of the lookup table.

Histogram equalisation is implemented in the MIPS software described in Appendix A. Two options are provided: use the entire histogram, or specify input histogram limits, to allow elimination of either the lower or the upper tail of the histogram, or both. For example, the histogram equalisation may be performed for the range of pixel values 100–199. In this case, image pixel values 0–99 and 200–255 are unaltered but values 100–199 are mapped on to the whole output brightness range. This process can lead to some unusual effects, and should therefore be used with care.

5.3.3   Gaussian stretch

A second method of contrast enhancement based upon the histogram of the pixel values is called a *Gaussian stretch* because it involves the fitting of the observed histogram to a normal or Gaussian histogram. A *normal distribution* that gives the probability of observing a value x given the mean $\bar{x}$ is defined by

$$p(x) = \frac{1}{\sigma\sqrt{2\pi}}\exp\left(\frac{-(x - \bar{x})^2}{2\sigma^2}\right)$$

The standard deviation, σ, is defined as the range of the variable for which the function $p(x)$ drops by a factor of $e^{-0.5}$ or 0.607 of its maximum value. Thus, 60.7% of the values of a normally distributed variable lie within one

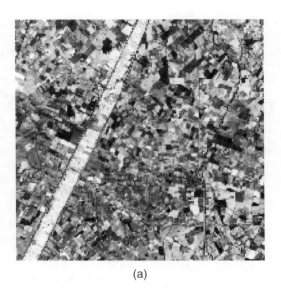

(a)

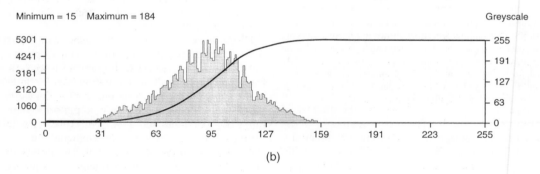

(b)

Figure 5.7   Histogram equalisation. (a) Landsat TM band 4 image shown in Figure 5.6(a) after histogram equalisation. (b) Plot of image histogram showing lookup table after histogram equalisation. Original data © ESA 1994, distributed by Eurimage.

standard deviation of the mean. For many purposes a *standard normal* distribution is useful. This is a normal distribution with a mean of zero and a unit standard deviation. Values of the standard normal distribution are tabulated in standard statistics texts, and formulae for the derivation of these values are given by Abramowitz and Stegun (1972). A Fortran subroutine listing is given by Hill (1973). The shape of the normal distribution, which will be familiar to many readers, is sketched in Figure 5.8. It is evident from this figure that the probability (or *y* value) decreases away from the mean in a systematic and symmetrical fashion.

An example of the calculations involved in applying the Gaussian stretch is shown in Table 5.3. The input histogram is the same as that used in the histogram equalisation example (Table 5.2). Again, 16 levels are

used for simplicity's sake. Since the normal distribution ranges from minus infinity to plus infinity, some delimiting points are needed to define the area of the distribution to be used for fitting purposes. The range ±3 standard deviations from the mean is used in the example. Level 1 is, in fact, the probability of observing a value of a normally distributed variable that is 3 standard deviations or more below the mean; level 2 is the probability of observing a value of a normally distributed variable that is between 2.6 and 3 standard deviations below the mean; and so on. These values are derived from an algorithm based on the approximation specified by Abramowitz and Stegun (1972). Column (i) of Table 5.3(a) shows the pixel values in the original, unenhanced image. Column (ii) gives the points on the standard normal distribution to which these pixel

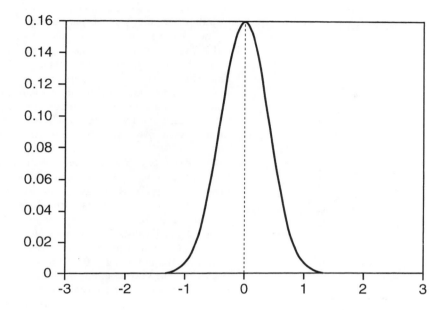

Figure 5.8 Normal (Gaussian) distribution with mean of zero and standard deviation of 2.5.

values will be mapped, while column (iii) contains the probabilities, as defined above, that are associated with the class intervals. Assume that the number of pixels in the image is $512 \times 512 = 262\,144$ and the number of quantisation levels is 16. The target number of pixels (that is, the number of pixels that would be observed if their distribution were normal) is found by multiplying the probability for each level by the value 262 144. These results are contained in column (iv) and, in cumulative form, in column (v). The observed counts for the input image are shown by class and in cumulative form in columns (vi) and (vii). The final column gives the level to be used in the Gaussian stretched image. These levels are determined by comparing columns (v) and (vii) in the following manner. The value in column (vii) at level 0 is 1311. The first value in column (v) to exceed 1311 is that associated with level 1, namely, 1398; hence, the input level 0 becomes the output level 1. Taking the input (cumulative) value associated with input level 1, that is, 3933, we find that the first element of column (v) to exceed 3933 is that value associated with level 3 (9595), so input level 1 becomes output level 3. This process is repeated for each input level. Once the elements of column (viii) have been determined, they can be written to the lookup table and the input levels of column (i) will automatically map to the output levels of column (viii).

The range ± 3 standard deviations as used in the example is not the only one that could have been used. A larger or smaller proportion of the total range of the standard normal distribution can be specified, depending on the requirements of the user. Usually, the limits chosen are symmetric about the mean, and the user can provide these limits from a terminal. An example of the Gaussian stretch is given in Figure 5.9. This image should be compared with the unenhanced version of the same image (Figure 5.6(a)) and the same image after a linear contrast stretch (Figures 5.6(b) and (c)) and histogram equalisation (Figure 5.7). Readers might like to select one of these images as the 'best' for visual interpretation, but it should be noted that the adequacy of any contrast stretch depends on the shape of the histogram of the unenhanced image. A Gaussian stretch option is included in the MIPS image display program (Appendix A).

If Tables 5.2 and 5.3 are compared it will be seen that the Gaussian stretch emphasises contrast in the tails of the distribution while the histogram equalisation method reduces contrast in this region. However, at the centre of the distribution the reverse may be the case, for the target number for a central class may well be larger for the Gaussian stretch than the histogram equalisation. In the worked example the target number for each class in the histogram equalisation was 16 384; note that

Table 5.3 The Gaussian stretch procedure.

(a) Fitting observed histogram of pixel values to a Gaussian histogram. See text for discussion. The column headings are described in the footnote and more fully in the text.

(i)	(ii)	(iii)	(iv)	(v)	(vi)	(vii)	(viii)
0	< −3.0	0.0020	530	530	1 311	1 311	1
1	−2.6	0.0033	868	1 398	2 622	3 933	3
2	−2.2	0.0092	2 423	3 821	5 243	9 176	3
3	−1.8	0.0220	5 774	9 595	9 176	18 352	4
4	−1.4	0.0448	1 175	21 346	13 108	31 460	5
5	−1.0	0.0779	20 421	41 767	24 904	56 364	6
6	−0.6	0.1156	30 303	72 070	30 146	86 510	7
7	−0.2	0.1465	38 401	110 471	45 875	132 385	8
8	0.2	0.1585	41 555	152 026	58 982	191 367	10
9	0.6	0.1465	38 401	190 427	48 496	239 863	11
10	1.0	0.1156	30 303	220 730	11 796	251 659	12
11	1.4	0.0779	20 421	241 151	3 932	255 591	13
12	1.8	0.0448	11 751	252 902	3 932	259 523	14
13	2.2	0.0220	5 774	258 676	2 621	262 144	15
14	2.6	0.0092	2 423	261 099	0	262 144	15
15	> 3.0	0.0040	1 045	262 144	0	262 144	15

Column headings:
(i) Original pixel value.
(ii) Standard deviations above (+) or below (−) the mean of the standard normal distribution.
(iii) Probability for each class from standard normal (Gaussian) distribution.
(iv) Target number of pixels at each level ($N = 262\,144$).
(v) Cumulative target number of pixels.
(vi) Observed number of pixels at each level.
(vii) Cumulative observed number of pixels.
(viii) Pixel value after transformation.

(b) Number of pixels at each level following transformation to Gaussian model.

Intensity	Number	Intensity	Number
0	0	8	45 875
1	1 311	9	0
2	0	10	58 782
3	7 865	11	48 496
4	9 176	12	11 796
5	13 108	13	3 932
6	24 904	14	3 932
7	30 146	15	2 621

the target numbers for classes 5 to 10 inclusive in the Gaussian stretch exceed 16 384. Input classes may well have to be amalgamated in order to achieve these target numbers. Tables 5.2(b) and 5.3(b) give the number of pixels allocated to each output class after the application of the histogram equalisation and Gaussian contrast stretches, respectively. In both cases, the range of levels allocated to the output image exceeds the range of

pixel values in the input image; this will result in an overall brightening of the displayed image.

The application of contrast enhancement techniques is discussed above in terms of a single grey-scale image. The techniques can be used to enhance a false-colour image by applying the appropriate process to the red, green and blue channels separately. Methods of simultaneously 'stretching' the colour components of a false-colour image are dealt with elsewhere (the HSI transform in section 6.5 and the decorrelation stretch in section 6.4.3).

If an image covers two or more spectrally distinctive regions, such as land and sea, then the application of the methods so far described may well be disappointing. In such cases, any of the contrast stretching methods described above can be applied to individual parts of the range of pixel values in the image; for instance, the histogram equalisation procedure could be used to transform the input range 0 − 60 to the output range 0–255, and the same could be done for the input range 61–255. The same procedure could be used whenever distinct regions occur if these regions can be identified by splitting the histogram at one or more threshold points. While the aesthetic appeal of images enhanced in this fashion may be increased, it should be noted that pixels with considerably different radiance values will be assigned to the same displayed or output colour value. The colour balance will also be quite different from that resulting from a standard colour-composite procedure.

5.4   PSEUDOCOLOUR ENHANCEMENT

In terms of the RGB colour model presented in section 5.2, a black-and-white image occupies only the diagonal of the RGB colour cube running from the 'black' to the 'white' vertex (Figure 5.3). In terms of the HSI model, grey values are ranged along the vertical or intensity axis (Figure 5.4). No hue or saturation information is present, yet the human visual system is particularly efficient in detecting variations in hue and saturation, but not so efficient in detecting intensity variations. Three methods are available for converting a grey-scale image to colour. The colour rendition in the output image is not true or natural, for the original (input) image does not contain any colour information, and no enhancement technique can generate information not present in the input image. Nor is the colour rendition correctly described as false colour, for a false-colour image is one composed of three bands of information, which are represented in visible red, green and blue. The name given to a colour rendition of a single band of imagery is a pseudocolour image. Two techniques are

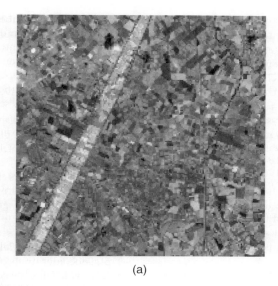

(a)

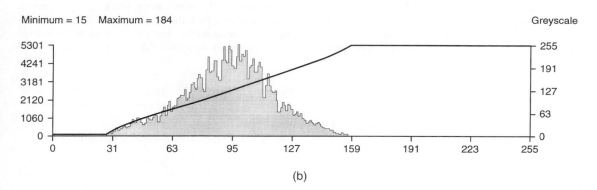

(b)

Figure 5.9   Gaussian stretch. (a) Landsat TM band 4 image shown in Figure 5.6(a) after Gaussian contrast stretch with $\sigma = 3.0$. (b) Plot of image histogram showing lookup table after Gaussian contrast stretch. Original data © ESA 1994, distributed by Eurimage.

available for converting from black-and-white to pseudocolour form. These are the technique of density slicing and the pseudocolour transform. Each provides a method for mapping from a one-dimensional grey scale to a three-dimensional (RGB) colour.

5.4.1   Density slicing

Density slicing is the mapping of a range of contiguous grey levels of a single-band image to a point in the RGB colour cube. The range of contiguous grey levels (such as 0 to 10 inclusive) is called a *slice*. The range 0–255 is normally converted to several colour slices. It is gen-

erally acknowledged that conversion of a single-band black-and-white image to pseudocolour is an effective way of highlighting different but internally homogeneous areas within an image, but at the expense of loss of detail. The loss of detail is due to the conversion from a 256-level image to an image represented in terms of many fewer colour slices. The effect is (i) to reduce the number of discrete levels in the image, for several grey levels are usually mapped on to a single colour, and (ii) to improve the visual interpretability of the image if the slice boundaries and the colours are carefully selected.

In most image processing systems the user is allowed to specify any colour for the current slice, and to alter slice boundaries in an upwards or downwards direction by means of a joystick or mouse. The slice boundaries are thus obtained by an interactive process, which allows the user to adjust the levels until a satisfactory result has been achieved. The choice of colour for each slice is important if information is to be conveyed to the viewer in any meaningful way, for visual perception is a psychological as well as a physiological process. Random colour selections may say more about the psychology of the perpetrator than about the information in the image. Consider, for example, a thermal infrared image of the heat emitted by the Earth. A colour scale ranging from light blue to dark blue, through the yellows and oranges to red would be a suitable choice, for most people have an intuitive feel for the 'meaning' of colours in terms of temperature. A scale taking in white, mauve, yellow, black, green and pink might confuse rather than enlighten.

Readers should use the MIPS software supplied with this book to experiment with the procedures of density slicing of a grey-scale image. A single band of a suitable image should be selected, loaded and displayed, then the density slicing option used to select grey-scale slices and convert them to different colours. Details are provided in Appendix A. Most readers will find that, although the technique is easily described in a textbook, it is less easy to select the 'right' colours and the 'right' number of slices.

5.4.2   Pseudocolour transform

A grey-scale image is displayed by setting the lookup tables for red, green and blue to the same values. Figure 5.10(a) shows, in graphical form, the default state of the lookup tables for grey-scale image display. A *pseudocolour transform* is carried out by setting the three lookup tables to the format shown in Figure 5.10(b). The settings shown in Figure 5.10(b) send different colour (RGB) information to the digital-to-analogue converter for the same grey-scale pixel value. The result is a pseudocolour image. Unlike the density slicing method, the pseudocolour transform method associates each grey level with a discrete colour, although the difference between 90% red and 89% red may not be physically discernible in reality. If the histogram of the grey-scale values in the black-and-white image is not approximately uniform then the resulting pseudocolour image will be dominated by one colour, and its usefulness thereby reduced. Analysis of the image histogram along

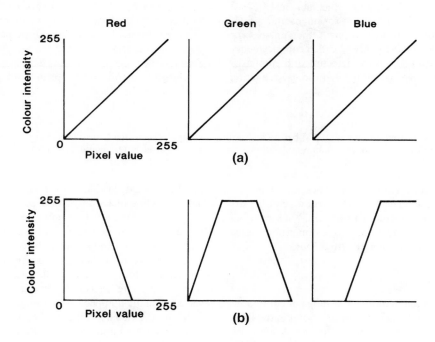

Figure 5.10   (a) Lookup table set for grey-scale image display. (b) Lookup table for pseudocolour display.

the lines of the histogram equalisation procedure (section 5.3.2) prior to the design of the pseudocolour LUTs would alleviate this problem.

The pseudocolour transform is implemented in the MIPS display program (Appendix A). The automatic mode sets the three lookup tables (red, green and blue) to the shapes shown in Figure 5.10(b) using the minimum and maximum pixel values in the image as the upper and lower limits rather than 0 and 255. A second option allows the user to set different upper and lower limits. Grey levels outside these limits are not converted to colour. Readers may like to experiment with the various settings and work out what is happening (the 'plot histogram' function can be used to check the shape of the lookup table plots).

The Fourier transform (section 6.6) can be used to split a grey-scale image into its low-, intermediate- and high-frequency components. These three components can subsequently be displayed as the red, green and blue inputs to a false-colour display. Such a display uses colour not only to enhance the visual interpretation of a grey-scale image, but also to transmit information about the spatial scale components of that image.

5.5 SUMMARY

Image enhancement techniques include, but are not limited to, those of contrast improvement and monochrome to colour transformations. Other image processing methods can justifiably be called enhancements. These include: (i) methods for detecting and emphasising detail in an image (section 7.4), (ii) noise reduction techniques, ranging from removal of banding (section 4.2.2) to filtering (Chapter 7), (iii) colour transforms based on principal components analysis, called the decorrelation stretch (section 6.4.3), and (iv) the hue–saturation–intensity (HSI) transform (section 6.5). All these methods alter the visual appearance of the image in such a way as to bring out or clarify some aspect or property of the image that is of interest to a user. The range of uses to which remotely-sensed images can be put is considerable and so, although there are standard methods of enhancement such as those described here, they should not be applied thoughtlessly but with due regard to the user's requirements and purpose.

5.6 QUESTIONS

1. Explain the meaning of the following terms: rod, cone, scotopic, dynamic range, pseudocolour, false colour.
2. What are the differences between the HSI and the RGB representations of colour? Illustrate your answer with reference to the RGB cube and the HSI hexcone.
3. Why is the 5–95% linear contrast stretch often preferred to the linear stretch based on the minimum and maximum pixel values present in the image?
4. Check that the calculations in Tables 5.2(a) and 5.3(a) are correct. What are the main differences in the effects achieved by the histogram equalisation and Gaussian contrast stretches?
5. What is the difference between density slicing and the pseudocolour transform? When would you use one rather than the other?

6

Image Transforms

6.1 INTRODUCTION

The image enhancement operations described in Chapter 5 are applied either to single-band images or separately to the individual bands of a multiband image set. With the exception of the Fourier transform, described in section 6.6, this chapter deals with operations on multiband image sets. Such image sets may consist of a single multispectral image of a particular area, or a number of images of the same area taken at different times (a *multitemporal* image set). The term 'transform' is used somewhat loosely in this chapter, for the arithmetic operators of addition, subtraction, multiplication and division are included, although they are not strictly transformations. These arithmetic operations, which are described in section 6.2, allow the generation of a new, composite, image from two or more bands of a multispectral or multitemporal image. The resulting image may well have properties that make it more suited to a particular purpose than the original. For example, the numerical difference between two images collected by the same sensor at different times may provide information about changes that have occurred between the two dates, while the ratio of the near-infrared and red bands of an image set is widely used as a *vegetation index*, which is often related to the attributes of the vegetation cover, particularly biomass and green leaf area index (LAI), of the area covered by the image.

Vegetation indices are discussed in section 6.3. They are based on a model of the distribution of data values on two or more spectral bands considered simultaneously. Two examples of these transformations are the perpendicular vegetation index (PVI), which uses a two-dimensional model of the relationship between vegetation and soil pixels, and the tasselled cap transformation, which is based on either the four bands of the Landsat MSS or the six reflective bands of Landsat TM.

Section 6.4 provides an introduction to the widely used technique of principal components analysis (PCA),

which is a method of re-expressing the information content of a multispectral set of m images in terms of a set of m principal components, which have two particular properties: zero correlation between the m principal components, and maximum variance. Because of the maximum variance property of principal components, it is usually found that much of the information in the original m correlated bands is expressible in terms of the first p of the full set of m principal components, where p is less than m. This property of principal components analysis is useful in generating a false-colour composite image. If the image set consists of more than three bands then the problem arises of selecting three bands for display in red, green and blue. Since the principal components of the image set are arranged in order of variance (which correlates with information) then the first three principal components can be used as the red, green and blue components of a false-colour composite image. No combination of the original bands can contain more information than is present in the first three principal components. Another use of principal components analysis is in reducing the amount of calculation involved in automatic classification (Chapter 8) by basing the classification on p principal components rather than on m spectral bands. In addition, the p principal component images require less storage space than the m-band multispectral image from which they were derived. Hence, principal components analysis can also be considered to be a data compression transformation.

Section 6.5 deals with a different transformation, which is concerned with the representation of the colour information in a set of three co-registered images. Theories of colour vision were summarised in section 5.2 where it was noted that the conventional red–green–blue (RGB) colour cube model is generally used to represent the colour information content of three images. The hue–saturation–intensity (HSI) hexcone model is considered in section 6.5 and its applications to image enhancement and to the problem of combining images from different sources (such as radar and optical images) are described.

The final section of this chapter (section 6.6) introduces a method of representing a single-band image in terms of its frequency content. The Fourier transform provides for the representation of image data in terms of a coordinate framework that is based upon spatial frequencies rather than upon distance from an origin (the conventional Cartesian x, y coordinate system). Image data that have been expressed in frequency terms are said to have been transformed from the image or *spatial domain* to the *frequency domain*. The frequency-domain representation of an image is useful in designing filters for special purposes (as described in Chapter 7) and in colour coding the scale components of the image.

6.2 ARITHMETIC OPERATIONS

The operations of addition, subtraction, multiplication and division are performed on two or more co-registered images of the same geographical area (see Chapter 4 for a survey of methods of geometric correction and image registration). These images may be separate spectral bands from a single multispectral data set or they may be individual bands from image data sets that have been collected at different dates. Addition of images is really a form of averaging for, if the dynamic range of the output image is to be kept equal to that of the input images, rescaling (usually division by the number of images added together) is needed. Averaging can be carried out on multiple images of the same area in order to reduce the noise component. Subtraction of pairs of images is used to reveal differences between those images and is often used in the detection of change if the images involved were taken at different dates. Multiplication of images is rather different from the other arithmetic operations for it normally involves the use of a single 'real' image and a binary image made up of ones and zeros. The binary image is used as a mask, for those image pixels in the real image that are multiplied by zero themselves become zero while those that are multiplied by one remain the same. Division or ratioing of images is probably the arithmetic operation that is most widely applied to images in geological, ecological and agricultural applications of remote sensing, for the division operation is used to detect the magnitude of the differences between spectral bands. These differences may be symptomatic of particular land cover types. Thus, a near-infrared: red ratio might be expected to be close to 1.0 for an object that reflects equally in both of these spectral bands (for example, a cloud-top), while the

value of this same ratio will be well above 1.0 if the near-infrared reflectance is higher than the reflectance in the visible red band, for example in the case of vigorous vegetation.

Image arithmetic operations can be carried out by the MIPS display program. First, load a false-colour (three-band) image, and choose `Utilities|Image Arithmetic`. This procedure is described in more detail in Appendix A.

6.2.1 Image addition

If multiple images of a given region are available for the same time and date of imaging then addition (averaging) of the multiple images can be used as a means of reducing the overall noise contribution. A single image might be expressed in terms of the following model:

$$G(x, y) = F(x, y) + N(x, y)$$

where $G(x, y)$ is the recorded image, $F(x, y)$ the true image and $N(x, y)$ the random noise component. $N(x, y)$ can be expected to be normally distributed with a mean of zero, since it is the sum of a number of small, independent errors or factors. The true signal, $F(x, y)$, is constant from image to image. Therefore, addition of two separate images of the same area taken at the same time might be expected to lead to the cancellation of the $N(x, y)$ term for, at any particular pixel position (x, y), the value $N(x, y)$ is as likely to be positive as to be negative. Image addition, as noted already, is really an averaging process. If two images $G_1(i, j)$ and $G_2(i, j)$ are added and if each has a dynamic range of 0–255 then the resulting image $G_{sum}(i, j)$ will have a dynamic range of 0–510. This is not a practicable proposition if the image display system has a fixed, 8-bit, resolution. Hence it is common practice to divide the sum of the two images by 2 to reduce the dynamic range to 0–255. The process of addition is carried out on a pixel-by-pixel basis as follows:

$$G_{sum}(i, j) = [G_1(i, j) + G_2(i, j)]/2$$

The result of the division is normally rounded to the nearest integer. If Fortran `INTEGER*1 (BYTE)` or `CHARACTER` data types are used to store the values of $G(x, y)$, then care must be taken to ensure that (i) the sum is stored in an `INTEGER*2` or `INTEGER*4` element before division and (ii) account is taken of the fact that the integer values 128 to 255 inclusive are treated as if they were -128 to -1 if single-byte, signed arithmetic is used.

If the algorithm described in the previous paragraph is applied then the dynamic range of the summed image will be approximately the same as that of the two input images. This may be desirable in some cases. However, it would be possible to increase the range by performing a linear contrast stretch (section 5.3.1) by subtracting a suitable offset o and using a variable divisor d:

$$G'(i,j) = [G_1(i,j) + G_2(i,j) - o]/d$$

The values of o and d might be determined on the basis of the user's experience, or by evaluating $G_{sum}(i,j)$ at a number of points systematically chosen from images $G_1 G$ and G_2. Image G' will have a stretched dynamic range in comparison with the result of the straight 'division by 2'.

6.2.2   Image subtraction

The subtraction operation is often carried out on a pair of co-registered images of the same area taken at different times. The purpose is to assess the degree of change that has taken place between the dates of imaging (see, for example, Dale *et al.*, 1996). Image differencing is performed on a pixel-by-pixel basis. The maximum negative difference (assuming both images have a dynamic range of 0–255) is $(0 - 255 =) - 255$ and the maximum positive difference is $(255 - 0 =) + 255$. The problem of scaling the result of the image subtraction operation on to a 0–255 range must again be faced. If the value 255 is added to the difference then the dynamic range is shifted to 0–510. Next, divide this range by 2 to give 0–255. Variable offsets and multipliers can be used as in the case of addition (section 6.2.1) to perform a linear contrast stretch operation. Formally, the image subtraction process can be written as:

$$G_{diff}(i,j) = [255 + G_1(i,j) - G_2(i,j)]/2$$

using the same notation as previously. If interest is centred on the magnitude rather than the direction of change then the following method could be used:

$$G_{absdiff}(i,j) = |G_1(i,j) - G_2(i,j)|$$

The vertical bars denote the absolute value (regardless of sign) of the difference. No difference is represented by the value 0 and the degree of difference increases towards 255.

A difference image $G_{diff}(i,j)$ tends to have a histogram that is normal in shape with the peak at 127 (if the standard scaling is used), tailing off rapidly in both

directions. The peak at 127 represents pixels that have not changed very much while the pixels in the histogram tails have changed substantially. The image $G_{absdiff}(i,j)$ has a histogram with a peak at or near zero and a long tail extending towards the higher values. It would be possible to design an interactive method of deciding on a threshold between 'change' and 'no change' pixels by using the colour lookup tables in the image display system (section 3.3) to transform selected parts of the dynamic range to black. If we had an image histogram such as that shown in Figure 6.1 then the areas of change could be highlighted if pixel values between 0 and 100 and between 150 and 255 were shown in white, with values in the range 101–149 being represented in black. Pixels showing only a small change will be black under this scheme, while those pixels with a large difference will be white. The lookup table to perform this operation is illustrated in Figure 6.2. Note that only one lookup table is shown. There are, in fact, three such lookup tables (one for each primary colour). All three must be identical if a greyscale image is to be produced on the monitor. The choice of threshold value (100 and 150) in this example is purely arbitrary. Normally these values would be fixed by a trial-and-error procedure.

An interesting use of the image subtraction operator is given by Jupp and Mayo (1982). They use a four-band Landsat MSS image, and first form a classified image (Chapter 8) in which all pixels are given a numerical label to indicate the land cover class to which they belong. For example, labels of 1, 2 and 3 could be used to indicate forest, grassland and water. Secondly, the mean pixel values in the four Landsat MSS bands are computed for each class. The pixels labelled as belong-

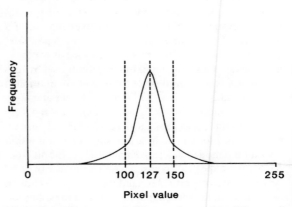

Figure 6.1   Histogram of difference image. No difference is coded 127. Thresholds at grey levels 100 and 150 are indicated.

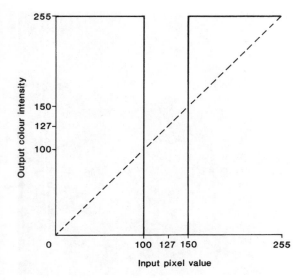

Figure 6.2 Graph of lookup table based on histogram shown in Figure 6.1. Input pixel values 0–100 and 150–255 are mapped to maximum white (output colour intensity 255) while values 101–149 are set to black (output colour intensity 0). Thus, areas of greatest change are displayed in white, and areas of lesser or no change are in black. The dashed line shows the graph for the default lookup table.

ing to class *i* are given the mean values for class *i* (one mean value per spectral band). This operation results in a set of four 'mean' images where Landsat MSS data are used. The third step is the generation of a set of four residual images, which are obtained by subtracting the actual pixel value recorded in a given MSS band from the corresponding value in the 'mean' image for that band. Residual images can be combined for colour composite generation. The procedure is claimed to assist in the interpretation and understanding of the classified image, as it highlights pixels that differ from the mean value of all pixels allocated the same class. This could be useful in answering questions such as: 'Have I omitted any significant land cover classes?' or 'Why is class *x* so heterogeneous?'

Other examples of the use of change detection techniques are Collins and Woodcock (1996), Johnson and Kasischke (1998), Lambin (1996), Lambin and Strahler (1994), Siljeström and Moreno (1995) and Varjo (1996). Gong *et al.* (1992) discuss the effects of mis-registration of images (Chapter 4) on the subtraction process, and present methods to reduce these effects, which will appear on the difference image as areas of change.

6.2.3 Image multiplication

Pixel-by-pixel multiplication of two images is rarely performed in practice. The multiplication operation is, however, a useful one if an image of interest is composed of two or more distinctive regions and if the analyst is interested only in one of these regions. Figure 6.3(a) shows a Landsat-2 MSS band 4 (green) image of part of the Tanzanian coast south of Dar-es-Salaam. Variations in the reflectance over the land area distract the eye from the more subtle variations in the radiance upwelling from the upper layers of the ocean. Variations over the distracting land region can be eliminated by a process of masking. The first step is the preparation of the mask that best separates land and water, using the near-infrared band (Figure 6.3(b)) since reflection from water bodies in the near-infrared spectral band is very low, while reflection from vegetated land areas is high. A suitable threshold is chosen by visual inspection of the image histogram of the near-infrared pixel values. A mask image is then generated from the near-infrared image by labelling with '1' those pixels that have values below the threshold (Figure 6.3(c)). Pixels whose values are above the threshold are labelled '0'. The second stage is the multiplication of the band 4 image and the mask image. Multiplication by 1 is equivalent to doing nothing, whereas multiplication by 0 sets the corresponding pixel to 0. Using the above procedure the pixels in the Tanzanian band 4 image that represent land are replaced by zero values, while 'ocean' pixels are unaltered. A linear contrast stretch (Chapter 5) produces the image shown in Figure 6.3(d). In practice, the two images (mask and band 4 in this example) are not multiplied but are processed by a simple logical function: if the mask pixel is zero then set the corresponding image pixel to zero, otherwise do nothing. Some software (for example, MIPS, which is described in the next paragraph) uses the two extremes (0 and 255) of the range of pixel values to indicate 'less than' and 'greater than' the threshold, respectively.

The masking procedure described in this section is implemented in the MIPS display program using the Utilities|Threshold operation. To create a mask, load as a grey-scale image the band on which the mask is to be based, then select Utilities|Threshold and choose an appropriate threshold value, and exit the procedure. Reset the lookup tables (using Utilities|Reset_LUTS) and view the masked image and its histogram. All values below the threshold are set to 0 and values above the threshold are given the value 255. If you load a false-colour image rather than a grey-scale image you can create a mask from one of the

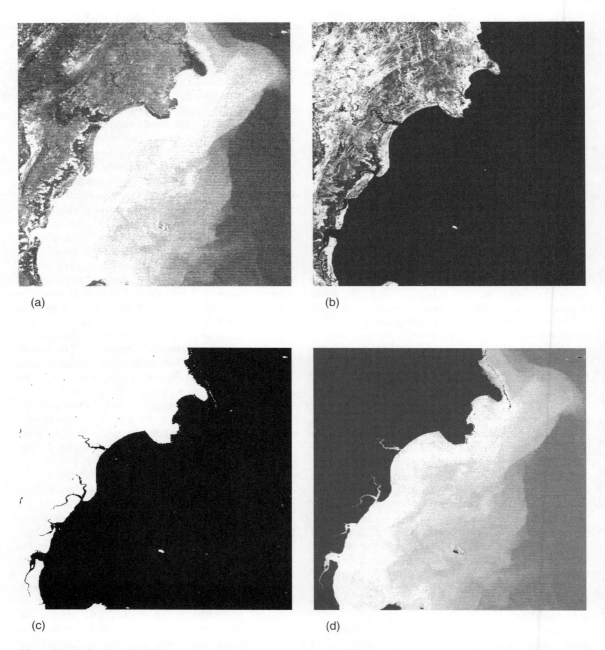

(a)

(b)

(c)

(d)

Figure 6.3   (a) Landsat-2 MSS band 4 (green) image of part of the coast of Tanzania, south of Dar-es-Salaam. (b) Landsat-2 MSS band 7 (near-infrared) image of the same area. (c) Mask derived by thresholding the near-infrared image. (d) Image shown in (a) after the application of the mask shown in (c). Data courtesy of USGS EROS Data Center, Sioux Falls, SD.

components of the false-colour image and then apply the mask to the displayed image (see Appendix A). A stand-alone program, mask, is also supplied (Appendix B), which applies the mask created by the MIPS image display program to any or all of the image files that are referenced in a given image dictionary file.

6.2.4   Image division

The process of dividing the pixels in one image by the corresponding pixels in a second image is known as *ratioing*. It is one of the most commonly used transformations applied to remotely-sensed images. There are two reasons why this is so. One is that certain aspects of the shape of spectral reflectance curves of different Earth-surface cover types can be brought out by ratioing. The second is that undesirable effects on the recorded radiances, such as that resulting from variable illumination (and consequently radiance) caused by variations in topography, can be reduced. Figure 6.4 shows the spectral reflectance curves for three cover types. The differences between the curves can be emphasised by looking at the gradient or slope between the red and the near-infrared bands, for example bands 5 (red) and 7 (near-infrared) in the

Landsat MSS, or bands 3 (near-infrared) and 2 (red) of the SPOT HRV. The curve for water shows a decline between these two points, while that for vegetation shows a substantial increase. The spectral reflectance curve for soil increases gradually between the two bands. If a pixel value in the vegetation curve in the near-infrared band is divided by the equivalent value in the red band then the result will be a positive real number that exceeds 1.0 in magnitude. The same operation carried out on the curve for water gives a result that is less than 1.0, while the soil curve gives a value somewhat higher than 1.0. The greater the difference between the pixel values in the two bands, the greater the value of the ratio.

The two images may as well be subtracted if this were the only result to be derived from the use of ratios. Figure 6.5 shows a hypothetical situation in which the irradiance at point B on the ground surface is only 50% of that at A due to the fact that one side of the slope is directly illuminated. Subtraction of the values for the two bands at point A will give a result that is double that which would be achieved at point B even though both points are located on the same ground-cover type. The ratios of the two bands at A and B are the same because the topographic effect has been largely cancelled out in

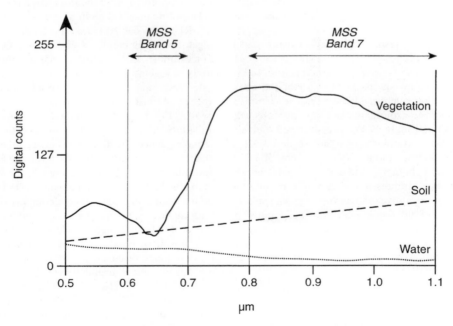

Figure 6.4   The ratio of pixel values in the near-infrared to red spectral bands emphasises the slope of the reflectance curve for vegetation between 0.8–1.1 μm and 0.6–0.7 μm and allows discrimination between vegetation (ratio greater than 1.0), soil (ratio around 1.0) and water (ratio less than 1.0).

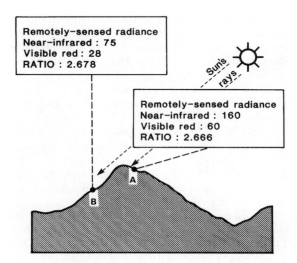

Remotely-sensed radiance
Near-infrared : 75
Visible red : 28
RATIO : 2.678

Sun's rays

Remotely-sensed radiance
Near-infrared : 160
Visible red : 60
RATIO : 2.666

A

B

Figure 6.5 Ratio of near-infrared to red spectral bands is virtually identical at points A and B despite different levels of solar irradiation.

this instance. This is not always the case, as shown by the discussion below.

The two properties of the ratio, that is, reduction of the 'topographic effect' and correlation between ratio values and the shape of the spectral reflectance curve between two given wavebands, have led to the widespread use of spectral ratios in, for example, bio-geography and geology (Rowan *et al.*, 1974; Curran, 1980, 1983; Tucker, 1979; Jackson, 1983; van der Meer *et al.*, 1995). One of the most common spectral ratios used in studies of vegetation status is the ratio of the near-infrared to the equivalent red band value for each pixel location. This ratio exploits the fact that vigorous vegetation reflects strongly in the near-infrared and absorbs radiation in the red waveband (section 1.3.2.1). The result is a grey-scale image that can be smoothed by a low-pass filter (Chapter 7) and density sliced (Chapter 5) to give a visual impression of variation in biomass (the amount of vegetative matter) and in green leaf area index (Bouman, 1992; Box *et al.*, 1989; Filella and Penuelas, 1994) as well as the state of health (physiological functioning) of plants. The theoretical justification of such interpretations is questioned by Myneni *et al.* (1995), who note that:

'A central question remains unanswered: what do vegetation indices indicate?'

An example of a near-infrared:red ratio image is shown in Figure 6.6. The near-infrared band in Figure 6.6(a) is a SPOT HRV band 3 image of the Nottingham

area (the city centre is in the upper right, and agricultural fields can be seen in the top left corner and lower centre). The near-infrared image shows more detail in the non-urban areas than in the urban areas of the image. Figure 6.6(b) is the corresponding SPOT HRV band 2 image, showing red reflectance. Features of the urban areas are more readily seen on the HRV band 2 image. The ratio of the near-infrared to the red image is shown in Figure 6.6(c) and the variation in grey level from 0 (black) to 255 (white) corresponds to the scale discussed above; water is dark and green fields are white. All three images have been contrast stretched using histogram equalisation for display purposes. The actual numerical values of the ratio could be used in any attempt to relate ratio values to biomass amounts or other indicators of vegetation health or productivity.

More complex ratios involve sums of and differences between spectral bands. For example, the *normalised difference vegetation index* (NDVI), defined in terms of the near-infrared (NIR) and red (R) bands as:

$$NDVI = \frac{NIR - R}{NIR + R}$$

is preferred to the simple red:near-infrared ratio by many workers because the ratio value is not affected by the absolute pixel values in the near-infrared (NIR) and red (R) bands. The NDVI has been used to study global vegetation using bands 1 and 2 of the NOAA AVHRR. For example, Justice *et al.* (1985) use the NDVI in a study of vegetation patterns on a continental scale, and data produced by this method from Global Area Coverage of AVHRR are commercially available from NOAA. Figure 6.6(d) shows the NDVI image calculated for the Nottingham SPOT sub-image. On the basis of visual evidence, the difference between the simple ratio and the NDVI is not great. However, the fact that sums and differences of bands are used in the NDVI rather than absolute values may make the NDVI more appropriate for use in studies where comparisons over time for a single area are involved, since the NDVI might be expected to be influenced to a lesser extent by variations in atmospheric conditions (but see below). Sensor calibration issues, discussed in Chapter 4, may have a significant influence on global NDVI calculations based on the NOAA AVHRR. Guttman and Ignatov (1995) show how the difference between pre- and post-launch calibrations lead to unnatural phenomena such as the 'greening of deserts'.

A range of different spectral band ratios is analysed by Jackson (1983), Logan and Strahler (1983) and Perry

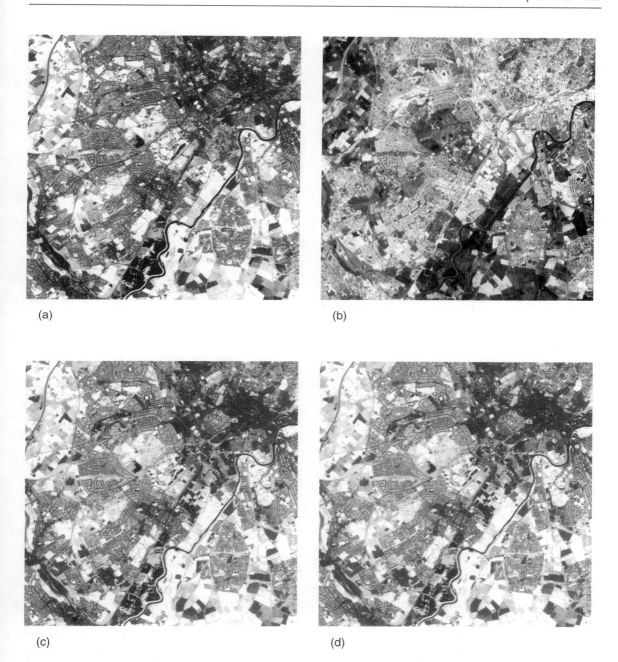

(a)

(b)

(c)

(d)

Figure 6.6   Illustrating the simple ratio and NDVI images. (a) SPOT HRV band 3 (near-infrared) image of the Nottingham area. The city centre is in the upper right corner, and the River Trent flows diagonally to the right of the image centre. Agricultural fields show up in the top left corner and lower centre of the image. (b) Corresponding SPOT HRV band 2 (red) image. (c) Ratio of near-infrared to red image, showing the way in which brightness increases from water (black) to vegetated fields (white). (d) NDVI (rescaled) image. All images have been subjected to a histogram equalisation contrast stretch (Chapter 5). Raw data © CNES – Spot Image Distribution. Permission to use the data was kindly provided by SPOT Image, 5 rue des Satelllites, BP 4359, F-31030 Toulouse, France (http://www.spotimage. fr).

and Lautenschlager (1984). The number of such ratios is considerable, and appears to be limited only by the imagination of the user and the degree of his or her determination to find an as-yet undiscovered combination. Fortunately, Perry and Lautenschlager (1984) find that most of these ratios are equivalent. Research into the derivation of *better* vegetation indices is a different matter. The simple ratio and NDVI, plus other band ratios, are all affected by external factors such as the state of the atmosphere, illumination and viewing angles, and soil background reflectance. This is one reason why NDVI-composite images derived from multiple NOAA AVHRR require careful analysis. These images are produced by selecting cloud-free pixels from a number of images collected over a short period (a week or a month) to produce a single image from these selected pixels. Because of the orbital characteristics of the NOAA satellite it is probable that the NDVI values for adjacent pixels have been collected at different illumination and viewing angles. See Chapter 1 for a discussion of the bidirectional reflectance properties of Earth-surface materials. The relationship between NDVI and vegetation parameters is discussed in Chilar *et al.* (1991), Hobbs and Mooney (1990) and Todd *et al.* (1998).

A class of indices called *soil-adjusted vegetation indices* has been developed, and there are quite a number to choose from: the original soil-adjusted vegetation index (SAVI), the transformed soil-adjusted vegetation index (TSAVI), the global environment monitoring index (GEMI), and a number of others. Bannari *et al.* (1995b), Rondeaux (1995) and Rondeaux *et al.* (1996) provide an extensive review. Steven (1998) discusses the optimised soil-adjusted vegetation index (OSAVI) and shows that the form

$$OSAVI = \frac{NIR - R}{NIR + R + 0.16}$$

minimises soil effects. Readers interested in pursuing this subject should refer to Baret and Guyot (1991), Huete (1989), Pinty *et al.* (1993) and Sellers (1989). Leprieur *et al.* (1996) assess the comparative value of the different vegetation indices using NOAA AVHRR data.

Figure 6.7 shows the range of values taken by (a) the simple red : near-infrared ratio, and (b) the NDVI. In the case of the simple ratio, the values range from zero (0/255) to infinity (255/0) in the case where image pixel values are recorded on an 8-bit scale. The corresponding values for 10-bit image pixels are 0/1023 and 1023/0. One of the problems involved in the use of ratios is the

mapping of the range of the ratios computed for a specific image on to the 0–255 range of the display system, given that we do not know beforehand the maximum and minimum values of the ratio for the image under analysis. There are two solutions to this problem. One is a two-pass solution. On the first pass the ratio values for all the pixels in the image are computed and the maximum and minimum values determined. On the second pass the ratios are calculated again and scaled on to the 0–255 range by the formula

$$R'(x, y) = \frac{R(x, y) - R_{min}}{R_{max} - R_{min}} \times 255$$

In this relationship, $R'(x, y)$ is the ratio between two wavebands for the pixel at location (x, y) after scaling, $R(x, y)$ is the corresponding raw ratio value, and R_{max} and R_{min} are the maximum and minimum raw ratio values for the whole image. The scaling factor 255 could, of course, be altered to suit the dynamic range of the display device being used. The amount of calculation involved is doubled in comparison with the one-pass approach, which will be considered next, though it could be reduced if R_{max} and R_{min} were estimated from a sample (say every nth pixel on every mth line, with n and m being selected so as to give a systematic 5% or 10% sample). If the sampling approach is followed then the output values $R'(x, y)$ must be checked to ensure that they lie in the required 0–255 range.

An alternative, one-pass approach uses the raw ratio value of 1.0 as a threshold. Ratio values equal to or less than 1.0 in magnitude are mapped on to the range 0–127 as follows:

$$R'(x, y) = R(x, y) \times 127$$

If the raw ratio value exceeds 1.0 then an alternative mapping function is used:

$$R'(x, y) = 255 - \frac{127}{R(x, y)}$$

A graph of the composite function is shown in Figure 6.8. Its advantage over the two-pass procedure is speed. Its disadvantage is that the full dynamic range of the display device is not utilised. In vegetated areas the raw ratio values may well occupy only a small part of the range, with water and shadow being the only cover types with ratio values of less than 1.0. The one-pass mapping function will assign the entire range of raw ratio values to a narrow range of counts on the 0–255

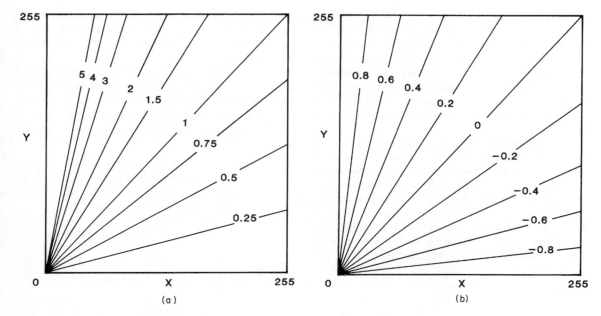

Figure 6.7   Distribution of (a) near-infrared: red ratio and (b) normalised difference vegetation index (NDVI) as a function of infrared (X) and red (Y) pixel values. The pixel values are assumed to be measured on a 0–255 scale.

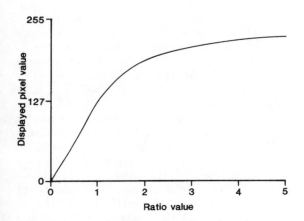

Figure 6.8   Function to map the IR/R ratio on to the 0–255 range.

scale. Once allocated, these grey-scale values cannot be subdivided by any subsequent enhancement procedure. A more sophisticated approach would use a mapping function that covered only part of the theoretical range of the raw ratio values, but then we are back to the problem of estimating the maximum and minimum ratio values in the image. A good review of the problem is provided by Wecksung and Breedlove (1977).

The MIPS software (Appendix A) computes the ratio by using floating-point arithmetic. The ratio is stored as a fractional value ranging from 1/255 to 255/1 (values of 0 are treated as if they are 1, to avoid division by zero). The floating-point image is then converted to the 0–255 range by searching for the minimum and maximum values, then scaling these to 0 and 255 respectively.

In the case of the NDVI the problem is similar. A two-pass approach could be adopted to determine the maximum and minimum NDVI values prior to scaling, or a mapping function could be specified *a priori*. The range of the NDVI is from -1 to $+1$ (again assuming that $y/0$ is equal to 0). A function to transform the range of the raw NDVI values to a 0–255 scale could be derived by adding 1.0 to the raw NDVI value (making the range 0–2) and multiplying the result by 127.

If the ratio values are to be correlated with field observations of, for example, leaf area index (LAI) or estimates of biomass, or if ratio values for images of the same area at different times of the year are to be compared, then some thought should be given to the scaling operation. Scaling is necessary only to convert the computed ratio values to integers on the range 0–255 for display purposes. If the user wishes to correlate field observations and ratio values from remotely-sensed images then unscaled ratio values should be used. For purposes of comparison of ratios through the year then a consistent scaling procedure should be adopted. This procedure should not rely, for example, on the minimum and maximum ratio values in an individual image, for this implies that the scaling is image-dependent. If each ratio image is projected on to a 0–255 scale using the same mapping function then changes in the ratios can be detected using image subtraction as described above.

Computational problems are not the only ones to bedevil ratio calculations. It was noted above that one of the reasons put forward to justify the use of ratios was the elimination of variable illumination effects in areas of topographic slope. This, of course, is not the case except in a perfect world. Assume that the variation in illumination at point (i, j) can be summarised by a variable $c(i, j)$ so that, for the pixel at location (i, j), the ratio between channels q and p is expressed as:

$$R(i, j) = \frac{c(i, j)v(q, i, j)}{c(i, j)v(p, i, j)} = \frac{v(q, i, j)}{v(p, i, j)}$$

where $v(q, i, j)$ is the radiance from pixel located at point (i, j) in channel q. The term $c(i, j)$ is constant for both bands q and p and therefore cancels out. If there were an additive as well as a multiplicative term then the following logic would apply:

$$R(i, j) = \frac{c(i, j)v(q, i, j) + r(q)}{c(i, j)v(p, i, j) + r(p)}$$

and it would be impossible to extract the true ratio $v(q, i, j)/v(p, i, j)$ unless the terms $r(p)$ and $r(q)$ were known or could be estimated. The terms $r(p)$ and $r(q)$ are the atmospheric path radiances for bands p and q (section 1.2.5), which are generally unknown. They are also unequal because the amount of scattering in the atmosphere increases inversely with wavelength. Switzer *et al.* (1981) consider this problem and show that the atmospheric path radiances must be estimated and subtracted from the recorded radiances before ratioing. A frequently used method, described in Chapter 4, involves the subtraction of constants $k(p)$ and $k(q)$, which are the minimum values in the histograms of channels p and q respectively; these values might be expected to provide a first approximation to the path radiances in the two bands. Switzer *et al.* (1981) suggest that this 'histogram minimum' method over-corrects the data. Other factors, such as the magnitude of diffuse irradiance (skylight) and reflection from cross-valley slopes, also confuse the issue. It is certainly not safe to assume that the 'topo-

graphic effect' is completely removed by a ratio operation. Atmospheric effects and their removal, including the 'histogram minimum' method, are considered in Chapter 4.

The effects of atmospheric haze on the results of ratio analyses are studied in an experimental context by Jackson *et al.* (1983). These authors find that, for turbid atmospheres, the near-infrared/red ratio was considerably less sensitive to variations in vegetation status and they conclude that:

'The effect [of atmospheric conditions] on the ratio is so great that it is questionable whether interpretable results can be obtained from satellite data unless the atmospheric effect is accurately accounted for on a pixel-by-pixel basis.'

Jackson *et al.* (1983, p. 195)

The same conclusion was reached for the normalised difference vegetation index (NDVI). Holben and Kimes (1986) also report the results of a study involving the use of ratios of NOAA AVHRR bands 1 and 2 under differing atmospheric conditions. They find that the NDVI is more constant than individual bands.

Other problems relate to the use of ratios where there is incomplete vegetation coverage. Variations in soil reflectivity will influence ratio values, as discussed above. The angle of view of the sensor and its relationship with solar illumination geometry must also be taken into consideration if data from off-nadir pointing sensors such as the SPOT HRV or from sensors with a wide angular field of view such as the NOAA AVHRR are used (Barnsley, 1983; Wardley, 1984; Holben and Fraser, 1984). In order to make his data comparable over time and space, Frank (1985) converted the Landsat MSS digital counts to reflectances (as described in section 4.6) and used a crude correction for solar elevation angle based on the Lambertian assumption (Chapter 1). Other useful references are Chalmers and Harris (1981), Elvidge and Lyon (1985), Gardner *et al.* (1985), Kowalik *et al.* (1983), Maxwell (1976) and Misra and Wheeler (1978). As noted above, extensions to the standard ratio for the reduction of soil reflectance effects and atmospheric effects have been developed in the past few years. The general review by Rondeaux (1995) is recommended.

6.3 EMPIRICALLY BASED IMAGE TRANSFORMS

Experience gained during the 1970s with the use of Landsat MSS data for identifying agricultural crops, together with the difficulties encountered in the use of ratio transforms (section 6.2) and principal components transforms (section 6.4), led to the development of image transforms based on the observations that (i) scatter plots of Landsat MSS data for images of agricultural areas show that agricultural crops occupy a definable region of the four-dimensional space based on the Landsat MSS bands, and (ii) within this four-dimensional space the region occupied by pixels that could be labelled as 'soil' is a narrow, elongated ellipsoid. Pairwise plots of Landsat MSS bands fail to reveal these structures fully because they give an oblique rather than a 'head-on' view of the subspace occupied by pixels representing vegetation. Kauth and Thomas (1976) propose a transformation that would, by rotating and scaling the axes of the four-dimensional space, give a more clear view of the structure of the data. They called their transform the *tasselled cap* since the shape of the region occupied by vegetation in different stages of growth appeared like a 'bobble hat'. Other workers have proposed other transforms; perhaps the best known is the *perpendicular vegetation index*, which is based on a similar idea to that of the tasselled cap, namely, that there is a definite axis in four-dimensional Landsat MSS space that is occupied by pixels representing soils, ranging from soils of low reflectance to those of high reflectance (see also Baret *et al.*, 1993). These two transformations are described briefly in the next two subsections.

6.3.1   Perpendicular vegetation index

A plot of radiance measured in the visible red band against radiance in the near-infrared for a partly vegetated area will result in a plot that looks something like Figure 6.9. Bare soil pixels will lie along the line S_1–S_2, with the degree of wetness of the soil being higher at the S_1 end of the 'soil line' than at the S_2 end. Vegetation pixels will lie below and to the right of the soil line, and the perpendicular distance to the soil line was suggested by Richardson and Wiegand (1977) as a measure that was correlated with the green leaf area index and with biomass. The formula used by Richardson and Wiegand (1977) to define the *perpendicular vegetation index* (PVI) is based on either Landsat-1 to -3 MSS band 7 or band 6, denoted by PVI7 and PVI6, respectively:

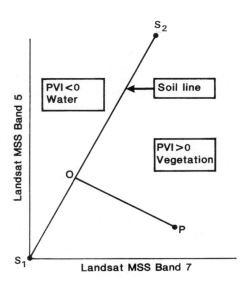

Figure 6.9   Illustrating the concept of the perpendicular vegetation index, which is proportional to the perpendicular distance from point P to the soil line S_1–S_2.

Neither of these formulae should be used without some forethought. First, the PVI is defined as the perpendicular distance from the soil line (Figure 6.9). Richardson and Wiegand's (1977) equation for the soil line is based on 16 points representing soils, cloud and cloud shadows in Hidalgo and Willacy Counties, Texas, for four dates in 1975. It is unlikely that such a small and geographically limited sample could adequately define the soil line on a universal basis. A locally valid expression relating 'soil' pixel values in Landsat MSS bands 5 and 7 (or 6 and 7) of the form $X_5 = c + X_7$ is needed. Secondly, the Richardson and Wiegand equation is based on the assumption that the maximum digital count in Landsat MSS bands 4, 5 and 6 is 127 with a maximum of 63 in Landsat MSS band 7. Images supplied by ESA–Earthnet are normally system-corrected and the pixel values in all four Landsat MSS bands are expressed on a 0–255 scale.

The PVI has been used as an index that takes into account the background variation in soil conditions which affect soil reflectance properties. Jackson *et al.* (1983) demonstrate that the PVI is affected by rainfall when the vegetation cover is incomplete. However, they

$$\text{PVI7} = \sqrt{(0.355\text{MSS7} - 0.149\text{MSS5})^2 + (0.355\text{MSS5} - 0.852\text{MSS7})^2}$$

$$\text{PVI6} = \sqrt{(0.498\text{MSS6} - 0.487\text{MSS5} - 2.507)^2 + (2.734 + 0.498\text{MSS5} - 0.543\text{MSS6})^2}$$

considered it to be 'moderately sensitive' to vegetation but was not a good detector of plant stress. The effects of atmospheric path radiance on the PVI was reported as reducing the value of the index by 10–12% from a clear to a turbid atmospheric condition. This is considerably less than the 50% reduction noted for the Landsat MSS band 7 to band 5 ratio.

6.3.2   Tasselled cap (Kauth–Thomas) transformation

The PVI (section 6.3.1) considers spectral variations in two of the four Landsat MSS bands, and uses distance from a soil line in the two-dimensional space defined by these two MSS bands as a measure of biomass or green leaf area index. Kauth and Thomas (1976) use a similar idea except that their model uses all four Landsat MSS bands. In the four-dimensional space defined by the Landsat MSS bands Kauth and Thomas suggest that pixels representing soils fall along an axis that is oblique with respect to each pair of the four MSS axes. A triangular region of the four-dimensional MSS space is occupied by pixels representing vegetation in various stages of growth. The *tasselled cap transform* is intended to define a new (rotated) coordinate system in terms of which the soil line and the region of vegetation are more clearly represented. The axes of this new coordinate system are termed 'brightness', 'greenness', 'yellowness' and 'nonesuch'. The brightness axis is associated with variations in the soil background reflectance. The greenness axis is correlated with variations in the vigour of green vegetation, while the yellowness axis is related to variations in the yellowing of senescent vegetation. The nonesuch axis has been interpreted by some authors as being related to atmospheric conditions. Owing to the manner in which these axes are computed they are statistically uncorrelated so that they can be represented by four orthogonal lines. However, the yellowness and nonesuch functions have not been widely used and the tasselled cap transformation has often been used to represent the four-band MSS data in terms of two functions, brightness and greenness. For further discussion of the model, see Crist and Kauth (1986).

The justification for this operation is that the tasselled cap axes provide a consistent, physically based coordinate system for the interpretation of images of an agricultural area obtained at different stages of the growth cycle of the crop. Since the coordinate transformation is defined *a priori* it will not be affected by variations in crop cover and stage of growth from image to image. The principal components transform (section 6.4) performs an apparently similar operation but the parameters of this transformation are defined by the statistical relationships among the spectral bands of the particular image being analysed. Consequently, the parameters of the principal components transform vary from one multispectral image set to another as the correlations among the bands depend upon the range and statistical distribution of pixel values in each band.

If the measurement for the jth pixel on the ith tasselled cap axis is given by u_j, the coefficients of the ith transformation by $\mathbf{R}_i$ and the vector of measurements on the four Landsat MSS bands for the same pixel by $\mathbf{x}_j$ then the tasselled cap transform is accomplished by:

$$u_j = \mathbf{R}'_i \mathbf{x}_j + c$$

In other words, the pixel values in the four MSS bands (the elements of $\mathbf{x}_j$) are multiplied by the corresponding elements of $\mathbf{R}_i$ to give the position of the jth pixel in the ith tasselled cap axis, $\mathbf{u}$. The constant c is an offset, which is added to ensure that the values of u are always positive. Kauth and Thomas (1976) use a value of 32.

The vectors of coefficients $\mathbf{R}_i$ are defined by Kauth and Thomas (1976) as follows:

$$\begin{aligned}
\mathbf{R}_1 &= \{\ \ 0.433, \quad\ \ 0.632, \quad\ \ 0.586, \quad 0.264\} \\
\mathbf{R}_2 &= \{-0.290, \quad -0.562, \quad\ \ 0.600, \quad 0.491\} \\
\mathbf{R}_3 &= \{-0.829, \quad\ \ 0.522, \quad -0.039, \quad 0.194\} \\
\mathbf{R}_4 &= \{\ \ 0.223, \quad\ \ 0.012, \quad -0.543, \quad 0.810\}
\end{aligned}$$

These coefficients assume that Landsat MSS bands 4 to 6 are measured on a 0–127 scale and band 6 is measured on a 0–63 scale. They are also calibrated for Landsat-1 data and slightly different figures may apply for other Landsat MSS data. The position of the $\mathbf{R}_1$ axis was based on measurements on a small sample of soils from Fayette County, Illinois. The representativeness of these soils as far as applications in other parts of the world is concerned is open to question.

Crist (1983) and Crist and Cicone (1984a, b) extend the tasselled cap transformation to data from the Landsat TM. Data from the thermal infrared channel are excluded. They found that the brightness function $\mathbf{R}_1$ for the MSS tasselled cap did not correlate highly with the TM tasselled cap equivalent, though MSS greenness function $\mathbf{R}_2$ did correlate with TM greenness. The TM data were found to contain significant information in a third dimension, which was identified as wetness. The coefficients for these three functions are given in Table 6.1.

The brightness function is simply a weighted average of the six TM bands, while greenness is a visible/near-infrared contrast, with very little contribution from bands 5 and 7. Wetness appears to be defined by a

Table 6.1 Coefficients for the tasselled cap functions 'brightness', 'greenness' and 'wetness' for Landsat Thematic Mapper bands 1–5 and 7.

TM band	1	2	3	4	5	7
Brightness	0.3037	0.2793	0.4343	0.5585	0.5082	0.1863
Greenness	−0.2848	−0.2435	−0.5436	0.7243	0.0840	−0.1800
Wetness	0.1509	0.1793	0.3299	0.3406	−0.7112	−0.4572

contrast between the mid-infrared bands (5 and 7) and the red/near-infrared bands (3 and 4). The three tasselled cap functions can be considered to define a three-dimensional space in which the positions of individual pixels can be computed using the coefficients above. The plane defined by the greenness and brightness functions is termed by Crist and Cicone (1984a) the 'plane of vegetation' while the functions brightness and wetness define the 'plane of soils' (Figure 6.10).

Several problems must be considered. The first is the now familiar problem of dynamic range compression, which in the case of the tasselled cap transform assumes an added importance. One of the main reasons for supporting the use of the tasselled cap method against, for example, the principal components technique (section 6.4) is that the coefficients of the transformation are defined *a priori*. However, if these coefficients are applied blindly, the resulting tasselled cap coordinates will not lie in the range 0–255 and will thus not be displayable on standard image processing equipment. The distribution of the tasselled cap transform values will vary from image to image, though. The problem is to define a method of dynamic range compression that will adjust the tasselled cap transform values on to a 0–255 range without destroying inter-image comparability. Because the values of the transform are scene-dependent it is unlikely that a single mapping function will prove satisfactory for all images, but if an image-dependent mapping function is selected then inter-image comparison will be made more difficult. Crist (personal communication) suggests that the range of tasselled cap function values met with in agricultural scenes will vary between 0 and 350 (brightness), between −100 and 125 (greenness) and from −150 to 75 (wetness). If inter-image comparability is important, then the calculated values should be scaled using these limits. Otherwise, the functions could be evaluated for a sample of pixels in the image (for example, 5% of the pixels could be selected) and the functions scaled using the extreme sample values according to the formula

$$y = \frac{x - y_{min}}{x_{max} - x_{min}} \times 255$$

where y is the scaled value (0–255) and x is the raw value. In order to prevent over- and under-shoots, a check should be made for negative scaled values (which are set to zero) or values greater than 255 (which are set to 255).

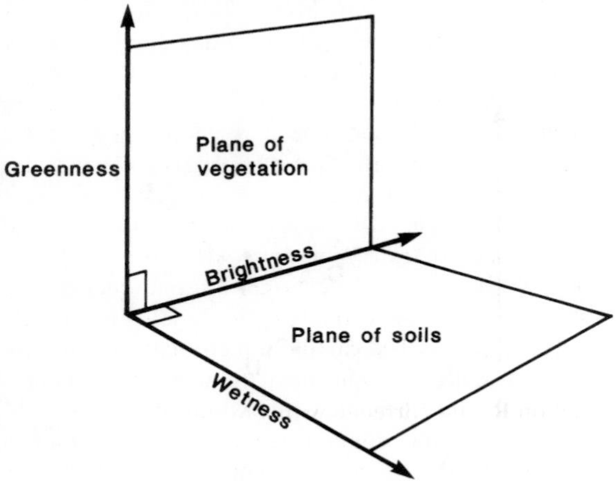

A second problem that interferes with the comparison of multi-date tasselled cap images is the problem of changes in illumination geometry and variations in the composition of the atmosphere. Both of these factors will influence the magnitude of the ground-leaving radiance from a particular point so that satellite-recorded radiances from a constant target will change even though the characteristics of the target do not change. These problems were addressed in a general way in Chapter 4 (sections 4.4 and 4.5); they are particularly

Figure 6.10 The tasselled cap brightness, greenness and wetness functions derived from Landsat Thematic Mapper data define two planes – the plane of vegetation and the plane of soils. Based on Crist, E.P. and Cicone, R.C., 1984b, figure 3; reproduced by permission of the American Society for Photogrammetry and Remote Sensing.

important in the context of techniques that endeavour to provide the means to carry out comparisons between multitemporal images.

Like the perpendicular vegetation index, the tasselled cap transform relies upon empirical data for the determination of the coefficients of the brightness axis (the soil line in the terminology of the perpendicular vegetation index). It was noted above that the Kauth and Thomas (1976) formulation of the MSS tasselled cap was based on a small sample of soils from Fayette County, Illinois. The TM brightness function is also based on a sample of North American soils; hence applications of the transformation to agricultural scenes in other parts of the world may not be successful if the position of the brightness axis as defined by the coefficients given above does not correspond to the reflectance characteristics of the soils in the local area. Jackson (1983) describes a method of deriving tasselled cap-like coefficients from soil reflectance data, using Gram–Schmidt orthogonal polynomials. He uses reflectance data in the four Landsat MSS bands for dry soil, wet soil, green and senesced vegetation to derive coefficients for three functions representing brightness, greenness and yellowness. This method is implemented for the Landsat TM bands in the program `specindex`. A second program, `tasscap` performs the tasselled cap transformation using user-defined coefficients. Both programs are supplied on the CD-ROM accompanying this book and are described in Appendix B. Where possible, the coefficients of the orthogonal, tasselled cap-like functions should be calculated for the area of study as they may differ from those provided by Crist and Cicone (1984a, b) and listed above.

6.4   PRINCIPAL COMPONENTS ANALYSIS

6.4.1   Standard principal components analysis

Adjacent bands in a multispectral remotely-sensed image are generally correlated. Multiband visible/near-infrared images of vegetated areas will show negative correlations between the near-infrared and visible red bands and positive correlations among the visible bands because the spectral characteristics of vegetation (section 1.3.2.1) are such that as the vigour or greenness of the vegetation increases the red reflectance diminishes and the near-infrared reflectance increases. The presence of correlations among the bands of a multispectral image implies that there is redundancy in the data. Some information is being repeated. It is the repetition of information between the bands that is reflected in their inter-correlations.

If two variables, x and y, are perfectly correlated then measurements on x and y will plot as a straight line sloping upwards to the right (Figure 6.11). Since the positions of the points shown along line AB occupy only one dimension then the relationship between these points could equally well be given simply by using the line AB as a single axis. Even if x and y are not perfectly correlated there may be a dominant direction of scatter or variability, such as that shown on Figure 6.12. If this dominant direction of variability (AB) is chosen as the major axis then a second, minor, axis (CD) could be drawn at right angles to it. A plot using the axes AB and

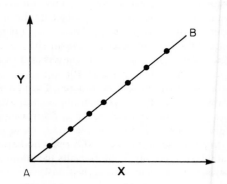

Figure 6.11   Plot of two variables, x and y, which are perfectly correlated ($r = 1.0$). The points (x_i, y_i) lie on the straight line AB passing through the origin. Although measurements are made on two variables, the scatter of points occupies only one dimension.

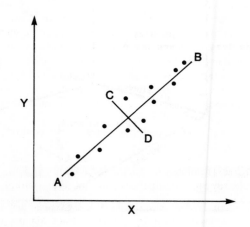

Figure 6.12   The two variables, x and y, show a high positive correlation. Most of the scatter lies in the direction of the line AB. There is much less scatter in the direction of the line CD, which is perpendicular to line AB.

CD rather than the conventional axes x and y might, in some cases, prove more revealing of the structures that are present within the data. Furthermore, if the variation in the direction CD in Figure 6.12 contains only a small proportion of the total variability in the data, then it may be ignored without too much loss of information.

This example shows that we must draw a basic distinction between the number of variables (spectral bands) in the image data set and the intrinsic dimensionality of that data set. In Figure 6.11 the number of variables is two (x and y) but the dimensionality of the data as shown in the scatter of points is one. In Figure 6.12 the dimensionality is again effectively one, although the number of observed variables is, in fact, two. In both examples the use of the single axis AB rather than the axes x and y accomplishes two aims: (i) a reduction in the size of the data set since a single coordinate on axis AB replaces the two coordinates on the axes x and y, and (ii) the information conveyed by the set of coordinates on AB is greater than the information conveyed by the measurements on either the x-axis or the y-axis individually. In this context information means variance or scatter about the mean; it can also be related to the range of states or levels in the data, as shown in the discussion of entropy in section 2.2.3.

Multispectral image data sets generally have a dimensionality that is less than the number of spectral bands. For example, it is shown in section 6.3.2 that the four-band Landsat MSS tasselled cap transform produces two significant dimensions (brightness and greenness) while the six-band Landsat TM tasselled cap defines three functions (dimensions). The purpose of principal (not 'principle') components analysis is to define the number of dimensions that are present in a data set and to fix the coefficients which specify the positions of the set of axes that point in the directions of greatest variability in the data (such as axes AB and CD in Figure 6.12). These axes are always uncorrelated. A principal components transform of a multispectral image (or of a set of registered multitemporal images, as used by Ribed and Lopez (1995)) might therefore be expected to perform the following operations:

- define the dimensionality of the data set, and
- identify the principal axes of variability within the data.

These properties of principal components analysis (sometimes also known as the Karhunen–Loève transform) might prove to be useful if the data set is to be compressed. Also, relationships between different groups of pixels representing different land cover types might become clearer if they are viewed in the principal axis reference system rather than in terms of the original spectral bands. The data compression property is useful if more than three spectral bands are available. A conventional RGB colour display system relates a spectral band to one of the three colour inputs (red, green and blue). The Landsat TM provides seven bands of data, hence a decision must be made regarding which three of these seven bands are to be displayed as a colour composite image. If the basic dimensionality of the TM data is only three then most of the information in the seven bands will be expressible in terms of three principal components. The principal components images could therefore be used to generate an RGB false-colour composite in which principal component number 1 was shown in red, number 2 in green and number 3 in blue. Such an image contains more information than any combination of three spectral bands.

The positions of the mutually perpendicular axes of maximum variability in the two-band data set shown in Figure 6.12 can be found easily by visual inspection to be the lines AB and CD. Where the number of variables is greater than two, a geometric solution is impracticable and an algebraic procedure must be followed. The direction of axis AB in Figure 6.12 is defined by the correlation between variables x and y; this high positive correlation results in the scatter of points being restricted to an elliptical region of the two-dimensional space defined by the axes x and y. Axis AB is, in fact, the major or principal axis of this ellipse and CD is the minor axis. In a multivariate context the shape of the ellipsoid enclosing the scatter of data points in a p-dimensional space is defined by the variance–covariance matrix computed from the p variables or spectral bands. The variance in each spectral band is proportional to the scatter of points in the direction parallel to the axis representing that variable, so that it can be deduced from Figure 6.12 that the variances of variables x and y are approximately equal. The covariance defines the shape of the ellipse enclosing the scatter of points. Figure 6.13 shows two distributions with the same variance. One (solid line) has a high positive covariance while the other (dashed line) has a covariance of zero. The mean of each variable gives the location of the centre of the ellipse (or ellipsoid in a space of dimensionality higher than two). Thus, the mean vector and the variance–covariance matrix define the location and shape of the scatter of points in a p-dimensional space. This information is also used in the definition of the maximum likelihood classification procedure in Chapter 8.

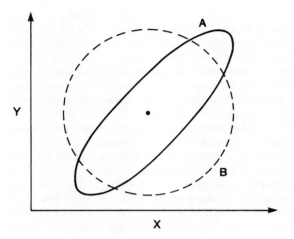

Figure 6.13 The ellipse A and the circle B show the distribution of two sets of points representing measurements on variables *x* and *y*. Both distributions have the same variance on *x* and *y*, but distribution A shows a positive covariance between *x* and *y*, whereas distribution B shows no covariance between *x* and *y*.

The relationship between the correlation matrix and the variance–covariance matrix sometimes leads to confusion. If the *p* variables making up the data set are measured on different and incompatible scales (such as metres above sea-level, barometric pressure in millibars or weight in kilograms) then the variances of these variables are not comparable. The importance of the variance in defining the scatter of points in a particular direction has already been stressed, so it is clear that if the variance is to be used in defining the shape of the ellipsoid enclosing the scatter of points in the *p*-dimensional space then the scales used to measure each variable must be comparable. To illustrate this point, consider what would be the outcome (in terms of the shape of the enclosing ellipsoid) if the three variables mentioned earlier were measured in feet above sea-level (rather than metres), inches of mercury (rather than millibars), and weight in ounces rather than kilograms. Not only would the variance of each of the variables be altered but the shape of the enclosing ellipsoid would change. The degree of change would not be constant in each dimension, and the shape of the second ellipsoid would not be related in any simple fashion to the shape of the first ellipsoid. It is in these circumstances that the correlation coefficient rather than the covariance is used to measure the degree of statistical association between the spectral bands. The correlation is simply the covariance measured for standardised variables. To standardise a variable the mean value is subtracted

from all the measurements and the result is divided by the standard deviation. This operation converts the raw measurements to standard scores or *z*-scores, which have a mean of zero and a variance of unity. The off-diagonal elements of the correlation matrix are the covariances of the standard scores and the diagonal elements are the variances of the standard scores, which are by definition always unity. Since the shape of the ellipsoid enclosing the scatter of data points is altered in a complex fashion by the standardisation procedure it follows that the orientation and lengths of the principal axes (such as AB and CD in Figure 6.12) will also change. The effects of the choice of variance–covariance matrix or correlation matrix on the results of the principal components operation are considered below, and an illustrative example is provided.

The principal components of a multispectral image set are found by algebraic methods which are beyond the scope of this book (see Mather (1976) or Richards (1993) for a more detailed discussion). In general terms, the variance–covariance matrix **S** or the correlation matrix **R** of the bands is computed. If there are *p* bands then each of these symmetric matrices will have *p* rows and *p* columns. A set of quantities called the *eigenvalues* are found for the chosen matrix using methods of linear algebra. The eigenvalues give the lengths of the principal axes of the ellipsoid whose shape is defined by **S** or **R**. The eigenvalues are measured in terms of units of variance. It follows from the previous paragraph that standardised units of variance must be used if the variables (bands) are measured on incompatible scales. If the variables are all measured in terms of the same units (such as milliwatts per square centimetre per steradian) then standardisation is unnecessary and, indeed, undesirable for it removes the effects of changes in variability between the bands.

Associated with each eigenvalue is a set of coordinates defining the direction of the associated principal axis. These coordinates are called the *eigenvectors* of the matrix **S** or **R**. The eigenvalues and eigenvectors therefore describe the lengths and directions of the principal axes. In effect the data are scaled and rotated so that the principal axes are the coordinate system in terms of which the principal component images are expressed. The eigenvectors, scaled by the square roots of the corresponding eigenvalues, can also be interpreted as correlations between the abstract principal components and the individual bands of the multispectral image. These correlations or *loadings* are used in the interpretation of the principal components.

Principal component images are computed from the eigenvectors in much the same way that the tasselled cap features are derived from their coefficients. The eigenvector for principal component 1 provides the co-efficients or weights that are used in conjunction with the observed pixel values to generate the first principal component image. The pixel values for the first principal component image are found from:

$$\text{pc1}_{ij} = a_1 x_{ij1} + a_2 x_{ij2} + a_3 x_{ij3} + \cdots + a_m x_{ijm}$$

In this expression pc1_{ij} is the value of the pixel at row i, column j in the first principal component image, the values $a_1, a_2, a_3, \ldots, a_m$ are the elements of the first scaled eigenvector and x_{ijk} is the observed pixel value at row i, column j of band k ($k = 1, \ldots, m$). The remaining principal component images are computed in the same way using the appropriate scaled eigenvectors. The scaling is performed by dividing the eigenvector by the square root of the corresponding eigenvalue.

If the principal component images are to be displayed on a monitor, as described in section 3.3.1, then the values pc1_{ij} (and their equivalents pc2_{ij} and pc3_{ij}) must be scaled to fit the range 0–255. This problem has already been discussed in the context of band ratios and the tasselled cap transform. Since the principal components are scene-dependent it is not necessary to seek a scaling that is consistent over a number of applications. The scaling factors can be computed separately for each application. This can be done by sampling the image data set (the x_{ijk}) to estimate the maximum and minimum of the corresponding component pixel values and then using a simple linear mapping function to translate from the observed sample range to integers on the range 0–255. This is generally adequate as the principal component pixel values, being a weighted sum of observed variables, are likely to be normally distributed. Where the component values are skewed then the sample mean component value can be used to derive a piecewise mapping function to convert the calculated scores (x) to counts on a 0–255 scale (y):

$$y = \frac{x - x_{\min}}{\bar{x} - x_{\min}} \times 127 \qquad \text{(computed value} \leq \text{mean)}$$

$$y = \frac{x - \bar{x}}{x_{\max} - \bar{x}} \times 127 + 127 \text{ (computed value} > \text{mean)}$$

Alternatively, a single function can be used to map the full range of principal component scores expressed as real (floating-point) numbers (x) to the range 0–255 (y) by locating the minimum and maximum image values and using the expression:

$$y = \frac{x - x_{\min}}{x_{\max} - x_{\min}} \times 255$$

When comparing principal component images produces by different computer programs, it is always wise to determine the nature of the scaling procedure used in the generation of each image.

If all available bands are input to the principal components procedure then, depending on whether the analysis is based on inter-band correlations or covariances, some bands may be under-represented. Siljeström *et al.* (1997) use a procedure that they call 'selective principal components analysis', which involves the division of the bands into groups on the basis of their inter-correlations. For TM data, these authors find that bands 1, 2 and 3 are strongly inter-correlated and are distinct from bands 5 and 7. Band 4 stands alone. They carry out principal components analysis separately on bands (1, 2, 3) and bands (4, 5) and use the first principal component from each group, plus band 4, to create a false-colour image, which is then used to help in the recognition of geomorphological features.

Principal components analysis is also used in the analysis of multitemporal images in order to identify areas of change (see above, section 6.2.2). For example, Siljeström and Moreno (1995) use covariance-based principal components analysis applied to a multitemporal image set and identify specific components with areas of change. Henebry (1997) also uses principal components analysis, applied to a multitemporal sequence of 11 ERS SAR images over a growing season, and was able successfully to identify individual principal components with specific land surface properties, such as terrain, burned/unburned prairie, and look angle variations between the 11 images.

Table 6.2(a) gives the correlation matrix for the six reflective (visible plus near- and mid-infrared) bands (numbered 1–5 and 7) of a Landsat TM image of the Littleport area shown in Figures 1.8 and 1.9. Correlations rather than covariances are used as the basis for the principal components procedure because the digital counts in each band do not relate to the same physical units. (That is, a change in level from 30 to 31, for example, in band 1 does not represent the same change in radiance as a similar change in any other band. See section 4.6 for details of the differences in calibration between TM bands.) High positive correlations among all reflective bands except band 4 (near-infrared) can be observed. The lowest correlation between any pair of bands (excluding band 4) is + 0.660. The correlations between band 4 and the other bands are negative. This

Table 6.2 Principal components analysis, based on correlation matrix.

(a) Correlations among Thematic Mapper reflective bands (1–5 and 7) for the Littleport TM image. The means and standard deviations of the six bands are shown in the rightmost two columns.

Band	1	2	3	4	5	7	Mean	Std dev.
1	1.000	0.916	0.898	−0.117	0.660	0.669	65.812	8.870
2	0.916	1.000	0.917	−0.048	0.716	0.685	29.033	5.652
3	0.898	0.917	1.000	−0.296	0.757	0.819	26.251	8.505
4	−0.117	−0.048	−0.296	1.000	−0.161	−0.474	93.676	24.103
5	0.660	0.716	0.757	−0.161	1.000	0.883	64.258	18.148
7	0.669	0.685	0.819	−0.474	0.883	1.000	23.895	11.327

(b) Principal component loadings for the six principal components of the Littleport TM image. Note that the sum of squares of the loadings for a given principal component (column) is equal to the eigenvalue. The percentage variance value is obtained by dividing the eigenvalue by the total variance (6 in this case because standardised components are used – see text) and multiplying by 100.

Band	PC 1	PC 2	PC 3	PC 4	PC 5	PC 6
1	0.899	0.242	−0.288	0.223	0.002	−0.002
2	0.914	0.303	−0.182	−0.143	−0.095	−0.103
3	0.966	0.033	−0.165	−0.117	0.086	0.134
4	−0.309	0.924	0.214	−0.006	0.076	−0.001
5	0.871	0.019	0.470	0.038	−0.124	0.059
7	0.904	−0.285	0.267	0.009	0.148	−0.094
Eigenvalue	4.246	1.086	0.481	0.085	0.059	0.041
% var.*	70.77	18.10	8.02	1.42	0.99	0.68
Cum. % var.†	70.77	88.88	96.90	98.32	99.31	100.00

*Percentage variance.
†Cumulative percentage variance.

generally high level of correlation implies that the radiances in each spectral band except band 4 are varying spatially in much the same way. The negative correlation between band 4 and the optical bands can be explained by the fact that the area shown in the image is an agricultural one, and the main land cover type is vegetation. The spectral reflectance curve for vegetation (Figure 1.18) shows the contrast between the near-infrared and red channels. The vegetation ratios exploit this fact, and as the ground cover of vegetation and the vigour (greenness) of the vegetation increase so the contrast between near-infrared and red reflectance increases as the near-infrared reflectance rises and the red reflectance falls. Hence, this negative correlation between band 4 and other bands can be partially explained. Table 6.2(a) also shows the means and standard deviations of the six bands. Note the significantly higher standard deviations for bands 4, 5 and 7, which show that variability of pixel values is greater in these bands than in bands 1, 2 and 3. The mean pixel values also differ. This is a result of the nature of the ground cover and the calibration of the TM sensor.

The eigenvalues and scaled eigenvectors, or principal component loadings, derived from the correlation matrix measure the concentration of variability in the data in six orthogonal directions (Table 6.2(b)). Over 70% of the variability in the data lies in the direction defined by the first principal component. Column 1 of Table 6.2(b) gives the relationship between the first principal component and the six TM bands; all bands except the near-infrared have entries more than 0.87, while the near-infrared band has an entry of − 0.309. This indicates (as was inferred from the correlation matrix) that there is considerable overlap in the information carried by the different channels, and that there is a contrast between the near-infrared and the other bands. The image produced by the first principal component (Figure 6.14(a)) summarises information that is

Figure 6.14 (a)–(f) respectively show the principal components 1–6 of TM bands 1–5 and 7 of the Littleport TM image shown in Figures 1.8 and 1.9. These principal component images are derived from the matrix of correlations among the TM bands. Compare with Figure 6.15(a)–(f). Original data © ESA 1994, distributed by Eurimage.

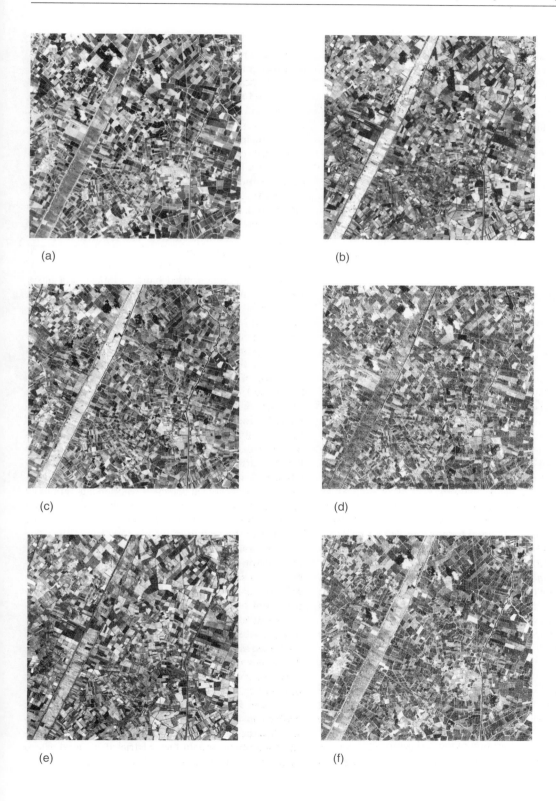

(a)

(b)

(c)

(d)

(e)

(f)

common to all channels. It can be seen to be a weighted average of five of the six TM bands contrasted with the near-infrared band.

The second principal component of the Littleport image set (Figure 6.14(b)) is dominated by the contribution of the near-infrared band. There is a small contribution from the three visible bands. Between them, principal components 1 and 2 account for over 88% of the variability in the original six-band data set. A further 8% is contributed by principal component 3 (Figure 6.14(c)), which appears to be highlighting a contrast between the infrared and the visible bands. Visual analysis of the principal component images shown in Figures 6.14(a)–(c) appears to indicate that principal components 1 and 2 may be picking out differences between different vegetation types in the area, while principal component 3 is related to water content of the soil.

Principal components 4–6 contain only 3.1% of the variation in the data. If the noise present in the image data set is evenly distributed among the principal components then the lower-order principal components might be expected to have a lower signal-to-noise ratio than the higher-order principal components. On these grounds it might be argued that principal components 4–6 are not worthy of consideration. This is not necessarily the case. While the contrast of these lower-order principal component images is less than that of the higher-order components, there may be patterns of spatial variability present that should not be disregarded, as Figures 6.14(d)–(f) show. The sixth principal component shown in Figure 6.14(f) is manifestly non-random. Townshend (1984) gives a good example of the utility of low-order principal components. His component 7 accounted for only 0.08% of the variability in the data yet a distinction between apple and plum orchards and between peaty and silty-clay soils was brought out. This distinction was not apparent on any other component or individual TM band, nor was it apparent from a study of the principal component loadings. It is important to check principal component images by eye, using one's knowledge of the study area, rather than rely solely upon the magnitudes of the eigenvalues as an indicator of information content, or on inferences drawn from the principal component loadings.

Do not be misled by the coherent appearance of the lower-order principal component images. Figure 6.14(f) shows the sixth principal component derived from the Littleport TM image data set. This principal component accounts for only 0.68% of the total (standardised) variance of the image data set, yet it is interpretable in terms of spatial variation in grey levels. It should be borne in mind that the information expressed by this principal component has been (i) transformed from an arbitrary scale on to the 0–255 range, as described above, and (ii) subjected to a histogram equalisation contrast stretch. Yet, if the first principal component were expressed on a 0–255 scale and the ranges of the other principal components adjusted according to their associated eigenvalue, then principal component 6 would have a dynamic range of 0–10. The use of lower-order components depends on the aims of the project. If the aim is to capture as much as possible of the information content of the image set in as few principal components as possible then the lower-order principal components should be omitted. The 'eigenvalue greater than one' (or 'eigenvalue-1') criterion could be used to determine how many principal components to retain. If an individual standardised band of image data has a variance of unity then it might be possible to argue that all retained principal components should have a variance of at least one. If this argument were used then only the first two principal components of the Littleport image set would be retained, and 11% of the information in the image data set would be traded for a reduction of 66.66% in the image data volume. On the other hand, one may wish merely to orthogonalise the data set (that is, express the data in terms of uncorrelated principal components rather than in terms of the original spectral bands) in order to facilitate subsequent processing. For example, the performance of the feed-forward artificial neural net classifier, discussed in Chapter 8, may perform better using uncorrelated inputs. In such cases, all principal components should be retained. It is important to realise that the aims of a project should determine the procedures followed, rather than the reverse.

There are additional complications. The example above showed that much of the variance in a multiband data set is concentrated in the first two or three principal components. If data compression is the primary aim of the exercise then these principal components can be used to summarise the total data set, as noted above. However, important information may be contained in the higher-order principal component. This is in part due to the fact that a multiband data set may not comprise a single statistical population the distribution of which can be approximated by an ellipsoid (as in Figure 6.13). Chang (1983) has studied the problem involved in using principal components analysis when the data set is made up of two or more statistical populations. He, in fact, restricted his analyses to mixtures of two multivariate-normal distributions, though his conclusions are claimed to be more general. He states:

'In using principal components analysis to reduce the dimensions it has always been the practice to select the components with the larger eigenvalues. We disprove this practice mathematically and by the use of hypothetical and real data.'

Although Chang's (1983) study points towards this general conclusion he does not consider either the effects of non-normal distributions or problems arising whenever the variance–covariance matrices of the populations are unequal. An earlier study by Lowitz (1978) also shows that high-order components can contain information relating to inter-group differentiation. This point should be borne in mind if principal components analysis is used as a feature-selection technique before image classification (section 8.9), and reinforces the conclusions presented in the preceding paragraph.

The example used earlier in this section is based on the eigenvalues and eigenvectors of the correlation matrix, $\mathbf{R}$. We saw earlier that the principal components of the covariance and correlation matrices are not related in a simple fashion. Consideration should therefore be given to the question of which of the two matrices to use. Superficially it appears that the image data in each band are comparable, all being recorded on a 0–255 scale. However, reference to section 4.6 shows that this is not so. The counts in each band can be referred to the calibration radiances for that band, and it will be found that:

• the same pixel value in two different bands relates to different radiance values, and
• if multi-date imagery is used then the same pixel value in the same band for two separate dates may well refer to different radiance values because of differences in sensor calibration.

The choice lies between the use of the correlation matrix to standardise the measurement scale of each band or conversion of the image data to radiances followed by principal components analysis of the variance–covariance matrix. The differences resulting from these alternative approaches have not as yet been fully explored (Singh and Harrison, 1985). However, if multi-date imagery is used the question of comparability is of great importance.

Table 6.3 and Figure 6.15 summarise the results of a principal components transform of the Littleport TM image data set based on the variance–covariance matrix rather than the correlation matrix. Compare the results shown with those for the analysis based on the correlation matrix, described above and summarised in Table 6.2 and Figure 6.14. The effects of the differences in the variances of the individual bands are very noticeable. Band 4 has a variance of 580.95, which is 47.56% of the total variance, and as a consequence band 4 dominates the two highest-order principal components. As in the correlation example, the first principal component is a contrast between band 4 and the other five bands, but this time the loading on band 4 is the highest in absolute terms, rather than the lowest, and the percentage of variance explained by the first principal component is 58.22 rather than 70.77. Principal component 2, accounting for 35.35% of the total variance rather than 18.10% in the correlation case, is now more like a weighted average. The low-order principal components are much less important even than they are when the analysis is based on correlations, and the corresponding principal component images are noisier. Which result is 'better' depends, of course, on the user's objectives. The aim of principal components analysis should be to generate a set of images that are more useful in some way than are the un-transformed images, rather than to satisfy the pedantic requirements of statistical theory. Nevertheless, the point is clear: the principal components of $\mathbf{S}$, the variance–covariance matrix, are quite different from the principal components of $\mathbf{R}$, the correlation matrix. Methods of dealing with the noise that affects the lower-order principal component images are considered below.

In section 6.3.2 the relationship between the principal components and the tasselled cap transformation is discussed. It was noted that the principal components transform is defined internally by the characteristics of the inter-band correlation or covariance structure of the image, whereas the tasselled cap transformation is based on external data, namely, the position of the soil brightness, greenness and wetness axes. The results of a principal components analysis are therefore image-specific and principal component-transformed data values cannot be directly compared between images. While principal components analysis is useful in finding the dimensionality of an image data set and in compressing data into a smaller number of channels (for display purposes, for example) it does not always produce the same components when applied to different image data sets. The differences between principal components analysis and the tasselled cap transform are illustrated in Figure 6.16.

6.4.2 Noise-adjusted principal components analysis

The presence of noise that tends to dominate lower-order principal component images is mentioned a

Table 6.3   Principal components analysis, based on variance–covariance matrix.

(a) Variance–covariance matrix for the Littleport TM image set. The last row shows the variance of the corresponding band expressed as a percentage of the total variance of the image set. The expected variance for each band is 16.66%, but the variance ranges from 2.61% for band 2 to 47.56% for band 4.

Band	1	2	3	4	5	7	Mean	Std dev.
1	78.67	45.90	67.71	−25.04	106.25	67.22	65.812	8.870
2	45.90	31.94	44.09	−6.51	73.45	43.87	29.033	5.652
3	67.71	44.09	72.34	−60.75	116.78	78.86	26.251	8.505
4	−25.04	−6.51	−60.75	580.95	−70.32	−129.52	93.676	24.103
5	106.25	73.45	116.78	−70.32	329.37	181.56	64.258	18.148
7	67.22	43.87	78.86	−129.52	181.56	128.30	23.895	11.327
% var.*	6.44	2.61	5.92	47.56	26.96	10.50	16.666	15.90

(b) Principal component loadings for the six principal components of the Littleport TM image, based on the covariance matrix shown in (a).

Band	PC 1	PC 2	PC 3	PC 4	PC 5	PC 6
1	0.531	0.593	0.575	−0.491	−0.168	−0.018
2	0.502	0.674	0.477	−0.034	0.139	0.215
3	0.707	0.530	0.421	0.057	0.178	−0.085
4	−0.834	0.552	0.000	0.019	−0.001	−0.002
5	0.664	0.714	−0.211	−0.070	0.006	−0.003
7	0.684	0.438	−0.064	0.233	−0.055	0.013
Eigenvalue	711.19	431.87	61.20	9.69	5.52	2.05
% var.*	58.22	35.35	5.01	0.79	0.45	0.17
Cum. % var.†	58.22	93.58	98.59	99.38	99.83	100.00

*Percentage variance.
†Cumulative percentage variance.

number of times in the preceding section. Some researchers, for example Green *et al.* (1988) and Townshend (1984), note that some airborne Thematic Mapper data do not behave in this way. Roger (1996) suggests that the principal components method could be modified so as to eliminate the noise variance. This 'noise-adjusted' principal components analysis would thus be capable of generating principal component images that are unaffected by noise. The method suggested by Roger (1996) is essentially similar to the method that unfortunately, in the present context, is called 'image factor analysis'. Image factor analysis is not simply a mathematical transformation, as is principal components analysis. It is a statistical method concerned with the estimation of the parameters of the factor analysis model. If we consider that a pixel vector **y** (for example, composed of the six values in the reflective bands of the Landsat TM) is generated by the additive effects of a set of factors x, then we could write:

$$\mathbf{y} = b_0 + b_1 x_1 + b_2 x_2 + \ldots + b_p x_p + e$$

where the b_i are weights to be estimated from the data

and e is random error. The original development of factor analysis occurred in the field of educational psychology. The **y** values were the test scores of an individual child (for example, in language, geometry, music, comprehension, etc.) and the x values were thought to be 'common factors' such as general intelligence, mathematical ability, spatial cognition, and so on. We need not be too concerned with the origins of the method, but the example helps to illustrate the motivation of those who developed the ideas. The term e encompasses errors of measurement plus the effects of influences specific to the child. To eliminate these effects, factor analysts developed procedures to estimate the communality or proportion of common variance (that is, variance that is

Figure 6.15   (a)–(f) respectively show the principal components 1–6 of TM bands 1–5 and 7 of the Littleport TM image shown in Figures 1.8 and 1.9. These principal component images are derived from the matrix of covariances among the TM bands. Compare with Figure 6.14(a)–(f). Original data © ESA 1994, distributed by Eurimage.

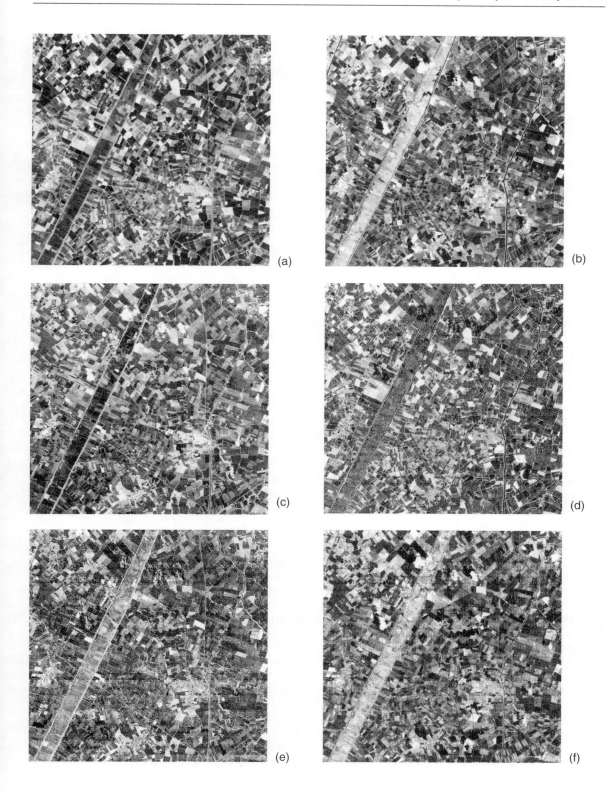

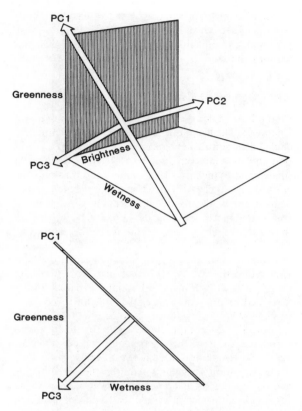

Figure 6.16 The TM tasselled cap transform defines three functions *a priori*: these functions are termed 'greenness', 'brightness' and 'wetness'. Principal components analysis identifies the number of dimensions in the TM data set, but the relationship between the principal component axes and interpretable concepts that identify the major sources of variability in the data set depends upon the inter-band correlations or covariances of the specific image being analysed. See text for explanation. Based on Crist, E.P. and Cicone, R.C., 1984b, figure 2; reproduced by permission of the American Society for Photogrammetry and Remote Sensing.

generated by the common factors x in the equation above). One such method is used in image factor analysis. The proportion of the total variance that is generated by the common factors x is held to be the proportion that is predicted for variable v_i from a linear combination of the remaining variables:

$$\hat{v}_i = \sum_{j=1}^{k} \alpha_j v_j, \qquad j \neq i$$

The α are regression coefficents; in fact, this equation

shows the regression of $v_j (j \neq i)$ on v_i. The proportion of the variance of v_i that is explained by each regression is the common variance of v_i, and this value is used in the principal diagonal of the correlation matrix instead of the value 1.0. In other words, the diagonal element r_{ii} of **R** (which is 1.0) is replaced by the proportion of the predictable variance of band i.

The effect of this operation is that only the coherent variance, or variance that is explained by the correlations among the spectral bands, is included in the analysis. Variations in grey levels that are peculiar to a specific band are eliminated, as is noise (which, in the case of optical imagery, is uncorrelated with the grey-level values in the image). The analysis then proceeds much as before, except that the common variance is partitioned among the principal components, rather than the total variance. Another development, which has not been widely applied, is to consider that the initial principal component solution merely defines the intrinsic dimensionality of the data space. For example, one might use the 'eigenvalue-1' rule, mentioned earlier, to conclude that the intrinsic dimensionality of the data space containing the six bands making up the Littleport TM image is two. The two principal components could be transformed or rotated in this two-dimensional space using a criterion other than maximum variance to define the location of the component axes. One possible criterion is interpretability, and factor analysts have defined a number of rotations of the initial solution that improve the interpretability of the results. The varimax rotation is an example. This rotation is based on the principle of simplicity. In effect, the varimax criterion is: the number of high (near $\pm$ 1) loadings should be small and the number of low (near zero) loadings should be large. A further possibility is to relax the requirement that the axes of this reduced-dimensionality space be oblique rather than orthogonal. Such esoteric considerations are discussed in detail in Mather (1976), which also contains a number of Fortran programs to implement these ideas.

Computer programs to carry out the procedures described in this section are provided on the CD-ROM that accompanies this book. The MIPS display program (Appendix A) calculates the first three principal components of an image set. First, select an image dictionary file (`*.inf`) and display any three bands. Choose the `Utilities` menu and select `PCA`. You can then choose some or all of the image bands that are referenced by the dictionary file and compute the first three principal components. Alternatively, for images larger than 512×512 in size, use the program `pca`. The noise-reduced method of principal components, called

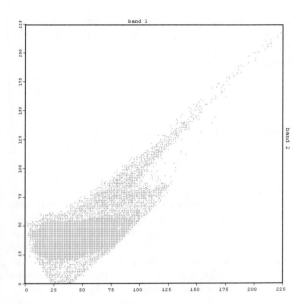

Figure 6.17 Plot of principal component 1 (*x*-axis) against principal component 2 (*y*-axis) for the Littleport TM image set. The principal components analysis was based on the image covariance matrix. This image was produced by the MIPS program scatter, which generates a colour-coded image in order to aid interpretation. Although the principal component axes are orthogonal, there appears to be a very high positive correlation between pixels that have values greater than 120.

image factor analysis, is implemented by the program imfact. Both pca and imfact allow the use of a mask. Recall from section 6.2.3 that a mask image can be created by thresholding one band of a multispectral image set. If such a mask file is referenced in the appropriate image dictionary file then both the pca and imfact programs allow the user to specify whether the principal components should be calculated for pixels that correspond to zero pixels in the mask or that correspond to non-zero pixels in the mask. The use of a mask may be useful in the analysis of images that cover two distinct Earth-surface cover types, for example land and water.

Program scatter, which is also included on the CD-ROM, plots the pixels in any two image bands (or principal component images) as an *x–y* scatter plot. Such plots are useful in the interpretation of principal components, and are also used in the identification of extreme (end-member) pixels in mixture modelling, a technique that is described in Chapter 8. Figure 6.17 shows a plot of the pixels making up the first two

principal component images based on the Littleport TM image set, which is used in earlier examples. These principal components are based on the image covariance matrix.

6.4.3 Decorrelation stretch

Methods of colour enhancement are the subject of Chapter 5. The techniques discussed in that chapter include linear contrast enhancement, histogram equalisation and the Gaussian stretch. All of these act upon a single band of the false-colour image and thus must be applied separately to the red, green and blue components of the image. As noted in section 6.4.1, principal components analysis removes the correlation between the bands of an image set by rotating the axes of the data space so that they become oriented with the directions of maximum variance in the data, subject to the constraint that these axes are orthogonal. If the data are transformed into principal components space, and are 'stretched' within this space, then the three contrast stretches will be at right-angles to each other. In RGB space the three colour components are likely to be correlated, so the effects of stretching are not independent for each colour (Gillespie *et al.*, 1986).

Decorrelation stretching requires the three bands making up the RGB colour composite images to be subjected to a principal components analysis, a stretch applied in the principal components space, and the result back-transformed to RGB space. The result is generally an improvement in the range of intensities and saturations for each colour, with the hue remaining unaltered (see section 6.5 for an extended discussion of these properties of the colour image). Poor results can be expected when the RGB images do not have approximately Gaussian histograms, or where the image covers large, homogeneous areas.

The decorrelation stretch, like principal components analysis, can be based on the covariance matrix **S** or the correlation matrix **R**, as discussed above. Use of **R** implies that all three bands are given equal weight. If the stretch is based on **S**, each band is weighted according to its variance. The following description uses the notation **R** to indicate *either* the correlation *or* the covariance matrix. Essentially, the principal component images **y** are calculated from the eigenvector matrix **E** and the raw images **x** using the relationship $\mathbf{y} = \mathbf{Ex}$, and the inverse transform is $\mathbf{x} = \mathbf{E}^{-1}\mathbf{y}$ (Alley, 1995; Rothery and Hunt, 1990).

1. Compute the eigenvectors **E** ($= e_{ij}$) and eigenvalues $\mathbf{\Lambda}(= \lambda_i)$ of **R**.

2. Obtain the inverse, $\mathbf{E}^{-1}$, of $\mathbf{E}$ by noting that $\mathbf{E}^{-1} = \mathbf{E}'$ for an orthogonal matrix. $\mathbf{E}'$ indicates the transpose of $\mathbf{E}$, that is, the rows of $\mathbf{E}$ are the columns of $\mathbf{E}'$.

3. Calculate a diagonal stretching matrix $\mathbf{S}$ which has elements s_{ii} equal to $50/\sqrt{\lambda_i}$. The value 50 is arbitrary and is meant to represent the desired standard deviation in the decorrelation stretched image. As $\mathbf{S}$ is a diagonal matrix it follows that $s_{ij} = 0$ for $i \neq j$.

4. Derive the transformation matrix $\mathbf{T} = \mathbf{E}^{-1}\mathbf{S}\mathbf{E}$.

5. Estimate the three correction factors forming the vector $\mathbf{g}$ required to convert the transformed data back to a 0–255 scale by using the vector of means $\mathbf{m}$ (the mean value of each raw image) to determine a vector $\mathbf{f}$ that is equal to $\mathbf{Tm}$ then calculating the values required to convert each f_i to 127, the midpoint of the 0–255 scale. If $\mathbf{f} = \mathbf{Tm}$ then simply divide each f_i by 127 to obtain the correction factors g_i.

6. Apply the transformation matrix to all the $\{r, g, b\}$ triples in the image. These triples, which are denoted here by the vector $\mathbf{x}$, are scaled on to the 0–255 range via the elements of $\mathbf{g}$ calculated at step 5, i.e. $\mathbf{y} = \mathbf{Tx}$ gives the 'raw' decorrelation stretched values for a given pixel with RGB values stored in $\mathbf{x}$. The raw $\mathbf{y}$ values are multiplied by the corresponding g_i to produce the scaled decorrelation stretched pixel.

The values used for the stretching parameter at step 3 require some thought (Campbell, 1996). The value of $50/\sqrt{\lambda_i}$ is suggested by Alley (1995). When combined with the use of the mean vector at step 5 to produce a vector of offsets, the result should be a distribution on each of the three bands that is centred at [127, 127, 127] and with a standard deviation of 50. This configuration does not necessarily produce optimum results. The description of the decorrelation stretch option in the MIPS display program (Appendix A) provides further discussion. As in the case of standard principal components analysis, and other image transforms that operate in the domain of real numbers, the scaling of the result back to the range 0–255 is logically difficult. The method suggested above for standard principal components analysis involves calculating the maximum and minimum of the raw transformed values, then scaling this range on to the 0–255 scale using offsets and stretching parameters. The program described in Appendix A uses a rather different method – the calculation of a sample of output values and the derivation from this output of scaling factors. These scaling factors are used to correct the final output, with separate transformations being used for values above and below the mean. The appearance of the final product depends crucially on the scaling method used, and there are few commonsense guidelines to help the user. The value of the decorrelation stretch is also a function of the nature of the image to which it is applied. The method seems to work best on images of semi-arid areas, and it seems to work least well where the image histograms are bimodal, for example, where the area covered by the image includes both land and sea.

The MIPS display program provides a decorrelation stretch function under the main window's `Enhance` menu. The appearance of the resulting image can be improved if a linear stretch, histogram equalisation or Gaussian stretch is applied following the decorrelation stretch. This may seem to defeat the object of the decorrelation stretch but essentially compensates for the inadequacy of the mapping of the decorrelation stretch output on to the range 0–255. Appendix A gives more details of the operation of this program. Further consideration of the procedure is provided by Guo and Moore (1996) and Campbell (1996). The latter author presents a detailed analysis of the decorrelation stretch process. Ferrari (1992) and White (1993) illustrate the use of the method in geological image interpretation.

6.5 HUE, SATURATION AND INTENSITY (HSI) TRANSFORM

Details of alternative methods of representing colours are discussed in section 5.2 where two models are described. The first is based on the red–green–blue colour cube. The different hues generated by mixing red, green and blue light are characterised by coordinates on the red, green and blue axes of the colour cube (Figure 5.3). The second representation uses the hue–saturation–intensity hexcone model (Figure 5.4) in which hue, the dominant wavelength of the perceived colour, is represented by angular position around the top of a hexcone, saturation or purity is given by distance from the central, vertical axis of the hexcone, and intensity or value is represented by distance above the apex of the hexcone. Hue is what we perceive as colour (such as mauve or purple). Saturation is the degree of purity of the colour, and may be considered to be the amount of white mixed in with the colour. As the amount of white light increases the colour becomes more pastel-like. Intensity is the brightness or dullness of the colour. It is sometimes useful to convert from RGB colour cube coordinates to HSI hexcone coordinates, and vice versa. The RGB coordinates will be considered to run from 0 to 1 (rather than 0–255) on each axis, while the coordinates for the hexcone model will consist of (i) hue expressed as an angle between 0 and 360°, and (ii) saturation and intensity on a 0–1 scale. Note that the acronym IHS (intensity–hue–saturation) is sometimes used in place of HSI.

The hue, saturation and intensity (HSI) transform is useful in two ways: first, as a method of image enhancement and, secondly, as a means of combining co-registered images from different sources. In the former case, the RGB representation is first converted to HSI, as described by Foley *et al.* (1990), Hearn and Baker (1994) and Shih (1995). Next, the intensity (I) and saturation (S) components are stretched independently and the HSI representation is converted back to RGB for display purposes. It does not make sense to stretch the hue (H) component, as this would upset the colour balance of the image. Gillespie *et al.* (1986) provide further details. This operation is implemented in the MIPS image display program (Appendix A) from the `Enhance` menu item. The forward transform (RGB to HSI) is carried out, and optionally the I and S components are subjected to a linear stretch using the 95% and 5% points of the corresponding histograms. The stretched HSI representation can be displayed or the inverse transform applied to convert back to RGB coordinates. In practice it is often necessary to apply a contrast enhancement procedure to the back-transformed image. See Chapter 5 for more details of enhancement methods.

The method can also be used to combine images from different sources (data fusion). For example, synthetic aperture radar (SAR) imagery from the Seasat satellite is combined with Landsat MSS imagery by Blom and Daily (1982); see also Harris *et al.* (1990). The steps used by Blom and Daily (1982) are as follows:

1. register the Seasat SAR and Landsat MSS images (Chapter 4),
2. convert the Landsat MSS image from RGB to HSI coordinates,
3. substitute the SAR image for the intensity coordinate, and
4. convert back to RGB space.

This transformation has been found to be particularly useful in geological applications; see for example Jutz and Chorowicz (1993) and Nalbant and Alptekin (1995). Further details of the HSI transform are given in Blom and Daily (1982), Foley *et al.* (1990), Green (1983), Hearn and Baker (1994), Pohl and van Genderen (1998) and Mulder (1980). Terhalle and Bodechtel (1986) illustrate the use of the transform in the mapping of arid geomorphic features, while Gillespie *et al.* (1986) discuss the role of the HSI transform in the enhancement of highly correlated images. Massonet (1993) gives details

of an interesting use of the HSI transform in which the amplitude, coherence and phase components of an interferometric image are allocated to hue, saturation and intensity, respectively, and the inverse HSI transform applied to generate a false-colour image, which highlights details of coherent and incoherent patterns. See Gens and van Genderen (1996) for a review of SAR interferometry. Schetselaar (1998) discusses alternative representations of the HSI transform, and Andreadis *et al.* (1995) give an in-depth study of the transform. Pitas (1993) lists *C* routines for the RGB to HSI colour transform.

6.6 FOURIER TRANSFORM

The coefficients of the tasselled cap functions and the eigenvectors associated with the principal components define coordinate axes in the multidimensional data space containing the multispectral image data. These data are re-expressed in terms of the new coordinate axes and the resulting images have certain properties, which may be more suited to particular applications. The Fourier transform operates on a single-band image. Its purpose is to break down the image into its scale components, which are defined to be sinusoidal waves with varying amplitudes, frequencies and directions. The coordinates of the two-dimensional space in which these scale components are expressed are given in terms of frequency (cycles per basic interval). This is called the *frequency domain* whereas the normal row/column coordinate system in which images are normally expressed is termed the *spatial domain* (Figure 6.18). The function of the Fourier transform is to convert a single-band image from its spatial-domain representation to the equivalent frequency-domain representation, and vice versa.

The idea underlying the Fourier transform is that the grey-scale values forming a single-band image can be viewed as a three-dimensional intensity surface, with the rows and columns defining two axes and the grey-level value at each pixel giving the third (z) dimension. A series of waveforms of increasing frequency are fitted to this intensity surface and the information associated with each such waveform is calculated. The Fourier transform therefore provides details of (i) the frequency of each of the scale components of the image and (ii) the proportion of information associated with each frequency component. Frequency is defined in terms of cycles per basic interval where the basic interval in the across-row direction is the number of pixels on the scan line and the basic interval in the down-column direction is the number of scan lines. Frequency could be

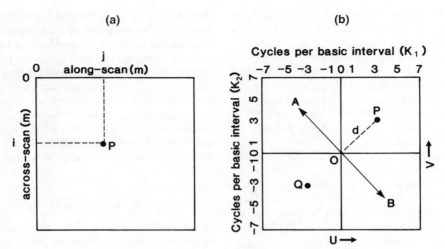

Figure 6.18 (a) Spatial-domain representation of a digital image. Pixel P has coordinates (i, j) with respect to the row (across-scan) and column (along-scan) coordinate axes, or $(j\Delta x, i\Delta y)$ metres, where Δx and Δy are the dimensions of the image pixels in the row and column direction, respectively. (b) Frequency-domain representation, showing the amplitude spectrum of the image. The value at P is the amplitude of a sinusoidal wave with frequency $K_1 = 3$ and $K_2 = 3$ cycles per basic interval in the *u* and *v* directions, respectively. The wavelength of this sinusoidal wave is proportional to the distance OP. The orientation of the waveform is along direction AB. Point Q $(-3, -3)$ is the mirror image of point P.

expressed in terms of metres by dividing the magnitude of the basic interval (in metres) by cycles per basic interval. Thus, if the basic interval is 512 pixels each 20 metres wide then the wavelength of the fifth harmonic component is $(512 \times 20)/5$ or 2048 metres. The first scale component, conventionally labelled zero, is simply the mean grey-level value of the pixels making up the image. The remaining scale components have increasing frequencies (decreasing wavelengths) starting with 1 cycle per basic interval, then $2, 3, \ldots, n/2$ cycles per basic interval where n is the number of pixels or scan lines in the basic interval.

This idea can be more easily comprehended by means of a simple example in one dimension. Figure 6.19(b) shows a complex waveform that might represent, for example, a cross-section along one scan line of an image. Figure 6.19(a) shows how this complex waveform can be broken down into a sequence of sine and cosine waves which collectively summarise the variation around the mean value (represented by the horizontal line). The first of these components is the mean value of the series and is represented by a horizontal line. Harmonic number 1 has a frequency of one cycle over the basic interval of the composite curve. The second harmonic has a frequency of two cycles per basic interval, the third has three, and so on. These frequencies are the harmonics of the basic interval. Figure 6.19(c) shows a plot of frequency (cycles per basic interval) against the

square of the amplitude (Figure 1.6) of each harmonic wave. The value shown as having a frequency of zero is the mean value. This is often referred to as the 'DC' value, following electrical engineering terminology. The plot is called an *amplitude spectrum* (be aware of possible confusion with the term electromagnetic spectrum, though the meaning should be clear from the context). It is clear from the amplitude spectrum that the major components of the composite curve shown in Figure 6.19(b) have frequencies of 1, 2, 3 and 4 cycles per basic interval – exactly as shown in Figure 6.19(a). So the Fourier transform has, in the example, broken down the composite waveform into its scale components and provided a measure of the relative importance of each scale component.

If this simple example were extended to a function defined over a two-dimensional grid then the only differences would be that (i) the scale components would be two-dimensional waveforms and (ii) each scale component would have an orientation as well as an amplitude. The squared amplitudes of the waves are plotted against frequency in the horizontal and vertical directions to give a two-dimensional amplitude spectrum, which is interpreted much in the same way as the one-dimensional amplitude spectrum in Figure 6.19(c), the major differences being as follows.

(i) The frequency associated with the point $[k_1, k_2]$ in the two-dimensional amplitude spectrum is given by

(a)

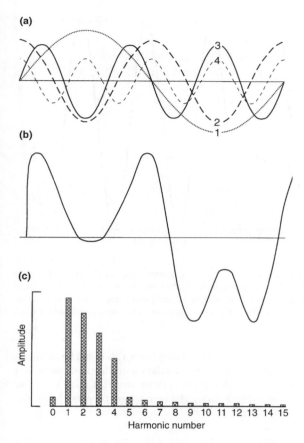

(b)

(c)

Figure 6.19 (a) Set of four sinusoidal waves having frequencies of 1, 2, 3 and 4 cycles over the basic interval, which is the length of the horizontal axis. (b) Composite curve derived by summing the four curves shown in (a) at 16 equally spaced points along the x-axis. (c) One-dimensional amplitude spectrum obtained from the Fourier transform of the curve shown in (b). The value at the harmonic numbered 0 is the mean value (which should be close to zero in this case). Harmonics 1–4 relate to the four curves shown in (a), and the amplitude corresponding to these four harmonics can be seen to be proportional to the amplitudes of the four curves shown in (a). Harmonics 5–15 are the result of rounding error and also of the fact that the curve in (b) is digitised only at 16 points.

$$k_{12} = \sqrt{(k_1^2 + k_2^2)}$$

where the basic intervals given by each axis of the spatial-domain image are equal, or by

$$k_{12} = \sqrt{(k_1/n_1\Delta t_1 + k_2/n_2\Delta t_2)}$$

where the basic intervals in the two spatial dimensions of the image are unequal. In the latter case, $n_1\Delta t_1$ and $n_2\Delta t_2$ are the lengths of the two axes, n_1 and n_2 are the number of sampling points along each axis, and Δt_1 and Δt_2 are the sampling intervals. This implies that frequency is proportional to distance from the centre of the amplitude spectrum, which is located at the point [0, 0] (Rayner, 1971).

(ii) The angle of travel of the waveform whose amplitude is located at point $[k_1, k_2]$ in the amplitude spectrum is perpendicular to the line joining the point $[k_1, k_2]$ to the centre (DC) point of the spectrum ([0, 0]). This point is illustrated in Figure 6.21(a), which shows the two-dimensional amplitude spectrum of an image made up of a set of equally spaced horizontal lines (represented digitally by rows of 1s against a background of 0s). The amplitude spectrum shows symmetric points located vertically above and below the DC point, which lies at the centre of the image. These points (shown in black against a white background for ease of viewing) represent the amplitudes of the waveforms reconstructed from the parallel, horizontal lines which could be considered to lie on the crests of a series of sinusoidal waveforms progressing down the image from top to bottom. Since the direction of travel could be top–bottom or bottom–top, the amplitude spectrum is symmetric and so the two points closest to the DC represent the amplitudes of the set of waves whose wavelength is equal to the spacing between the horizontal lines. The two points further out from the DC represent a spurious waveform, which has a wavelength equal to double the distance between the horizontal lines. Such spurious waveforms represent a phenomenon termed *aliasing* (Rosenfeld and Kak, 1982; Gonzales and Woods, 1992; Figure 6.20). Figure 6.21(b) shows a second example using a set of vertical lines as input. The amplitude spectrum is rotated through 90° in comparison with Figure 6.21(b) but can be interpreted in a similar way.

Figure 6.22 may help to illustrate some of these ideas. Figure 6.22(a) shows a two-dimensional sine wave oriented in a direction parallel to the x-axis of the spatial-domain coordinate system. Figure 6.22(b) shows another sine wave, this time oriented parallel to the y-axis, and Figure 6.22(c) shows the result of adding these two sine waves together to generate an undulating surface, which might be considered to represent the grey values of a digital image. Higher grey values are lighter, lower grey values are darker. The two-dimensional amplitude spectrum of Figure 6.22(c) would show two high values located at a distance from the DC that is proportional to the frequency of the sine wave (i.e. the number

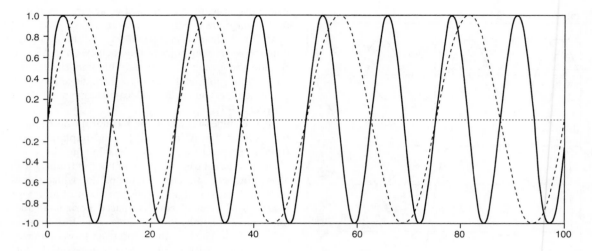

Figure 6.20 Illustrating the concept of aliasing. The sine wave shown as a solid line has a period that is one-half of that of the sine wave shown as a dashed line. Both have the same amplitude. The solid line passes through all of the data points, while the dashed line is an alias. It passes through every second data point and has a frequency twice that of the 'true' frequency. It would be possible to add other sine waves that have frequencies of three, four, . . . times the 'true' frequency. All would be *aliases* of the first ('true') sine wave.

of times that the complete sine wave is repeated in the x direction (or the y direction, since the two axes are equal). One of these high values will be located on axis K_1 of Figure 6.18, and the other on axis K_2. The mirror images of these two points will also show high amplitude values. In effect, the two high values in the spectrum show that Figure 6.22(c) is made up of two frequency components. The high value on axis K_1 indicates that one of these frequency components is parallel to the x-axis of the spatial-domain (image) representation. The high value on axis K_2 of Figure 6.18 shows that the second component is parallel to the y-axis. The distances of these two points along axes K_1 and K_2 are proportional to the frequencies of the two curves shown in Figures 6.22(a) and (b).

The calculation of the amplitude spectrum of a two-dimensional digital image involves techniques and concepts that are too advanced for this book. A simplified account will be given here. Fourier analysis is so called because it is based on the work of Jean Baptiste Joseph, Baron de Fourier, who was Governor of Lower Egypt and later Prefect of the Department of Grenoble during the Napoleonic era. In his book *Theorie Analytique de la Chaleur*, published in 1822, he set out the principles of the Fourier series, which has since found wide application in a range of subjects other than Fourier's own, the analysis of heat flow. The principle of the Fourier series is that a single-valued curve (i.e. one that has only a

single y value for each separate x value) can be represented by a series of sinusoidal components of increasing frequency. This point is illustrated in Figure 6.19. The form of these sinusoidal components is given by

$$f(t) = a_0 + \sum_i a_i \cos(i\omega t) + \sum_i b_i \sin(i\omega t)$$

in which $f(t)$ is the value of the function being approximated at point t, ω is equal to $2\pi/T$ and T is the length of the series. The term a_0 represents the mean level of the functions and the summation terms represent the contributions of a set of sine and cosine waves of increasing frequency. The fundamental waveform is that with a period equal to T, or a frequency of ω. The second harmonic has a frequency of 2ω, while 3ω is the frequency of the third harmonic, and so on.

The a_i and b_i terms are the cosine and sine coefficients of the Fourier series. It can be seen in the formula above that the a_i are the multipliers of the cosine terms and the b_i are the multipliers of the sine terms. Sometimes the sine and cosine terms are expressed as a complex number, that is, a number that has 'real' and 'imaginary' parts. Hence, the a_i and b_i are also called the real and imaginary coefficients. We will not use complex-number notation here. The coefficients a_i and b_i can be calculated by a least-squares procedure. Owing to its con-

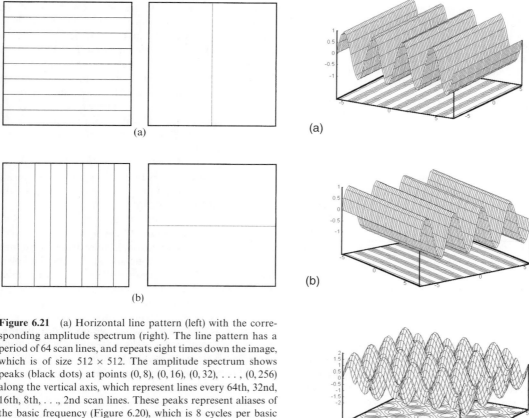

(a)

(b)

(a)

(b)

(c)

Figure 6.21 (a) Horizontal line pattern (left) with the corresponding amplitude spectrum (right). The line pattern has a period of 64 scan lines, and repeats eight times down the image, which is of size 512 × 512. The amplitude spectrum shows peaks (black dots) at points $(0, 8)$, $(0, 16)$, $(0, 32)$, . . . , $(0, 256)$ along the vertical axis, which represent lines every 64th, 32nd, 16th, 8th, . . ., 2nd scan lines. These peaks represent aliases of the basic frequency (Figure 6.20), which is 8 cycles per basic interval (the line repeats eight times over the length of the image, which is 512 scan lines). (b) Vertical line pattern and corresponding amplitude spectrum. The same pattern is seen as in (a) except that it is horizontal rather than vertical. Again, aliasing is proposed as the explanation for the peaks in the amplitude spectrum at frequencies of 16, 24, . . ., 256 cycles per basic interval. Both images have been enhanced using a 5–95% linear contrast stretch.

Figure 6.22 (a) Two-dimensional sine wave running parallel to the x-axis. (b) Two-dimensional sine wave running parallel to the y-axis. (c) Sum of the two sine waves shown in (a) and (b).

siderable computing time requirements this method is no longer used in practice. In its place, an algorithm called the fast Fourier transform (FFT) is used (Bergland, 1969; Gonzales and Woods, 1992; Lynn, 1982; Pavlidis, 1982; Pitas, 1993; Ramirez, 1985). The advantage of the FFT over the older method can be summarised by the fact that the number of operations required to evaluate the coefficients of the Fourier series using the older method is proportional to N^2 where N is the number of sample points (length of the series) whereas the number of operations involved in the FFT is proportional to $N \log_2 N$. The difference is brought out by a comparison of columns (ii) and (iii) of Table 6.4. However, in its normal implementation the FFT requires that N (the length of the series) should be a power of 2. Singleton (1979a) gives an algorithm that evaluates the FFT for a series of length N which need not be a power of 2, while Bergland and Dolan (1979) provide a Fortran program listing of an algorithm that computes the FFT in a very efficient manner.

Once the coefficients a_i and b_i are known the amplitude of the ith harmonic is computed from

$$A_i = \sqrt{a_i^2 + b_i^2}$$

Table 6.4 Number of operations required to compute the Fourier transform coefficients a and b for a series of length N (column (i)) using least-squares methods (column (ii)) and the fast Fourier transform (FFT) (column (iii)). The ratio of column (ii) to column (iii) shows the magnitude of the improvement shown by the FFT. If each operation took 0.01 second then, for the series of length $N = 8096$, the least-squares method would take 7 days, 18 hours and 26 minutes. The FFT would accomplish the same result in 17 minutes and 45 seconds.

(i) N	(ii) N^2	(iii) $N \log_2 N$	(iv) (ii)/(iii)
2	4	2	2.00
4	16	8	2.00
16	256	64	4.00
64	4 096	384	10.67
128	16 384	896	18.29
512	262 144	4 608	56.89
8096	67 108 864	106 496	630.15

and the phase angle (the displacement of the first crest of the sinusoid from the origin, measured in radians or degrees) is defined as

$$\theta = \tan^{-1}(b_i/a_i)$$

Normally, only the amplitude information is used.

The procedure to calculate the forward Fourier transform (it is, in fact, the forward discrete Fourier transform) for a two-dimensional series, such as a grey-scale image, involves the following steps:

1. Compute the Fourier coefficients for each row of the image, storing the coefficients a_i and b_i in separate two-dimensional arrays. The coefficients form two-dimensional arrays of real numbers (in the Fortran sense) or they can be considered to form a single two-dimensional complex array.
2. Compute the Fourier transform of the columns of these a_i and b_i matrices to give the Fourier coefficients of the two-dimensional image. There are two sets of coefficients, corresponding to the a_i and b_i above.

Step 2 requires that the two coefficient matrices are transposed; this is a very time-consuming operation if the image is large. Singleton (1979b) describes an efficient algorithm for the application of the FFT to two-dimensional arrays.

Figure 6.23(b) shows the amplitude spectrum of a Landsat TM band 7 image of an area of the Kenyan Rift Valley in East Africa (Figure 6.21(a)). Concentric circles

spaced at 32 units apart are superimposed on the amplitude spectrum to assist interpretation. Most of the power is seen to be concentrated in the region around the DC, showing that the main frequency components are low and mid-frequency. The pronounced horizontal and vertical lines in the amplitude spectrum are the result of edge effects – the sides of the TM image are oriented either horizontally or vertically. In most cases, the amplitude spectrum is not interpreted directly but used during the frequency-domain filtering process (section 7.5). Nevertheless, an ability to interpret the amplitude spectrum is an important skill.

As already noted, the main use of the Fourier transform in remote sensing is in frequency-domain filtering (Chapter 7). For example, Lei *et al.* (1996) use the Fourier transform to identify and characterise noise in MOMS-02 panchromatic images in order to design filters to remove the noise. Other applications include the characterisation of particular terrain types by their Fourier transforms (Leachtenauer, 1977), and the use of measures of heterogeneity of the grey levels over small neighbourhoods based on the characteristics of the amplitude spectra of these neighbourhoods. If an image is subdivided into 32×32 pixel sub-images and if each sub-image is subjected to a Fourier transform then the sum of the amplitudes in the area of the amplitude spectrum closest to the origin gives the low-frequency or smoothly varying component while the sum of the amplitudes in the area of the spectrum furthest away from the origin gives the high-frequency, rapidly changing component. These characteristics of the amplitude spectrum have also been used as measures of image texture, which is considered further in section 8.7.1. Fourier-based methods have also been used to characterise topographic surfaces (Brown and Scholz, 1985), in the calculation of image-to-image correlations during the image registration process (section 4.3.1.4), in analysis of the performance of resampling techniques (Shlien, 1979; section 4.3.1.3) and in the derivation of pseudocolour images from single-band images (Chapter 5). De Souza Filho *et al.* (1996) use Fourier-based methods to remove defects from optical imagery acquired by the Japanese JERS-1 satellite. Temporal sequences of vegetation indices (section 6.2.4) are analysed using one-dimensional Fourier transforms by Olsson and Ekhlund (1994), and a similar approach is used by Menenti *et al.* (1993).

In the derivation of a pseudocolour image the low-frequency component (see Figure 7.9) is extracted from the amplitude spectrum and an inverse Fourier transform applied to give an image that is directed to the red monitor input. The intermediate or mid-range frequencies are dealt with similarly and directed to the green

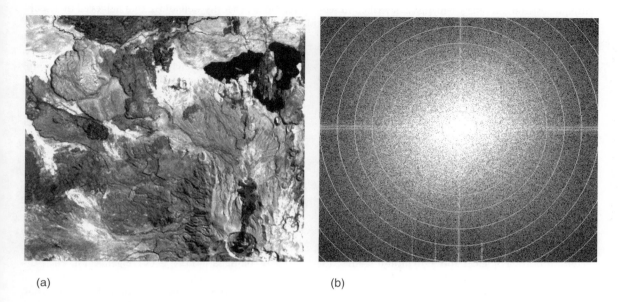

(a)   (b)

Figure 6.23   (a) Landsat Thematic Mapper band 7 image of part of the Kenyan Rift Valley. This image is 512×512 pixels in size. (b) Logarithm of the amplitude spectrum of (a). Both images have been enhanced using a 5–95% linear contrast stretch. The concentric circles on (b) are drawn at a spacing of 32 units. Most of the information in (a) is seen to be low and mid-frequency. The strong horizontal and vertical features in the amplitude spectrum are edge effects, with some contribution from the 16-line effect caused by the manner in which a Thematic Mapper image is scanned.

input. The blue input is derived from the high-frequency components, giving a colour rendition of a black-and-white image in which the three primary colours (red, green and blue) represent the low-, intermediate- and high-frequency or scale components of the original monochrome image. Blom and Daily (1982) and Daily (1983) describe and illustrate the geological applications of a variant of this method using Seasat synthetic aperture radar images. Their method is based on the observation that grey-level (tonal) variations in synthetic aperture radar images can be attributed to two distinct physical mechanisms. Large-scale features (with low spatial frequency) are produced by variations in surface backscatter resulting from changes in ground-surface cover type. High spatial frequencies correlate with local slope variations, which occur on a much more rapid spatial scale of variability. The amplitude spectrum of a Seasat SAR image is split into these two components (high and low frequencies) using frequency-domain filtering methods as described in Chapter 7. The result is two filtered amplitude spectra, each of which is subjected to an inverse Fourier transform to convert from the frequency back to the spatial domain. The low-pass

filtered image is treated as the hue component in HSI colour space (section 6.5) and the high-pass filtered image is treated as the intensity component. Saturation is set to a value that is constant over the image; this value is chosen interactively until a pleasing result is obtained. The authors claim that the pseudocolour image produced by this operation is significantly easier to interpret.

Other developments in the field of image transforms include the *wavelet transform*, which was discovered in the late 1980s at the AT&T Bell Telephone Laboratories. A readable introduction is provided by Strang (1994). Ranchin and Wald (1993) and Prasad and Iyengar (1997) present applications of wavelets in image processing. Yocky (1996) and Zhou *et al.* (1998) discuss the use of the wavelet transform in merging SPOT panchromatic (10 m resolution) and Landsat TM (30 m resolution) data.

A computer program, `fourier`, is provided on the CD accompanying this book. It will take a square image with a side length that is a power of 2 up to 512, and compute its amplitude spectrum, which is placed on the Windows clipboard. The program also provides a range of frequency-domain filters, which are described in section 7.5. Once the program has terminated, the amplitude spectrum can be pasted into a word processor or into an image processing application for further enhancement, or for printing. Further details are given in Appendix B.

6.7 SUMMARY

A range of image transforms is considered in this chapter. Arithmetic operations (addition, subtraction, multiplication and division) have a utilitarian use – for example, image subtraction is used routinely in the separation of the high-frequency component of an image during the filtering process (section 7.3.1) while addition is used in the method of lineament detection described in section 7.4. Image division, or ratioing, is one of the most common transformations applied to remotely-sensed images in both geological and agricultural studies, for simple band ratios reflect differences in the slopes of the spectral reflectance curves of Earth-surface materials. Problems experienced with the use of ratios include the difficulty of separating the effect of atmospheric path radiances, and the choice of dynamic range compression technique. Nevertheless ratio images are a widely used and valuable tool. The empirical transformations considered in section 6.3 have been developed for use with images of agricultural areas. One of the special problems here is that the database on which these transformations (the perpendicular vegetation index and the tasselled cap transformation) are based is limited and site-specific. Their unrestricted use with images from parts of the world other than those regions of the United States where they were developed is questionable.

The principal components analysis (Karhunen-Loève) transformation has a long history of use in multivariate statistics. Even so it is not as well understood by the remote-sensing community as it might be. Like most parametric statistical methods it is based on a set of assumptions, which must be appreciated, if not completely satisfied, if the methods are to be used successfully. The final transformation technique covered in this chapter is the Fourier transform. The level of mathematics required to understand a formal presentation of this method is generally well above that achieved by undergraduate students in geography, geology and other Earth sciences. The intuitive explanation given in section 6.6 might serve to introduce such readers to the basic principles of the method and allow a fuller understanding of the frequency-domain filtering techniques described in section 7.5.

It will be realised that the presentation in this chapter is largely informal and non-mathematical. Many pitfalls and difficulties are not covered in this chapter. These will no doubt be discovered serendipitously by the reader in the course of project work.

6.8 QUESTIONS

1. Explain the following terms: multitemporal, LAI, correlation matrix, variance–covariance matrix, spatial domain, frequency domain, mask, topographic effect, biomass, NDVI, SAVI, eigenvector, basic interval, DC.
2. Review the methods for measuring change over time from a multitemporal image set.
3. What is the motivation underlying the use of spectral band ratios? Give examples from the literature to show the application of band ratios in studies of (a) vegetation and (b) geology. What are the problems involved in relating values of vegetation ratios and characteristics of vegetation cover such as LAI and biomass?
4. What was the rationale behind the development of the tasselled cap transform? How does this transform differ from the principal components transform?
5. Distinguish between principal components analysis based on correlations, principal components analysis based on covariances, and 'noise-adjusted' principal components analysis.
6. Why should the decorrelation stretch produce a 'better' result than histogram equalisation applied separately to the red, green and blue components of a false-colour image?
7. Explain how to interpret the amplitude spectrum of (a) a one-dimensional series and (b) a two-dimensional image. Use simple examples to demonstrate your knowledge.

7

Filtering Techniques

7.1 INTRODUCTION

The enhancement techniques discussed in Chapter 5 change the way in which the information content of an image is presented to the viewer, either by altering image contrast or by coding a grey-scale image in pseudocolour so as to emphasise or amplify some property of the image that is of interest to the user. In these procedures, the values held in the image display memory are unchanged. All that is done is to modify the lookup tables (LUTs) in such a way that the appearance of the image on the screen is altered. In other words, the way in which the numerical data held in the image memory are translated to viewable form as a television picture is altered in order to change the perceived balance of information in the image. The data themselves are not changed. On the other hand, image transforms, such as principal components analysis, the hue–saturation–intensity transform and the decorrelation stretch (all of which are reviewed in Chapter 6), alter the way in which the spectral information in the image set is expressed and viewed.

Methods for selectively emphasising or suppressing information at different *spatial scales* over an image are described in this chapter. For example, we may wish to suppress the noise pattern caused by detector imbalance that is sometimes seen in Landsat MSS and TM images and which results from the fact that the image is mechanically scanned in groups of six lines (MSS) or 16 lines (TM). The origin of this banding phenomenon is described in Chapter 2. Methods other than filtering for the removal of banding are described in Chapter 4 under the heading of pre-processing. On the other hand, we may wish to emphasise some spatial feature or features of interest, such as curvilinear boundaries between areas that are relatively homogeneous in terms of their tone or colour, in order to sharpen the image and reduce blurring. Unlike the methods described in Chapters 4 and 5, these techniques require that the image data

(rather than the lookup tables) be altered. This is because the techniques operate selectively on the image data, which are considered to contain information at various spatial scales. The idea that a spatial (two-dimensional) pattern, such as the variation of grey levels in a grey-scale image, can be considered as a composite of patterns at different scales superimposed upon each other was introduced in section 6.6. Large-scale background or regional patterns, such as land and sea, are the basic components of the image. These large-scale patterns can be thought of as 'background' with 'detail' being added by small-scale patterns. Noise, either random or systematic, is normally also present. Random noise, such as the speckle prevalent on radar images, is a high spatial frequency phenomenon.

The representation of the spatial variability of a feature in terms of a regional pattern with local information and noise superimposed has been widely used in disciplines that deal with spatially distributed phenomena. Patterns of variation are often summarised in a generalisation. For example, a geographer might say that, in Great Britain, 'Mean annual rainfall declines from west to east' in the knowledge that such a statement describes only the background pattern, upon which is superimposed the variations attributable to local factors. In both geography and geology, the technique of trend surface analysis has been found useful in separating the regional and local components of such spatial patterns (Davis, 1973; Mather, 1976).

By analogy with the procedure used in chemistry laboratories to separate the components of a suspension, the techniques described in this chapter are known as *filtering*. A digital filter can be used to extract a particular spatial-scale component from a digital image. The slowly varying background pattern in the image can be envisaged as a two-dimensional waveform with a long wavelength or low frequency, hence a filter that separates this component from the remainder of the information present in the image is called a *low-pass filter*. Conversely, the more rapidly varying detail is like a two-dimensional waveform with a short wavelength or high frequency. A filter to separate out this component is called a *high-pass filter*. These two types of filter will be considered separately. Low-frequency information allows the identification of the background pattern, and produces an output image in which the detail has been smoothed or removed from the original (input) image (hence low-pass filtering can be thought of as a form of blurring the image). High-frequency information allows us either to isolate or to amplify the local detail. If the high-frequency detail is amplified by adding back to the image some multiple of the high-frequency component extracted by the filter, then the result is a sharper, de-blurred image.

Two complementary methods exist for the separation of the scale components of the spatial patterns exhibited in a remotely-sensed image. The first is based upon the transformation of the image into its scale or spatial frequency components using the Fourier transform (section 6.6), while the second method is applied directly to the image data in the spatial domain. Fourier-based methods are summarised in section 7.5. In the following two sections the most common spatial-domain filtering methods are described. There is generally a one-to-one correspondence between spatial- and frequency-domain filters. However, specific filters may be easier to design in the frequency domain but may be applied more efficiently in the spatial domain. The concept of spatial- and frequency-domain representations is shown in Figure 6.18. Whereas spatial-domain filters are generally classed as either high-pass (sharpening) or as low-pass (smoothing), filters in the frequency domain can be designed to suppress, attenuate, amplify or pass any group of spatial frequencies. The choice of filter type can be based either on spatial frequency or on direction, for both these properties are contained in the Fourier amplitude spectrum (section 6.6).

7.2   LOW-PASS (SMOOTHING) FILTERS

Before the topic of smoothing a two-dimensional image is considered, we will look at a simpler expression of the same problem, which is the smoothing of a one-dimensional pattern. Figure 7.1 shows plots of grey levels along the cross-section from the top left corner (0, 0) to the bottom right corner (511, 511) of the Littleport TM band 7 image shown in Figure 1.9(b). Figure 7.1(a) shows the cross-section for the unfiltered image, while Figure 7.1(b) shows the same cross-section after the application of a low-pass (smoothing) filter. Clearly, the

level of detail has been reduced and the cross-section curve is more generalised, though the main peaks are still apparent. Figure 7.2 is also a plot showing grey-level value (vertical axis) against position across a scan line of a digital image. The underlying pattern is partially obscured by the presence of local patterns and random noise. If the local variability, and the random noise, were to be removed then the overall pattern would become more clearly apparent and a general description of trends in the data could then be more easily made. The solid line in Figure 7.2 is a plot of the observed pixel values against position along the scan line, while the dotted line and the dashed line represent the output from median and moving-average filters respectively. These filters are described below. Both produce smoother plots than the raw data curve, and the trend in the data is more easily seen. Local sharp fluctuations in value are removed. These fluctuations represent the high-frequency component of the data and may be the result of local anomalies or of noise. Thus, low-pass filtering is used by Crippen (1989), Eliason and McEwen (1990) and Pan and Chang (1992) to remove banding effects on remotely-sensed images, while Dale *et al.* (1996) use a low-pass filter to smooth away the effects of image-to-image mis-registration.

Cross-sectional plots such as those shown in Figures 7.1 and 7.2 can be produced using the MIPS display program, which is supplied with this book. To use this procedure, load an image and select the `Plot|Cross-sections` menu item. The 'raw' pixel values are plotted, whether or not an image enhancement procedure has been applied. Readers are encouraged at this point to familiarise themselves with various cross-sections in order to gain a 'feel' for the way in which the shape of a grey-level curve can be estimated visually from the image. As individual filters are introduced in the following pages, cross-sections of filtered images can be compared to the original cross-section, as is done in Figures 7.1(a) and (b).

7.2.1   Moving-average filter

If the coordinate on the horizontal axis of Figure 7.2 is denoted by the index j then the moving-average filtered value at any point j is x'_j. The calculation of x'_j depends on the number of local values around the filtered point that are used in the calculation of the moving average. This number is always an odd, positive integer so that there is a definite central point (thus, the central value in the sequence 1, 2, 3 is 2, whereas there is no specific central value in the sequence 1, 2, 3, 4). The dashed line

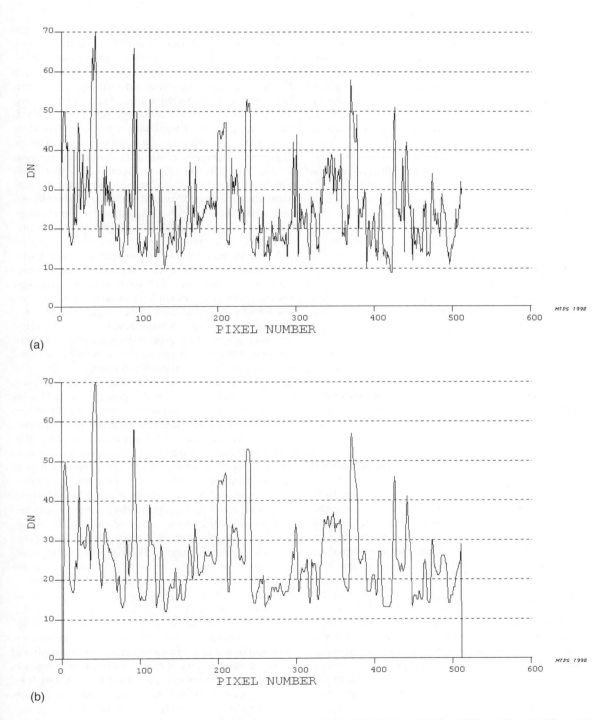

(a)

(b)

Figure 7.1   (a) Cross-section from the top left to the bottom right corner of the Littleport TM band 7 image shown in Figure 1.9 b). (b) Cross-section between the same points as used in (a) after the application of a smoothing filter (a 5 × 5 median filter was used to generate this cross-section, as described in section 7.2.2). The reduction in detail is clearly apparent.

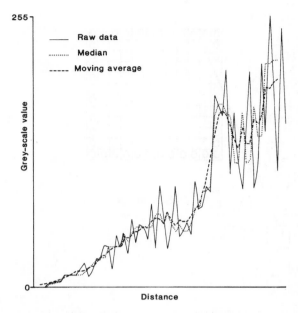

Figure 7.2 One-dimensional data series showing the effect of median (low-pass) filtering and moving-average (low-pass) filtering.

in Figure 7.2 is based on a five-point moving average, defined by

$$x'_j = (x_{j-2} + x_{j-1} + x_j + x_{j+1} + x_{j+2})/5$$

Five raw data values (x) centred on point j are summed and averaged to produce one output value (x'_j). If a three-point moving average had been used then three raw data values centred on point j would be summed and averaged to give one output value x'_j. If the number of data elements included in the averaging process is n, then $[n/2]$ values at the beginning of the input series and $[n/2]$ values at the end of the input series do not have output values associated with them, because some of the input terms x_{j-1}, x_{j-2} and so on will not exist for $j < [n/2]$, just as some of the terms $x_{j+1}, x_{j+2}, \ldots$ will not exist for $j > N - [n/2]$. (Here the symbol $[\ \ ]$ indicates the integer part of the given expression and N is the number of raw data values.) The filtered (output) series is thus shorter than the input series by $n - 1$ elements, where n is the length of the filter (three-point, five-point, etc.). Thus, a 5×5 filter will leave an unfiltered margin of two pixels around the four sides of the image. These marginal pixels are usually set to zero (see Figure 7.3).

In calculating a five-point moving average for a one-

dimensional series the following algorithm might be used: add input (x) values 1–5 and divide their sum by 5 to give x'_3. Note that filtered values x'_1 and x'_2 cannot be calculated; the reason is given above. Next, add raw data values x_2 to x_6 and divide their sum by 5 to give x'_4. This procedure is repeated until output value x'_{N-2} has been computed, where N is the number of input values (again, x'_{N-1} and x'_N are left undefined). This algorithm is rather inefficient, for it overlooks the fact that the sum of x_2 to x_6 is easily obtained from the sum of x_1 to x_5 simply by subtracting x_1 from the sum of x_1 to x_5 and adding x_6. The terms x_2, x_3 and x_4 are present in both summations, and need not be included in the second calculation. If the series is a long one then this modification to the original algorithm will result in a more time-efficient program.

A two-dimensional moving-average filter is defined in terms of its horizontal (along-scan) and vertical (across-scan) dimensions. Like the one-dimensional moving-average filter, these dimensions must be odd, positive and integral. However, they need not be equal. A two-dimensional moving average is described in terms of its size, such as 3×3. Care is needed when the filter dimensions are unequal to ensure that the order of the dimensions is clear; the Cartesian system uses (x, y) where x is the horizontal and y the vertical coordinate, with an origin in the lower left of the positive quadrant. In matrix (image) notation, the position of an element is given by its row (vertical) and column (horizontal) coordinates, and the origin is the upper left corner of the matrix or image. The central element of the filter, corresponding to the element x'_j in the one-dimensional case described earlier, is located at the intersection of the central row and column of the $n \times m$ filter window. Thus, for a 3×3 window, the central element lies at the intersection of the second row and column. To begin with, the window is placed in the top left corner of the image to be filtered (Figure 7.3) and the average value of the elements in the input image that are covered by the window is computed. This value is placed in the output image at the point in the output image corresponding to the location of the central element of the window. In effect, the window can be thought of as a matrix with all its elements equal to 1; the output from the convolution of the window and the image is the sum of the products of the corresponding window and image elements divided by the number of elements in the window. If **F** ($=f_{ij}$) is the filter matrix, **G** ($=g_{ij}$) is the input image and **O** ($=o_{ij}$) the output (filtered) image then

$$o_{ij} = \left\{ \sum_{p=-b}^{b} \sum_{q=-c}^{c} g_{p+i,q+j} f_{p+r,q+s} \right\} / mn$$

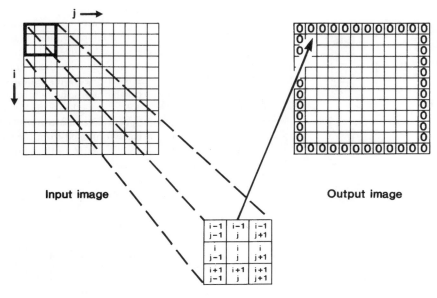

Figure 7.3   Moving-window calculations. Average of 3×3 filter window is placed in cell of output image corresponding to cell (i, j) of window.

where p and q are indices, b = integer part of $n/2$, c = integer part of $m/2$, n = number of rows in filter matrix, m = number of columns in filter matrix, r = central row in filter matrix $(= [n/2])$, s = central column in filter matrix $(= [m/2])$, i, j = image pixel underlying element (r, s) of filter matrix (coordinates in row/column order) and $[e]$ = the integer part of expression e.

For example, given a 5×3 filter matrix the value of the pixel in the filtered image at row 8, column 15 is given by:

$$o_{8,15} = \left\{ \sum_{p=-2}^{2} \sum_{q=-1}^{1} g_{p+8,q+15} f_{p+3,q+2} \right\} / 15$$

with $b = 2$, $c = 1$, $r = 3$ and $s = 2$. Notice that the indices i and j must be in the range $b < i < (N - b + 1)$ and $c < j < (M - c + 1)$ if the image has N rows and M columns numbered from 1 to N and 1 to M respectively. This means that there are b empty rows at the top and bottom of the filtered image and c empty columns at either side of the filtered image. This unfiltered margin can be filled with zeros (as in Figure 7.3) or the unaltered pixels from the corresponding cells of the input image can be placed there.

The initial position of the filter window with respect to the image is shown in Figure 7.3. Once the output value from the filter has been calculated, the window is moved one column (pixel) to the right and the operation is repeated (Figure 7.4). The window is moved rightwards and successive output values are computed until the right-hand edge of the filter window hits the right margin of the image. At this point, the filter window is moved down one row (scan line) and back to the left-hand margin of the image. This procedure is repeated until the filter window reaches the bottom right-hand corner of the input image. The output image values form a matrix that has fewer rows and columns than the input image; the unfiltered margin corresponds to the rows and columns of the input matrix that the filter window cannot reach. Generally, these missing rows and columns are filled with zeros in order to keep the input and output images the same size. The effect of the moving-average filter is to reduce the overall variability of the image and lower its contrast. At the same time those pixels which have larger or smaller values than their neighbourhood average are respectively reduced or increased in value so that local detail is lost. Noise components, such as the banding patterns evident in line-scanned images, are also reduced in magnitude by the averaging process, which can be considered as a smearing or blurring operation. In cases where the overall pattern of grey-level values is of interest, rather than the details of local variation, neighbourhood grey-level averaging is a useful technique. The moving-average filter described in this section is an example of a general class

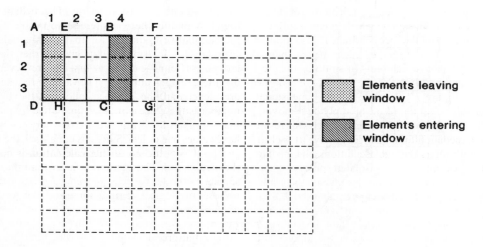

Figure 7.4 Updating filter window calculations. The window size is 3 × 3. If the histogram of area ABCD is known then, to find the histogram of area EFGH, adjust for cells in column 1 that are leaving the window and for cells in column 4 that are entering the window.

of filters called box filters; refer to McDonnell (1981) for a survey.

The moving-average filter can be implemented in the MIPS display program (Appendix A) using the `filter|userdefined` menu option. Examples of 3 × 3 and 5 × 5 moving-average filter weights are given in Figure 7.5. Note that the weights are given as integers – the actual weights are 1/9 or 1/25 rather than 1.0. Hence a divisor is needed. For the 3 × 3 filter the divisor is 9. For the 5 × 5 filter the divisor is 25. If you apply these two filters you will see that increasing the window size for a moving average results in a greater degree of smoothing since more pixels are included in the averaging process.

7.2.2 Median filter

An alternative type of smoothing filter utilises the median of the neighbourhood rather than the mean. The median filter is generally thought to be superior to the moving-average filter, for two reasons. First, the median of a set of n numbers (where n is an odd integer) is always one of the data values present in the set. Secondly, the median is less sensitive to errors or to extreme data values. This can be demonstrated by a simple, one-dimensional, example. If the nine pixel values in the neighbourhood of, and including, the point (x, y) are $\{3, 1, 2, 8, 5, 3, 9, 4, 27\}$ then the median is the central value (the fifth in this case) when the data are ranked in ascending or descending order of magnitude. In this example the ranked values are $\{1, 2, 3, 3, 4, 5, 8, 9, 27\}$

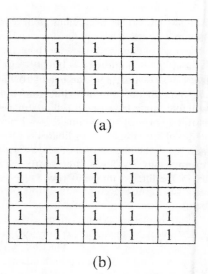

Figure 7.5 Weights for the two-dimensional moving-average filter: (a) 3 × 3 filter, (b) 5 × 5 filter. A divisor of 9 should be specified for the 3 × 3 filter (i.e. $1/3^2$) and a divisor of 25 ($1/5^2$) should be specified for the 5 × 5 filter.

giving a median value of 4. The mean is 6.89, which would be rounded up to a value of 7. The value 7 is not present in the original data, unlike the median value of 4. Also, the mean value is larger than six of the nine observed values, and may be thought to be unduly influenced by the extreme data value 27, which is three times larger than the next highest value in the set. Thus, isolated extreme pixel values, which might represent noise, are removed by the median filter. It also follows that the median filter preserves edges better than a moving-average filter, which blurs or smooths the grey levels in the neighbourhood of the central point of the filter window. Figure 7.2 shows (a) a one-dimensional sequence of values, (b) the result of applying a moving average of length five to the given data, and (c) the result of applying a median filter also of length five. It is clear that, while both filters remove oscillations, the median filter more successfully removes isolated spikes and better preserves edges, defined as pixels at which the gradient or slope of grey-level value changes markedly. Synthetic aperture radar (SAR) images often display a noise pattern called *speckle* (Chapter 2). This is seen as a random pattern of bright points over the image. The median filter has been used by a number of researchers in order to eliminate this speckle without unduly blurring the sharp features of the image (MacFarlane and Thomas, 1984; Blom and Daily, 1982).

The mean is relatively easily computed; it involves summation and division, as explained earlier, and considerable savings of computer time can be achieved by methods described in section 7.2.1. In contrast, the median requires the ranking of the data values lying within the $n \times m$ filter window centred on the image point that is being filtered. The operation of ranking or sorting is far slower than that of summation, for $[n/2] + 1$ passes through the data are required to determine the median ($[n/2]$ indicates 'the integer part of the result of dividing n by 2'). At each pass the smallest value remaining in the data must be found by a process of comparison. Using the values given in the preceding paragraph, the value 1 would be picked out after the first pass, leaving eight values. A search of these eight values gives 2 as the smallest remaining value, and so on for $[n/2] + 1 = [9/2] + 1 = 4 + 1 = 5$ passes. The differences between the summation and ranking methods are amplified by the number of times the operation is carried out; for a 3×3 filter window and a 512×512 image the filter is evaluated $510 \times 510 = 260\,100$ times. Thus, although the median filter might be preferred to the moving-average filter for the reasons given earlier, it might be rejected if the computational cost was too high.

Fortunately, a less obvious but faster method of computing the median value for a set of overlapping filter windows is available. This fast algorithm begins as usual with the filter window located in the top left-hand corner of the image, as shown in Figure 7.3. A histogram of the $n \times m$ data points lying within the window is computed and the corresponding class frequencies are

stored in a one-dimensional array. The median value is found by finding that class (grey-level) value such that the cumulative frequency for that class equals or exceeds $[n/2] + 1$. Using the data given earlier, the cumulative class frequencies for the first few grey levels are (0) 0, (1) 1, (2) 2, (3) 4 and (4) 5. The value in brackets is the grey level and the number following the bracketed number is the corresponding cumulative frequency, so that no pixels have values of zero, whereas four pixels have values of 3 or less. Since n is equal to 9, the value of $[n/2] + 1$ is 5 and no further calculation is necessary; the median is 4. This method is faster than the obvious sorting method because histogram calculation does not involve any logical comparisons, and the total histogram need not be checked in order to find the median. Further savings are achieved if the histogram is updated rather than recalculated when the filter window is moved to its next position, using a method similar to that illustrated in Figure 7.4. First, the cumulative frequencies are reduced as necessary to take account of the left-hand column of the window, which is moving out of the filter, and then the frequencies are incremented according to the values of the new right-hand column of pixel values, which is moving into the window. This part of the procedure is similar to the updating of the sum of pixel values as described in section 7.2.1 in connection with the moving-average filter. If the fast algorithm is used then the additional computational expense involved in computing the median is not significant.

The concept of the median filter was introduced by Tukey (1977) and its extension to two-dimensional images is discussed by Pratt (1978). The fast algorithm described above was reported by Huang *et al.* (1979), who also provide a number of additional references. See also Brownrigg (1984) and Danielsson (1981). Blom and Daily (1982) and Rees and Satchell (1997) illustrate the use of the median filter applied to SAR images.

The MIPS display program includes a median filtering option. Click filter on the main menu bar and select menu item median. You can choose any filter size as long as (i) the filter dimension is an odd positive integer and (ii) the filter dimension is not greater than 9×9. The median filtered Littleport TM band 7 image (Figure 1.9(b)) is shown in Figure 7.6.

7.2.3   Adaptive filters

Both the median and the moving-average filter apply a fixed set of weights to all areas of the image, irrespective of the variability of the grey levels underlying the filter window. Several authors have considered smoothing methods in which the filter weights are calculated for

Figure 7.6   Result of 5 × 5 median filter applied to the Little-port TM band 7 image shown in Figure 1.9(b). A plot of the cross-section from the top left corner to the bottom right corner of this image is shown in Figure 7.1(b). Original data © ESA 1994, distributed by Eurimage.

each window position, the calculations being based on the mean and variance of the grey levels in the area of the image underlying the window. Such filters are termed *adaptive filters*. Their use is particularly important in the attenuation of the multiplicative noise effect known as speckle, which affects synthetic aperture radar images. As noted above, the median filter has been used with some success to remove speckle. However, more advanced filters will, in general, produce superior results in the sense that they are capable of removing speckle without significantly degrading the high-frequency component of the SAR image.

One of the best-known speckle suppression filters is the *sigma filter*, proposed by Lee (1983a, b). This filtering method is based on the concept of the normal distribution. Approximately 95% of the values of observations belonging to a normal distribution with mean μ and standard deviation σ fall within $\pm 2\sigma$ of the mean value. Lee's method assumes that the grey-level values in a single-band SAR image are normally distributed and, for each overlapping, rectangular window, computes estimates of the local mean $\bar{x}$ and standard deviation s from the pixels falling within the window. A local average is computed from those pixels which lie within $\pm 2s$ of the window mean. Pixels outside this range are not included in the calculation. The method

breaks down when only a few of the pixels in the window are within $\pm 2s$ of the window mean. A parameter k is used to act as a threshold. If fewer than k pixels are to be included in the calculation then the procedure is aborted, and the filtered pixel value to the left of the current position is used. Alternatively, the average of the four neighbouring pixels replaces the window centre pixel. This can cause problems in the first case if the pixel concerned is on the left margin of the image. In the second case, the filtered values for the pixels to the right of and below the current pixel will need to be calculated before the average can be obtained. Other developments in the use of the sigma filter are summarised by Smith (1996), who describes two simple modifications to the standard sigma filter to improve its computational efficiency and preserve fine features. Reviews of speckle filtering of SAR images are provided by Desnos and Matteini (1993) and Lee *et al.* (1994). Wakabayeshi and Arai (1996) discuss a new approach to speckle filtering using a chi-square test. Martin and Turner (1993) consider a weighted method of SAR speckle suppression, while Alparone *et al.* (1996) present an adaptive filter using local order statistics to achieve the same objective. Order statistics are based on the local grey-level histogram, for example the median. More advanced methods of speckle filtering using simulated annealing are described by White (1993). Other references are Beauchemin *et al.* (1996), who use a measure of texture (the contrast feature derived from the grey-level co-occurrence matrix, described in Chapter 8) as the basis of the filter, and Lopes *et al.* (1993).

A similar idea, that of edge-preserving smoothing, was put forward by Nagao and Matsuyama (1979). Their method attempts to avoid averaging pixel values that belong to different 'regions'. The boundary between two regions contained within a window area might be expected to be represented by an 'edge' or sharp discontinuity in the grey-level values. Hence, Nagao and Matsuyama suggest that a bar be rotated around the centre of the window and the bar at the position with the smallest standard deviation of the pixels' grey-scale values be selected as the 'winner', since a small standard deviation indicates the absence of any edges. The centre pixel value is replaced by the average of the pixel values in the winning bar.

A modified form of the sigma filter is incorporated into the MIPS image display software. Select Filter|Sigma and provide details of the window size. Instead of using the standard deviation of the window to control filtering, the MIPS implementation simply uses a user-supplied threshold to include or exclude pixels from the averaging process. Finally, the parameter k mentioned above (the minimum number of pixels that need to pass the sigma test) is supplied.

7.3   HIGH-PASS (SHARPENING) FILTERS

The process of imaging or scanning involves blurring, as noted in the discussion of the point spread function (PSF) in section 2.2.1. High frequencies are suppressed relative to the low-frequency components of the image. It would seem intuitive that the visual quality of an image might be improved by selectively increasing the contribution of its high-frequency components. Since the low-pass filters discussed in section 7.2 involve some form of averaging (or spatial integration) then the use of a spatial derivative function might appear suitable to perform the opposite function, that of sharpening or de-blurring an image. A simpler way of performing a high-pass filtering is considered before the derivative-based methods are discussed.

7.3.1   Image subtraction method

According to the model described in sections 6.6 and 7.1 an image can be considered to be the sum of its low- and high-frequency components. The low-frequency part can be isolated by the use of a low-pass filter as explained in section 7.2. This low-frequency image can be subtracted from the original, unfiltered, image, leaving behind the high-frequency component. The resulting image can be added back to the original, thus effectively doubling the high-frequency part or, as in the case of Thomas *et al.* (1981), a proportion of the low-pass filtered image is subtracted from the original image to give a high-frequency image. Thomas *et al.* (1981) use the model

$$R^* = R - fR' + C$$

in which R^* is the filtered pixel value at the centre of a 3×3 window, R is the original value, R' is the average of the 3×3 window, f is a proportion between 0 and 1 (Thomas *et al.* (1981) use 0.8) and C is a constant whose function is to ensure that R^* is always positive. This subtractive box filter is used by Thomas *et al.* to enhance a circular geological feature. The addition and subtraction operations must be done with care (section 6.2). The sum of any two pixel values drawn from images each having a dynamic range of 0–255 can range from 0 to 510, so division by 2 is needed to keep the sum within the 0–255 range. The difference between two pixel values can range from -255 to $+255$; if the result is to be expressed

on the range 0–255 then (i) 255 is added to the result and (ii) this value is divided by 2. In a difference image, therefore, 'zero' has a grey-scale value of 127.

7.3.2   Derivative-based methods

Other methods of high-pass filtering are based on the mathematical concept of the derivative. The derivative of a continuous function is the rate of change of that function at a point. For example, the first derivative of position with respect to time (the rate of change of position over time) is velocity, assuming direction is constant. The greater the velocity of an object, the more rapidly it changes its position with respect to time. The velocity can be measured at any time after motion commences. The velocity at time t is the first derivative of position with respect to time at time t. If the position of an object were to be graphed against time then the velocity (and hence the first derivative) at time t would be equal to the slope of the curve at the point time $= t$. Hence, the derivative gives a measure of the rate at which the function is increasing or decreasing at a particular point in time or, in terms of the graph, it measures the gradient of the curve.

In the same way that the rate of change of position with time can be represented by velocity so the rate of change of velocity with time can be found by calculating the first derivative of the function relating velocity and time. The result of such a calculation would be acceleration. Since acceleration is the first derivative of velocity with respect to time and, in turn, velocity is the first derivative of position with respect to time, then it follows that acceleration is the second derivative of position. It measures the rate at which velocity is changing. When the object is at rest its acceleration is zero. Acceleration is also zero when the object reaches a constant velocity. A graph of acceleration against time would be useful in determining those times when velocity was constant or, conversely, the times when velocity was changing.

In terms of a continuous grey-scale image, the analogue of velocity is the rate of change of grey-scale value over space. This derivative is measured in two directions – one with respect to x, the other with respect to y. The overall first derivative (with respect to x and y) is the square root of the sum of squares of the two individual first derivatives. These three derivatives (in the x direction, y direction and overall) tell us (i) how rapidly the grey-scale value is changing in the x direction, (ii) how rapidly it is changing in the y direction and (iii) the maximum rate of change in any direction, and the direction of this maximum change. All these values are calculable at any point in the interior of a continuous image. In those areas of the image that are homogeneous, all three derivatives (x, y and overall) will be small. Where there is a rapid change in the grey-scale values, for example at a coastline in a near-infrared image, the gradient (first derivative) of the image at that point will be high. These lines or edges of sharp change in grey level can be thought of as being represented by the high-frequency component of the image for, as mentioned earlier, the local variation from the overall background pattern is due to high-frequency components (the background pattern is the low-frequency component). The first derivative or gradient of the image therefore identifies the high-frequency portions of the image.

What does the second derivative tell us? Like the first derivative it can be calculated in both the x and y directions and also with respect to x and y together. It identifies areas where the gradient (first derivative) is constant, for the second derivative is zero when the gradient is constant. It could be used, for example, to find the top and the bottom of a 'slope' in grey-level values.

Images are not continuous functions. They are defined at discrete points in space, and these points are usually taken to be the centres of the pixels. It is therefore not possible to calculate first and second derivatives using the methods of calculus. Instead, the derivatives are estimated by differences between the values of adjacent pixels in the x and y directions, though diagonal or corner differences are also used. Figure 7.7 shows the relationship between a discrete, one-dimensional function (such as the values along a scan line of a digital image, as shown in Figure 7.1) and its first and second derivatives estimated by the method of differences. The first differences (equivalent to the first derivatives) are

$$\Delta x\, p(i,j) = p(i,j) - p(i-1,j)$$
$$\Delta y\, p(i,j) = p(i,j) - p(i,j-1)$$

in the x (along-scan) and y (across-scan) directions respectively, while the second difference in the x direction is:

$$\Delta x_2\, p(i,j) = \Delta x\, p(i+1,j) - \Delta x\, p(i,j)$$
$$= [p(i+1,j) - p(i,j)] - [p(i,j) - p(i-1,j)]$$
$$= p(i+1,j) + p(i-1,j) - 2p(i,j)$$

Similarly, the second difference in the y direction is:

$$\Delta y_2\, p(i,j) = p(i,j+1) + p(i,j-1) + 2p(i,j)$$

The calculation and meaning of the first and second

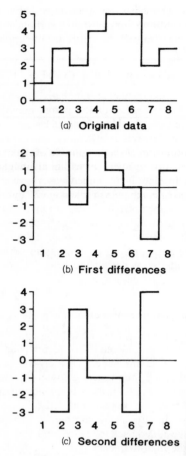

Figure 7.7   Graphical representation of (a) a one-dimensional data series with corresponding (b) first and (c) second differences.

differences in one dimension are illustrated in Table 7.1. A discrete sequence of values, which can be taken as pixel values along a scan line, is shown in the top row of Table 7.1, and the first and second differences are shown in rows two and three. The first difference is zero where the rate of change of grey-scale value is zero, positive when 'going up' a slope and negative when going down. The magnitude of the first difference is proportional to the steepness of the 'slope' of the grey-scale values, so steep 'slopes' (where grey-scale values are increasing or decreasing rapidly) are characterised by first differences that are large in absolute value. The second difference is zero where the first difference values are constant, negative at the foot of a 'slope' and positive at the top of the 'slope'. The extremities of a 'slope' are thus picked out by the second difference.

Table 7.1 Relationship between discrete grey-level values (f) across a scan line, and the first and second differences ($\Delta(f)$, $\Delta_2(f)$). The first difference (row 2) indicates the rate of change of the values of f, shown in row 1. They are calculated from $\Delta(f) = f_i - f_{i-1}$. The second difference (row 3) gives the points at which the rate of change itself alters. The second difference is calculated from $\Delta_2(f) = \Delta(\Delta f)) = f_{i+1} + f_{i-1} - 2f_i$.

$f0$		0	0	0	1	2	3	4	5	5	5	5	4	3	2	1	0
$\Delta(f)$			0	0	1	1	1	1	1	0	0	0	-1	-1	-1	-1	-1
$\Delta_2(f)$			0	1	0	0	0	0	-1	0	0	-1	0	0	0	0	

The computation of the magnitude of the maximum first difference or gradient of a digital image can be carried out by finding Δx and Δy as above and then determining the composite gradient, given by

$$\Delta xy\, p(i,j) = \sqrt{\{[\Delta x\, p(i,j)]^2 + [\Delta y\, p(i,j)]^2\}}$$

and the direction of this composite gradient is

$$\theta = \tan^{-1}[\Delta y\, p(i,j)/\Delta x\, p(i,j)]$$

Other gradient measures exist; one of the most common is termed the *Roberts gradient* ΔR. It is computed in the two diagonal directions rather than in the horizontal and vertical directions from:

$$\Delta R = \sqrt{\{[p(i,j) - p(i+1,j+1)]^2 + [p(i,j+1) - p(i+1,j)]^2\}}$$

or

$$\Delta R = |p(i,j+1) - p(i+1,j+1)| + |p(i,j+1) - p(i+1,j)|$$

The second form is sometimes preferred for reasons of efficiency; the absolute value is more quickly computable than the square root, and raising the inter-pixel difference values to the power of 2 is avoided. The Roberts gradient function is implemented in the MIPS display program, via the `Filter` main menu item (Appendix A). Figure 7.8(a) shows the Littleport TM band 7 image (Figure 1.9(b)) after the application of the Roberts gradient operator. The grey-level values are proportional to ΔR.

In order to emphasise the high-frequency components of an image a multiple of the gradient values at each pixel location (except those on the first and last rows and columns) can be added back to the original image. Normally the absolute values of the gradient are used in this operation. The effect is to emphasise those areas where the grey-scale values are changing rapidly. Another possibility is to define a threshold value by inspection of the histogram of gradient values. Where the gradient at a pixel location exceeds this threshold the pixel value is set to 255, otherwise the gradient is added back as before. This will overemphasise the areas of greatest change in grey level.

The second difference function of a digital image is given by

$$\begin{aligned}\Delta xy_2\, p(i,j) &= \Delta x_2\, p(i,j) + \Delta y_2\, p(i,j) \\ &= [p(i+1,j) + p(i-1,j) + p(i,j+1) \\ &\quad + p(i,j-1)] - 4p(i,j)\end{aligned}$$

This function is called the *Laplacian operator*. Like its one-dimensional analogue shown in Table 7.1, this operator takes on a negative value at the foot of a grey-scale 'slope' and a positive value at the crest of a 'slope'. The magnitude of the value is proportional to the gradient of the 'slope'. If absolute values are taken then the Laplacian will pick out the top and the bottom of 'slopes' in grey-scale values (see Figure 7.8(b)). Alternatively, the signed values (negative at the foot, positive at the crest) can be displayed by adding 127 to all values, thus making 127 the 'zero' point on the grey scale. Negative values of the Laplacian will be shown by darker shades of grey, positive values by lighter grey tones. Like the gradient image, the Laplacian image can be added back to the original image though, as noted below, it is more sensible to subtract the Laplacian. The effect is often quite dramatic, though much depends on the 'noisiness' of the image. Any adding-back of high-frequency information to an already noisy image will inevitably result in disappointment. Figure 7.8(c) shows the effects of subtracting the Laplacian (shown in Figure 7.8(b)) from the Littleport TM band 7 image (Figure 1.9(b)). The result is much less 'hazy' than the original.

Rosenfeld and Kak (1982, pp. 241–244) give reasons why this should be so. If the discussion of the point spread function (Chapter 2) is recalled, it will be realised that the effect of imaging through the atmosphere and the use of lenses in the optical system is to diffuse the radiance emanating from a point source so that the image of a sharp point source appears as a circular blob.

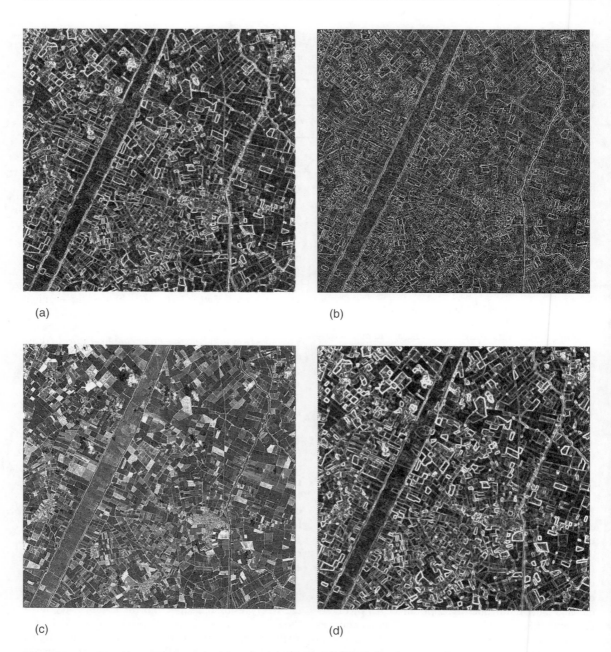

(a)   (b)

(c)   (d)

Figure 7.8   (a) Littleport TM band 7 image shown in Figure 1.9(b) after application of the Roberts gradient operator, followed by a 5–95% linear stretch. (b) Same image after application of the Laplacian operator. (c) Result of subtracting the Laplacian of the Littleport TM band 7 image (shown in (b)) from the original image (Figure 1.9(b)). (d) Output from the Sobel filter for the Littleport TM band 7 image. Original data © ESA 1994, distributed by Eurimage.

Rosenfeld and Kak show that the Laplacian operator approximates in mathematical terms to the equation known as Fick's law, which describes the two-dimensional diffusion process. Thus, subtracting the Laplacian from the original image is equivalent to removing the diffused element of the signal from a given pixel. Another possible explanation is that the value recorded at any point contains a contribution from the neighbouring pixels. This is a reasonable hypothesis, for the contribution could consist of the effects of diffuse radiance, that is, radiance from other pixels that has been scattered into the field of view of the sensor. The Laplacian operator effectively subtracts this contribution.

The weight matrix to be passed across the image to compute the Laplacian is:

0	1	0
1	-4	1
0	1	0

while the 'image-minus-Laplacian' operation can be performed directly using the following weight matrix:

0	-1	0
-1	5	-1
0	-1	0

The absolute values of the Laplacian of an image can be computed in the MIPS display software (Appendix A) using the `User-Defined` option in the `Filter` menu and entering the weights shown above (using a divisor of 1). The 'image-minus-Laplacian' operation can be performed directly from the `Filter` menu. Other forms of the weight matrix are conceivable; for example, the diagonal differences rather than the vertical and horizontal differences could be used, or the diagonal differences plus the vertical/horizontal differences. A wider neighbourhood could be used, with fractions of the difference being applied. There seems to be little or no reason why such methods should be preferred to the basic model unless the user has some motive based upon the physics of the imaging process.

The basic model of a high-pass image-domain filter involves the subtraction of the pixel values within a window from a multiple of the central pixel. The size of the window is not limited to 2×2 or 3×3, which are used in the derivative-based filters described above. Generally, if the number of pixels in a window is k then the weight given to the central pixel is $(k-1)$ while all other pixels have a weight of -1. The product of the window weights and the underlying image pixel values is subsequently divided by k. The size of the window is proportional to the wavelengths allowed through the filter. A low-pass filter will remove more of the high-frequency components as the window size increases (i.e. the degree of smoothing is proportional to the window size). A high-pass filter will allow through a broader range of wavebands as the window size increases. Unless precautions are taken, the use of very large window sizes will cause problems at the edge of the image; for instance, if the window size is 101×101 then the furthest left that the central pixel can be located is at row 51, giving a margin of 50 rows that cannot be filtered. For 3×3 filters this margin would be one pixel wide, and it could be filled with zeros. A zero margin 50 pixels wide at the top, bottom, left and right of an image might well be unacceptable. One way around this problem is to ignore those window weights which overlap the image boundary and compute the filtered value from those weights which fall inside the image area. The value of the central weight will need to be modified according to the number of weights that lie inside the image area. This implies that the bandwidth of the filter will vary from the edge of the image until the point at which all the window weights lie inside the image area.

High-pass filters are used routinely in image processing, especially when high-frequency information is the focus of interest. For instance, Ichoku *et al.* (1996) use the 'image-minus-Laplacian' filter as part of a methodology to extract drainage-pattern information from satellite imagery. Krishnamurthy *et al.* (1992) and Nalbant and Alptekin (1995) demonstrate the value of high-frequency enhancement and directional filtering in geological studies. Al-Hinai *et al.* (1991) use a high-pass filter to enhance images of sand dunes in the Saudi Arabian desert.

7.4 EDGE DETECTION

A high-pass filtered image that is added back to the original image is a high-boost filter and, as shown in Figure 7.8(c), the result is a sharpened or de-blurred image. The high-pass filtered image can be used alone, particularly in the study of the location and geographical distribution of 'edges'. An edge is a discontinuity or sharp change in the grey-scale value at a particular pixel point and it may have some interpretation in terms of cultural features, such as roads or field boundaries, or in terms of geological structure or relief. We have already noted that the first difference can be computed for the horizontal, vertical and diagonal directions, and the magnitude and direction of the maximum spatial gradient can also be used. Other methods include the subtraction of a low-pass filtered image from the original (section 7.3.1) or the use of the Roberts gradient. A method not so far described is the *Sobel* non-linear edge operator (Gonzales and Woods, 1992), which is applied to a 3 × 3 window area. The value of this operator for the 3 × 3 window defined by

A	B	C
D	E	F
G	H	I

is given for the pixel underlying the central window weight E by the function

$$S = \sqrt{(X^2 + Y^2)}$$

where

$$X = (C + 2F + I) - (A + 2D + G)$$
$$Y = (A + 2B + C) - (G + 2H + I)$$

This operation can also be considered in terms of two sets of filter weight matrices. X is given by the following weight matrix, which determines horizontal differences in the neighbourhood of the centre pixel:

−1	0	1
−2	0	2
−1	0	1

while Y is given by a weight matrix that involves vertical differences:

−1	−2	−1
0	0	0
1	2	1

Some thought should be given to the implementation of the Sobel operator, for its calculation involves 16 additions/subtractions, two squarings and one square-root operation at each interior pixel point. If sufficient physical memory is available then a table of 256 × 256 bytes can be built up; the row is indexed by X and the column by Y. Element (X, Y) of the table contains the value of the Sobel operator (after division by 4 to convert to the 0–255 range. The value of element (X, Y) is inserted in the table the first time it is calculated. Thereafter the evaluation of the Sobel operator can be achieved by 16 additions and one table reference. The overhead is the need to check the table each time X and Y are calculated. Nevertheless the savings in computing time are considerable. The Sobel operator is implemented in the MIPS display software (Appendix A) and an example of the output from the application of this operator is shown in Figure 7.8(d), which should be compared to the results achieved by the Roberts gradient and Laplacian operators, which are shown in Figures 7.8(a) and (b).

Shaw *et al.* (1982) provide a comparative assessment of these techniques of edge detection. They conclude that first-differencing reveals local rather than regional boundaries, and that increasing the size of a high-pass filter window increases the amount of regional-scale information. The Roberts and Sobel techniques produced a too-intense enhancement of local edges but did not remove the regional patterns.

One of the main uses of edge-detection techniques has been in the enhancement of images for the identification and analysis of geological *lineaments*, which are defined as:

> 'Mappable, simple or composite linear features whose parts are aligned in a rectilinear or slightly curvilinear relationship and which differ distinctly from the pattern of adjacent features and which presumably reflect a subsurface phenomenon.'
> O'Leary *et al.* (1976, p. 1467)

The subsurface phenomena to which the definition refers are presumed to be fault and joint patterns in the underlying rock. However, linear features produced from remotely-sensed images using the techniques described in this section should be interpreted with care. An example of the use of an edge-detection procedure to highlight linear features for geological interpretation is to be found in Moore and Waltz (1983).

Other applications of edge-detection techniques include the determination of the boundaries of homogeneous regions (segmentation) in synthetic aperture radar images in order to segment those images (Quegan and

Wright, 1984). The topic of image segmentation is considered within the context of image understanding in a comprehensive review article by Brady (1982). Gurney (1980) discusses the problem of threshold selection for line-detection algorithms. Chavez and Bauer (1982) consider the choice of filter window size for edge enhancement, and Nasrabadi *et al.* (1984) present a new line-detection algorithm for use with noisy images. Chittenini (1983) and Drewniok (1994) look at the problem of edge and line detection in multidimensional noisy images. Eberlein and Wezka (1975), Haralick (1980) and Peli and Malah (1982) consider the general problem of edge and region analysis. A widely used line detector is the one presented by Vanderbrugg (1976). Algorithms for edge detection and region segmentation are discussed by Farag (1992), Pavlidis (1982) and Pitas (1993). Reviews of edge detection and linear feature extraction methodologies are provided by Budkewitsch *et al.* (1994), Peli and Malah (1982) and Wang (1993). Riazanoff *et al.* (1990) describe ways of thinning (*skeletonising*) lines that have been identified using edge-detection techniques. Such lines are generally defined by first applying a high-pas filter, then thresholding the resulting image on the basis of edge magnitude or strength to give a binary image. The definitive reference is Marr and Hildreth (1980).

7.5 FREQUENCY-DOMAIN FILTERS

The Fourier transform of a two-dimensional digital image is discussed in section 6.6. The Fourier transform of an image, as expressed by the amplitude spectrum, is a breakdown of the image into its frequency or scale components. Filtering can be viewed as a technique for separating these components, hence it would seem logical to consider the use of frequency-domain filters in remote sensing. Such filters operate on the amplitude spectrum of an image and remove, attenuate or amplify the amplitudes in specified wavebands. A simple filter might set the amplitudes of all frequencies less than a selected threshold to zero. After the amplitude spectrum has been converted back to the spatial domain by an inverse Fourier transform, the result would be a low-pass filtered image. Any wavelength or waveband can be operated upon in the frequency domain, but three general categories of filter will be considered here – low-pass, high-pass and band-pass. The terms low-pass and high-pass are defined in section 7.1. A *band-pass* filter removes high and low frequencies and allows an intermediate range of frequencies through, as shown in Figure 7.9. In addition, directional filters can be developed, for the amplitude spectrum contains information

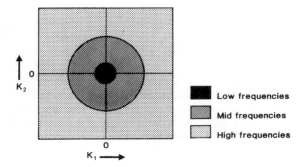

Figure 7.9 Location of low-, mid- and high-frequency components of the two-dimensional amplitude spectrum.

on the frequencies and orientations as well as the amplitudes of the scale components of the image.

The different types of high-, low- and band-pass filters are distinguished on the basis of their transfer functions. The *transfer function* is a graph of frequency against filter weight, though the term filter weight should, in this context, be interpreted as 'proportion of input amplitude that is passed by the filter'.

Figure 7.10 shows a cross-section of a transfer function that passes all frequencies up to f_1 without alteration. Frequencies higher than f_1 are subjected to increasing attenuation until the point f_2. All frequencies beyond f_2 are removed completely. Care should be taken in the design of filter transfer functions. As noted earlier, the amplitude spectrum is multiplied by the two-dimensional filter transfer function (Figure 7.11) and the resulting filtered amplitude spectrum is converted to its spatial-domain representation by means of an inverse Fourier transform. Any sharp edges in the

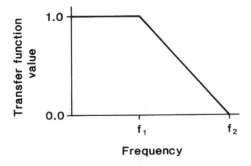

Figure 7.10 Filter transfer function (one-dimensional slice) that passes unchanged all spatial frequencies lower than f_1, attenuates frequencies in the range f_1 to f_2, and suppresses spatial frequencies higher than f_2.

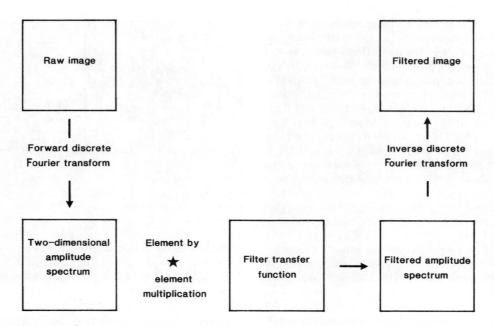

Figure 7.11   Steps in the frequency-domain filtering of a digital image.

filtered amplitude spectrum will convert to a series of concentric circles in the spatial domain, producing a pattern of light and dark rings on the filtered image. This phenomenon is termed *ringing*, for reasons that are self-evident. Gonzales and Woods (1992) discuss this aspect of filter design.

A cross-section through the transfer function of a *low-pass ideal filter* is shown in Figure 7.12(a). The degree of smoothing achieved by the low-pass ideal filter depends on the position of the cut-off frequency, f_0. The lower the value of f_0, the greater the degree of smoothing, as more intermediate and high frequencies are removed

by the filter. The transfer functions for band-pass and high-pass ideal filters are also shown in Figures 7.12(b) and (c). Their implementation in software is not difficult as the cut-off frequencies form circles of radius f_0 and f_1 around the centre of the transform (the DC point). Figures 7.13(a)–(c) illustrate the results of the application of increasingly severe low-pass ideal filters to the Littleport TM band 7 image, using D_0 values of 100, 50 and 5. The degree of smoothing increases as the cut-off frequency decreases. Figure 7.13(c) shows very little real detail but is, nevertheless, one of the frequency components of the Littleport TM image.

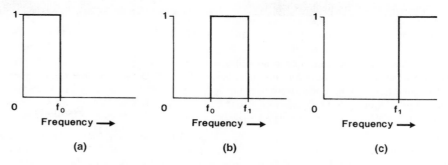

Figure 7.12   Cross-sections of transfer functions for (a) low-pass ideal, (b) band-pass ideal, and (c) high-pass ideal filters.

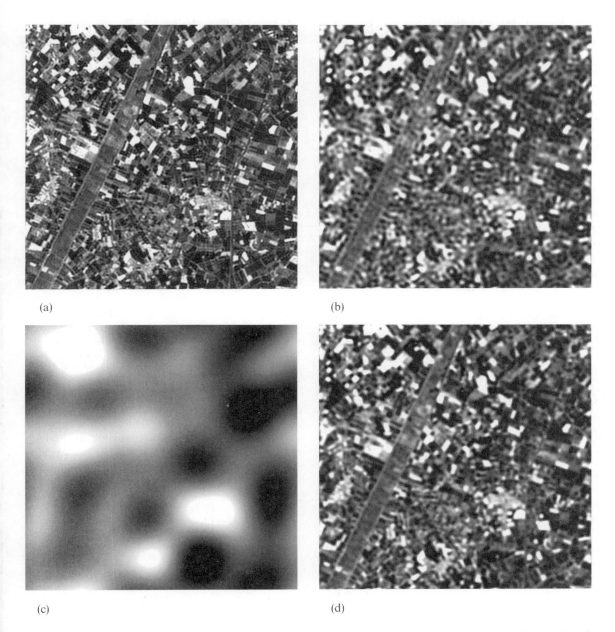

(a)

(b)

(c)

(d)

Figure 7.13 Low-pass ideal frequency-domain filters applied to the Littleport TM band 7 image shown in Figure 1.9(b). The cut-off frequencies are (a) 100, (b) 50 and (c) 5. Part (d) shows the effects of a Butterworth low-pass filter with D_0 equal to 50. Compare the result with that shown in (b). The degree of smoothing is slightly less, as the Butterworth low-pass filter transfer function has a 'tail' of higher frequencies. A 5–95% linear contrast stretch has been applied to the images. The degree of smoothing increases as the cut-off frequency D_0 decreases (see text). Original data © ESA 1994, distributed by Eurimage.

Because of their sharp cut-off features, ideal filters tend to produce a filtered image that can be badly affected by ringing, as discussed earlier. Other filter transfer functions have been designed to reduce the impact of ringing by replacing the sharp edge of the ideal filter with a sloping edge or with a function that decays exponentially from the cut-off frequency. An example of this latter type is the Butterworth filter (Figure 7.14), which is defined by:

$$H(u,v) = \frac{1.0}{1.0 + 0.414[D(u,v)/D_0]^2}$$

$H(u,v)$ is the value of the filter transfer function for frequencies u and v, $D(u,v)$ is the distance from the origin to the point on the amplitude spectrum with coordinates (u,v) and D_0 is the cut-off frequency, as shown in Figure 7.14, which is a plot of the value of the transfer function $H(u, v)$ against frequency (the image size to which this filter was applied was 512 × 512 pixels). This form of the Butterworth filter ensures that $H(u,v) = 1.0\sqrt{2}$ when $D(u,v)$ equals D_0. Gonzales and Woods (1992) describe other forms of filter transfer function. The result of applying a low-pass Butterworth filter with D_0 equal to 50 to the Littleport TM band 7 image (Figure 1.9(b)) is shown in Figure 7.13(d). Compare this result with the output from a low-pass ideal filter (Figure 7.13(b)). It is clear that the Butterworth filter has retained more of the high-frequency information.

Directional filters can be implemented by making use of the fact that the amplitude spectrum contains scale

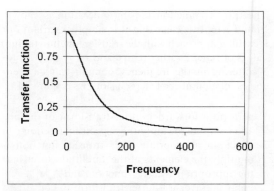

Figure 7.14   Plot of transfer function value against frequency for a Butterworth low-pass filter with a cut-off frequency of 50 for application to an image of size 512 × 512. The transfer function takes the value 0.5 when the frequency is equal to the cut-off frequency.

and orientation information (section 6.6, Figure 6.18(b)). A filter such as the one illustrated in Figure 7.15 removes all those spatial frequencies corresponding to sinusoidal waves oriented in a north–south direction. Such filters have been used in the filtering of binary images of geological fault lines (McCullagh and Davis, 1972).

High-frequency enhancement is accomplished by first defining the region of the amplitude spectrum containing 'high' frequencies and then adding a constant, usually 1.0, to the corresponding amplitudes before carrying out an inverse Fourier transform to convert from the frequency- to the spatial-domain representation.

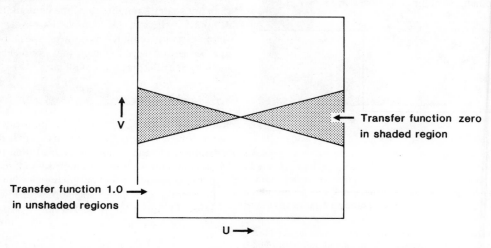

Figure 7.15   Directional filter. The value of the transfer function is equal to zero in the shaded region and to 1.0 in the unshaded area. All spatial frequency components of the image with a north–south orientation in the image (spatial) domain are suppressed.

The transfer function for this operation is shown in Figure 7.16. It is clear that it is simply a variant of the ideal filter approach with the transfer function taking on values of 1 and 2 rather than 0 and 1.

Filtering in the frequency domain can be seen to consist of a number of steps, as follows (Figure 7.11):

1. perform a forward Fourier transform of the image and compute the amplitude spectrum (section 6.6),
2. select an appropriate filter transfer function and multiply the elements of the amplitude spectrum by the appropriate transfer function, and
3. apply an inverse Fourier transform to convert back to spatial-domain representation.

Although frequency-domain methods are far more flexible that the spatial-domain filtering techniques, the cost of computing the forward and inverse Fourier transforms is a limiting factor. As noted in section 6.6, the two-dimensional Fourier transform requires the transposition of two large matrices holding the intermediate sine and cosine coefficients. This is a time-consuming operation particularly if a computer with a small random-access memory (Chapter 3) is used. The cost factor has therefore limited the use of frequency-based methods to specialist applications. However, the more widespread availability of computers with access to several megabytes of physical memory and the availability of special hardware units termed array processors has meant that the computational cost of performing frequency-domain filtering has fallen considerably in recent years and will no doubt continue to fall as developments in computing take place.

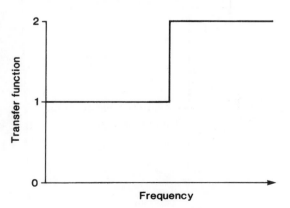

Figure 7.16 Transfer function for high-frequency enhancement. Low spatial frequencies are unaltered, but high spatial frequencies, representing detail, are doubled. The effect is to sharpen the image by enhancing detail.

Examples of the use of frequency-domain filtering include de Souza Filho *et al.* (1996), who describe a method to remove noise in JERS-1 imagery. Lei *et al.* (1996) also use frequency-domain filtering methods to clean up MOMS-02 images. Aeromagnetic data are analysed using frequency-domain techniques by Hussein *et al.* (1996).

A program, `fourier`, to carry out the filtering operations described in this section is provided on the accompanying CD, and details of its operation are given in Appendix B. The range of filters implemented by this program is wider than those described in this section, and interested readers should consult a more advanced text, such as Gonzales and Woods (1992) and Pitas (1993) for a more detailed account.

7.6   SUMMARY

Filtering of digital images is used to remove, reduce or amplify specific frequency components of an image. The most commonly used filters operate in the spatial domain and can be divided into low-pass or smoothing filters and high-pass or sharpening filters. Uses of smoothing filters include the suppression of noise and other unwanted effects, such as the six-line banding phenomenon that afflicts Landsat MSS images. Sharpening filters are used to improve the visual interpretability of the image by, for example, de-blurring the signal. Edge and line detection is seen as an extension of the technique of image sharpening. Filtering in the frequency domain is achieved via the application of the principles of the Fourier transform, discussed in section 6.6. While these methods are inherently more flexible than are spatial-domain filters, the computational cost of applying them is considerable. However, recent developments in computer hardware, especially memory and processor speed, mean that frequency-domain methods may become more popular.

7.7   QUESTIONS

1. Give extended definitions of the following terms: filter (low-pass, band-pass, high-pass), first derivative, median, average, speckle filter, box filter, frequency-domain filter, Butterworth high-pass filter.
2. Explain the origin of the horizontal banding phenomenon that is commonly seen over low-radiance areas of Landsat MSS and TM images. Compare the effects of removing this banding using (a) low-pass filters (spatial- and frequency-domain) and (b) statistical methods as described in section 4.2.2.
3. Compare and contrast the effects of low-pass filtering

of an image using the median and the moving-average filters.

4. Explain the operation of a low-pass 'moving window' filter. Give a worked example using a 3×3 median filter and a 15×15 set of fictitious data. What happens at the image margins?

5. Using the MIPS display program (Appendix A) examine the effects of (a) a 5×5 median filter, and (b) the 'image-minus-Laplacian' operator, using one of the test images contained on the CD. Write an explanation of the ways in which these operations modify the appearance of the original image, giving some example applications in which these operations might be used.

6. Describe the effects of a low-pass Butterworth filter (program `fourier` – Appendix B) on one of the test images contained on the CD. Use D_0 values of 25, 50 and 75.

7. What is meant by 'high-frequency enhancement'? How would you carry out this operation (give examples of both spatial- and frequency-domain methods).

8

Classification

8.1 INTRODUCTION

The process of classification has two stages. The first is the recognition of categories of real-world objects. In the context of remote sensing of the land surface these categories could include, for example, woodlands, water bodies, grassland and other land cover types, depending on the geographical scale and nature of the study. The second stage in the classification process is the labelling of the entities (normally pixels) to be classified. In digital image classification these labels are numerical, so that a pixel that is recognised as belonging to the class 'water' may be given the label '1', 'woodland' may be labelled '2', and so on. The process of image classification requires the user to perform the following steps:

1. determine *a priori* the number and nature of the categories in terms of which the land cover is to be described, and
2. assign numerical labels to the pixels on the basis of their properties using a decision-making procedure, usually termed a classification rule or a decision rule.

Sometimes these steps are called *classification* and *identification* (or *labelling*), respectively.

In contrast, the process of *clustering* does not require the definition of a set of categories in terms of which the land surface is to be described. Clustering is a kind of exploratory procedure, the aim of which is to determine the number (but not initially the identity) of distinct land cover categories present in the area covered by the image, and to allocate pixels to these categories. Identification of the clusters or categories in terms of the nature of the land cover types is a separate stage that follows the clustering procedure. Several clusters may correspond to a single land cover type. Methods of relating the results of clustering to real-world categories are described by Lark (1995), while Gong and Howarth (1990) discuss factors influencing land cover classification by remote sensing.

These two approaches to pixel labelling are known in the remote sensing literature as *supervised* and *unsupervised* classification procedures, respectively. They can be used to segment an image into regions with similar attributes. Although land cover classification is used above as an example, similar procedures can be applied to clouds, water bodies and other objects present in the image. In all cases, however, the properties of the pixel to be classified are used to label that pixel. In the simplest case, a pixel is characterised by a vector whose elements are its grey levels in each spectral band. This vector represents the spectral properties of that pixel.

A set of grey-scale values for a single pixel measured in a number of spectral bands is known as a *pattern*. The spectral bands (such as the seven Landsat TM bands) or other, derived, properties of the pixel (such as context and texture, which are described in later sections of this chapter) that define the pattern are called *features*. The classification process may also include features such as land-surface elevation or soil type that are not derived from the image. A pattern is thus a set of measurements on the chosen features for the individual that is to be classified. The classification process may therefore be considered as a form of pattern recognition, that is, the identification of the pattern associated with each pixel position in an image in terms of the characteristics of the objects or materials that are present at the corresponding point on the Earth's surface.

Pattern recognition methods have found widespread use in fields other than Earth observation by remote sensing; for example, military applications include the identification of approaching aircraft and the detection of targets for cruise missiles. Robot or computer vision involves the use of mathematical descriptions of objects 'seen' by a television camera representing the robot eye and the comparison of these mathematical descriptions with patterns describing objects in the real world. In every case, the crucial steps are: (i) selection of a set of features that best describe the pattern and (ii) choice of a suitable method for the comparison of the pattern describing the object to be classified and the target patterns. In remote-sensing applications it is usual to include a third stage, that of assessing the degree of accuracy of the allocation process.

A geometrical model of the classification or pattern recognition process is often helpful in understanding the procedures involved; this topic is dealt with in section 8.2. The more common methods of unsupervised and supervised classification are covered in sections 8.3 and 8.4. Supervised methods include those based on statistical concepts and those based on artificial neural networks. The methods described in these sections have generally been used on spectral data alone (that is, on the individual vectors of pixel values). This approach is called 'per-point' or 'per-pixel' classification based on spectral data. The addition of features that are derived from the image data has been shown to improve the classification in many cases. *Texture* is a measure of the homogeneity of the neighbourhood of a pixel, and is widely used in the interpretation of aerial photographs. Objects on such photographs are recognised visually not just by their grey-scale value (tone) alone but also by the variability of the tonal patterns in the region or *neighbourhood* that surrounds them. Texture is described in section 8.7.1.

Visual analysis of a photographic image often involves assessment of the context of an object as well as its tone and texture. *Context* is the relationship of an object to other, nearby, objects. Some objects are not expected to occur in a certain context; for example, jungles are not observed in polar regions. Conversely glacier ice is unlikely to be widespread in southern Algeria. In the same vein, a pixel labelled 'wheat' may be judged to be incorrectly identified if it is surrounded by pixels labelled 'snow'. The decision regarding the acceptability of the label might be made in terms of the pixel's context rather than on the basis of its spectral reflectance values alone. Contextual methods are not yet in widespread use, though they are the subject of on-going research. They are described in section 8.8.

The number of spectral bands used by satellite and airborne sensors ranges from the single band of the SPOT HRV in panchromatic to several hundred bands provided by hyperspectral remote-sensing instruments. The methods considered in this chapter are, however, most effective when applied to multispectral image data in which the number of spectral bands is less than 12 or so. The addition of other 'bands' or features such as texture descriptors or external data such as land-surface elevation or slope derived from a digital elevation model can increase the number of features available for

classification. The effect of increasing the number of features on which a classification procedure is based is to increase the computing time requirements but not necessarily the accuracy of the classification. Some form of feature selection process to allow a trade-off between classification accuracy and the number of features is therefore desirable (section 8.9). The assessment of the accuracy of a thematic map produced from remotely-sensed data is considered in section 8.10.

8.2 GEOMETRICAL BASIS OF CLASSIFICATION

One of the easiest ways to perceive the distribution of values measured on two features is to plot one feature against the other. Figure 8.1 is a plot of catchment area against stream discharge for a hypothetical set of river basins. It is apparent that there are two basic types of river basin. The first type has a small catchment area and a low discharge whereas the second type has a large area and a high discharge. This example might appear trivial but it demonstrates two fundamental ideas. The first is the representation of the selected features of the objects of interest (in this case the catchment area and discharge) by the axes of a Euclidean space (the *feature space*), and the second is the use of measurements of closeness or resemblance of points (representing river basins) in this space as the basis of decisions to classify particular river basins as large-area/high-discharge or small-basin/low-discharge. The axes of the graph in Figure 8.1 are the *x*- and *y*-axes of a Cartesian coordinate system. They are orthogonal (at right angles) and

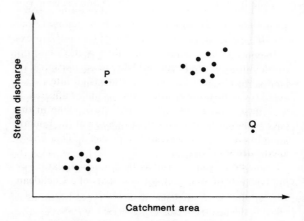

Figure 8.1 Plot of catchment area (*x*) against stream discharge (*y*). Two distinct groups of river basins are apparent, but basins P and Q are not seen as members of either group.

define a two-dimensional Euclidean space. Variations in basin area are shown by changes in position along the *x*-axis and variations in river discharge are shown by position along the *y*-axis of this space. Thus, the position of a point in this two-dimensional space is directly related to the magnitude of the values of the two features (area and discharge) measured on the particular drainage basin represented by that point.

The eye and brain combine to provide what is sometimes disparagingly called a 'visual' interpretation of a pattern of points such as that depicted in Figure 8.1. If we analyse what the eye–brain combination does when faced with a distribution such as that shown in Figure 8.1 we realise that a 'visual' interpretation is not necessarily a simple one, though it might be intuitive. The presence of two clusters of points is recognised by the existence of two regions of feature space that have a relatively dense distribution of points, with more or less empty regions between them. A point is seen as being in cluster 1 if it is closer to the centre of cluster 1 than it is to the centre of cluster 2. Distance in feature space is being used as a measure of similarity (more correctly 'dissimilarity', as the greater the inter-point distance the less the similarity). Points such as those labelled P and Q in Figure 8.1 are not allocated to either cluster, as their distance from the centres of the two clusters is too great. We can also visually recognise the compactness of a cluster using the degree of scatter of points (representing members of the cluster) around the cluster centre. We can also estimate the degree of separation of the two clusters by looking at the distance between their centres and the scatter of points around those centres. It seems as though a visual estimate of distance (closeness and separation) in a two-dimensional Euclidean space is used to make sense of the distribution of points shown in the diagram. However, we must be careful to note that the scale on which the numerical values are expressed is very important. If the values of the *y*-coordinates of the points in Figure 8.1 were to be multiplied or divided by a scaling factor, then our visual interpretation of the inter-point relationships would be affected. If we wished to generalise, we could draw a line in the space between the two clusters to represent the boundary between the two kinds of river basin. This line is called a *decision boundary*.

The same concepts – the association of a feature or characteristic of an object with one axis of a Euclidean space and the use of inter-point distance as the basis of a decision rule – can easily be extended to three dimensions. Figure 8.2 shows the same hypothetical set of river basins, but this time they are represented in terms of altitude above sea-level as well as area and discharge.

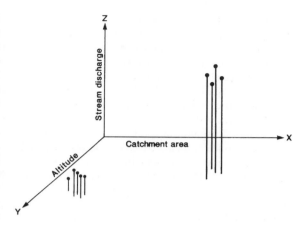

Figure 8.2 Plot of catchment area (*x*), altitude above sea-level (*y*) and stream discharge (*z*). Two groups of river basins are identifiable in this three-dimensional feature space.

Two groupings are evident as before, though it is now clear that the small basins with low discharge are located at higher altitudes than the large basins with high discharge. Again, the distance of each point from the centres of the two clusters can be used as the basis of an allocation or decision rule but, in this three-dimensional case, the decision boundary is a plane rather than a line.

Many people find great difficulty in extending the concept of inter-point distance to situations in which the objects of interest have more than three characteristics. There is no need to try to visualise what a four-, five- or even seven-dimensional version of Figure 8.2 would look like; just consider how straight-line distance is measured in one-, two- and three-dimensional Euclidean spaces in which *x*, *y* and *z* represent the axes:

$$d_{12} = \sqrt{(x_1 - x_2)^2} \qquad \text{(one-dimensional)}$$

$$d_{12} = \sqrt{(x_1 - x_2)^2 + (y_1 - y_2)^2} \quad \text{(two-dimensional)}$$

$$d_{12} = \sqrt{(x_1 - x_2)^2 + (y_1 - y_2)^2 + (z_1 - z_2)^2}$$
$$\text{(three-dimensional)}$$

The squared differences on each axis are added and the square root of the sum is the Euclidean distance from point 1 to point 2. This is a simple application of the theorem of Pythagoras (Figure 8.3). If we replace the terms x_j, y_j and z_j (where j is an index denoting the individual point) by a single term x_{ij}, where i is the axis number and j the identification of the particular point, then the three expressions above can be seen to be

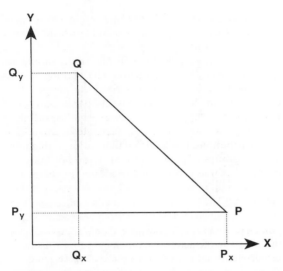

Figure 8.3 Pythagoras's theorem is used to calculate the distance QP in a two-dimensional feature space, given the coordinates (measurements) of the two points on features *x* and *y*. The distance PQ is given by:

$$PQ = \sqrt{(Q_x - P_x)^2 + (Q_y - P_y)^2} \,.$$

particular instances of the general case, in which the distance from point *a* to point *b* is:

$$d_{ab} = \sqrt{\sum_{i=1}^{p} (x_{ia} - x_{ib})^2}$$

where d_{ab} is the Euclidean distance between point *a* and point *b* measured on *p* axes or features. There is no reason why *p* should not be any positive integer value – the algebraic formula will work equally well for *p* = 4 as for *p* = 2 despite the fact that most people cannot visualise the *p* > 3 case. The geometrical model that has been introduced in this section is thus useful for the appreciation of two of the fundamental ideas underlying the procedure of automated classification, but the algebraic equivalent is preferable in real applications because (i) it can be extended to beyond three dimensions and (ii) the algebraic formulae are capable of being used in computer programs.

It may help to make things clearer if an example relating to a remote-sensing application is given at this point. The discussion of the spectral response of Earth-surface materials in section 1.3.2 showed that deep, clear water bodies have a very low reflectance in the near-

infrared waveband, and their reflectance in the visible red waveband is not much higher. Vigorous vegetation, on the other hand, reflects strongly in the near-infrared waveband whereas its reflectance in the visible red band is relatively low. The red and near-infrared wavebands might therefore be selected as the features on which the classification is to be based. Estimates can be made of the pixel grey-scale values in each spectral band for sample areas on the image that can be identified *a priori* as 'water' and 'vigorous vegetation' on the basis of observations made in the field, or from maps or aerial photographs, and these estimates used to fix the mean position of the points 'VV' and 'W' in Figure 8.4. The two axes of the figure represent near-infrared and red reflectance, respectively, and the mean position is found by finding the average red reflectance (*y*-coordinate) and near-infrared reflectance (*x*-coordinate) of the sample values for each of the two categories.

The points labelled *a*, *b*, *c* and *d* in Figure 8.4 represent unclassified pixels. We might choose a decision rule such as 'points will be labelled as members of the group whose centre is closest in feature space to the point concerned'. The distance formula given above could then be used on the points taken one at a time to give the Euclidean straight-line distance from that point (representing a pattern associated with a particular pixel) to each of the centres 'VV' and 'W'. Those points which are closer to 'VV' are labelled 'vigorous vegetation' (or '1')

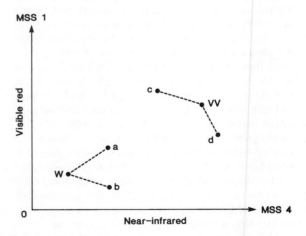

Figure 8.4 Plot of visible red and near-infrared spectral bands. Point W represents average reflectance for water in these wavebands. Point VV represents average reflectance for vigorous vegetation. Points *a* and *b* are more similar (closer) to point W than they are to point VV and are therefore labelled as 'water', whereas points *c* and *d* are labelled 'vigorous vegetation'.

while those closer to 'W' are labelled 'water' (or '2'). If this procedure is applied to a two-band image, as shown in Figure 8.4, the end-product is a matrix of the same dimensions as the image being classified. The elements of this new matrix are numerical pixel labels, which in this example are either '1's or '2's. If the colour green is associated with the value '1' and the colour blue with the value '2' then a colour-coded thematic map of the image area would result, in which water would be blue and vigorous vegetation would be green, assuming, of course, that the classification procedure was a reliable one. The position of the decision boundary is given by the set of points that are equidistant from both 'VV' and 'W'.

It will be shown later that the decision rule used in this example ('allocate an unknown pixel to the closest class centroid') is not the only one that can be applied. However, the process of image classification can be understood more clearly if the geometrical basis of the example is clearly understood.

8.3   UNSUPERVISED CLASSIFICATION

It is sometimes the case that insufficient observational or documentary evidence of the nature of the land cover types is available for the geographical area covered by a remotely-sensed image. In these circumstances, it is not possible to estimate the mean centres of the classes, as described above. Even the number of such classes might be unknown. In this situation we can only 'fish' in the pond of data and hope to come up with a suitable catch. In effect, the automatic classification procedure is left largely to its own devices – hence the term *unsupervised clustering*. The relationship between the labels allocated by the classifier to the pixels making up the multispectral image and the land cover types existing in the area covered by the image is determined after the unsupervised classification has been carried out. Identification of the spectral classes picked out by the classifier in terms of information classes existing on the ground is achieved using whatever information is available to the analyst. The term *exploratory* might be used in preference to *unsupervised* because a second situation in which this type of analysis might be used can be envisaged. The analyst may well have considerable ground data at his or her disposal but may not be certain (i) whether the spectral classes he or she proposes to use can, in fact, be discriminated, and/or (ii) whether the proposed spectral classes are 'pure' or 'mixed'. As we will see in section 8.4, some methods of supervised classification require that the frequency distribution of points belonging to a single spectral class in the p-dimensional feature space

has a single mode or peak. In either case, exploratory or unsupervised methods could be used to provide answers to these questions.

An exploratory classification algorithm should require little, if any, user interaction. The workings of such a technique will now be described by means of an example. Figure 8.5 shows two well-separated groups of points in a two-dimensional feature space. The members of each group are drawn from separate bivariate-normal distributions. It is assumed that we know that there are two groups of points but that we do not know the positions of the centres of the groups in the feature space. Points '1_0' and '2_0' represent a first guess at these positions. The 'shortest-distance-to-centre' decision rule, as described earlier, is used to label each unknown point (shown by the dots) with a '1' or a '2' depending on the relative Euclidean distance of the point from the respective initial cluster centres, labelled '1_0' and '2_0'. Thus, the (squared) Euclidean distances to cluster centres 1 and 2 (d_{q1}^2 and d_{q2}^2) are computed for each point q, and q is allocated the label '1' if d_{q1}^2 is less than d_{q2}^2 or the label '2' if d_{q2}^2 is less than d_{q1}^2. If the two squared distances are equal, then the point is arbitrarily allocated the label '1'.

At the end of this labelling sequence the mean of the

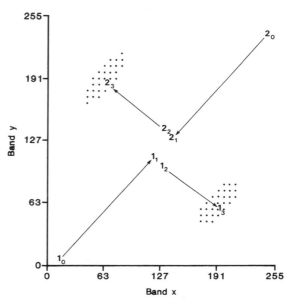

Figure 8.5   Iterative calculation of centroid positions for two well-separated groups of points. Each point represents a pixel in a two-dimensional feature space. The axes (features) are band x and band y of a multispectral image set.

---

values of all points labelled '1' is computed for each of the axes of the feature space, and the same is done for all points labelled '2' to give the coordinates in the feature space of the centroids of the two groups of points. These new centroids are shown in the diagram as '1_1' and '2_1'. The points are relabelled again using the 'shortest-distance-to-mean' decision rule, based this time on the new positions of the centroids. Again, the position of the centroid of the points labelled '1' at this second iteration is computed and is shown as '1_2'. The centroid of the set of points labelled '2' is found in a similar fashion and is shown as '2_2'. Distances from all points to these new centres are calculated and another pair of new centroids are found ('1_3' and '2_3'). These centroids are now at the centres of the two groups of points that were artificially generated for this example, and relabelling of the points does not cause any change in the position of the centroids, hence this position is taken as the final one.

To show that the technique still works even when the groups of points are not so well separated, as in Figure 8.5, a second pair of groups of points was generated. This time the coordinates of the points in a two-dimensional feature space were computed by adding random amounts to a pre-selected pair of centre points to give the distribution shown in Figure 8.6. The start positions of the migrating centroids were selected randomly and are shown on the figure as '1_0' and '2_0' respectively. The same relabelling and recalculation process as that used in the previous example was carried out and the centroids again migrate towards the true centres of the point sets, as shown in Figure 8.6. However, this time the decision boundary is not so clear-cut and there may be some doubt about the class membership (label) of points that are close to the decision boundary.

Since the relabelling procedure involves only the rank orders of the distances between point and centroids, the squared Euclidean distances can be used, for the squares of a set of distance measures have the same rank order as the original distances. Also, it follows from the fact that the squared Euclidean distances are computed algebraically that the feature space can be multidimensional. The procedures in the multidimensional case involve only the additional summations of the squared differences on the feature axes as shown in section 8.2; no other change is needed.

The procedure described above is called *k-means clustering*. The program `kmeans` that is contained on the accompanying CD (Appendix B) carries out *k*-means clustering. The program accepts up to 100 cluster centres. In unsupervised mode the coordinates of these centres are generated randomly, and the iterative allocation procedure described above continues until the labels of fewer than `minrel` pixels are changed at a given iteration, where `minrel` is a value specified by the user.

In the examples used so far in this section, it has been assumed that the number of clusters of points is known in advance. More elaborate schemes are needed if this is not the case. The basic assumption on which these schemes are based is that the clusters present in the data are 'compact' (that is, the points associated with each cluster are tightly grouped around the cluster centre and thus occupy a spheroidal region of feature space). A measure of the compactness of a cluster can be taken as the set of standard deviations for the cluster measured separately for each feature (Figure 8.7). If any of these standard deviations for a particular cluster is larger than a user-specified value, then that cluster is considered to be elongated in the direction of the axis concerned.

A second assumption is that the clusters are well separated in that their inter-centre distances are greater than a pre-selected threshold. If the *p*-space coordinates of a trial number of cluster centres are generated randomly (call the number of centres k_0) then the closest-distance-to-centre decision rule can be used to label the pixels which, as before, are represented by points in feature space. Once the pixels have been labelled then (i) the standard deviation for each feature axis is computed for each of the non-null k_0 clusters and (ii) the Euclidean distances between the k_0 cluster centres are found. Any

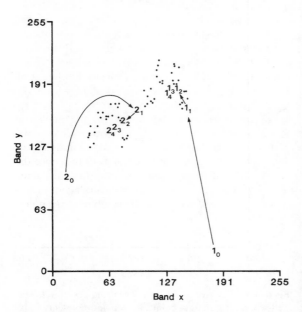

Figure 8.6 Iterative calculation of centroid positions for two diffuse groups of pixels. See text for discussion.

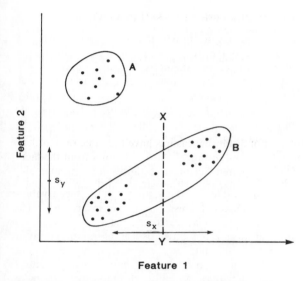

Feature 1

Figure 8.7 Compact cluster (A) and elongated cluster (B) in a two-dimensional feature space. Cluster B has standard deviations s_x and s_y on the two features. Since s_x is larger than a user-specified threshold value, cluster B is split along the line XY.

cluster that has one or more 'large' standard deviations is split in half along a line perpendicular to the feature axis concerned (Figure 8.7) while any clusters that are closer together than a second user-supplied threshold (in terms of their inter-centre distance) are amalgamated. The application of this split and merge routine results in a number k_1 of new cluster centre positions, and the pixels are relabelled with respect to these k_1 centroids. The split and merge function is again applied, and a new number k_2 of centres is found. This process iterates until no clusters are split or merged and no pixels change cluster. This gives the number k_p of clusters and the positions of their centroids in the feature space. At each cycle, any cluster centres that are associated with less than a pre-specified number of pixels are eliminated. The corresponding pixels are either ignored as unclassifiable in subsequent iterations or else are put back into the pool for relabelling at the next iteration.

This split and merge procedure forms the ISODATA algorithm (ISODATA is an acronym derived from Iterative Self-Organising Data Analysis Technique, with a terminal 'A' added for aesthetic reasons). It can be surprisingly voracious in terms of computer time if the data are not cleanly structured (i.e. do not possess clearly separated and spheroidal clusters). Unless precautions are taken it can easily enter an endless loop when clus-

ters that are split at iteration i are merged again at iteration $i + 1$, then are split at iteration $i + 2$. Little general guidance can be given on the choice of the number of initial cluster centres k_0 or the elongation and closeness threshold values to be used. It is often sensible to experiment with small subsets of the image to be classified to get a 'feel' for the data. This algorithm has been in use for many years. A full description is given by Tou and Gonzales (1974) while the standard reference is Duda and Hart (1973). A more recent account is Bow (1992), who provides a flow-chart of the procedure as well as a lengthy example.

Appendix B contains details of a program (isodata) to carry out the procedures described above. The coordinates of the initial centres in feature space can be estimated by the user or they can be generated randomly. Input includes the 'desired' number of clusters, which is usually fewer than the initial number of clusters. The stopping criteria are (i) when the average inter-centre Euclidean distance falls below a user-specified threshold or (ii) when the change in average inter-centre Euclidean distance between iteration i and iteration $i - 1$ is less than a user-specified amount.

A second, less complicated, method of estimating the number of separable clusters in a data set involves a modification of the k-means approach to allow merging of clusters. An overestimate of the expected number of cluster centres (k_{max}) is provided by the user, and pixels are labelled as before using the closest-distance-to-centre decision rule. Once the pixels have been labelled the centroids of the clusters are calculated and the relabelling procedure is employed to find stable positions for the centroids. Once such a position has been located, a measure of the compactness of cluster i is found by summing the squared Euclidean distances from the pixels belonging to cluster i to the centroid of cluster i. The square root of this sum divided by the number of points in the cluster gives the root-mean-square deviation for that cluster. It is now necessary to find the pair of clusters that can be combined so as (i) to reduce the number of cluster centres by one, and (ii) at the same time cause the least increase in the overall root-mean-square deviation. This is done by computing the quantity

$$\frac{n_i n_j}{n_i + n_j} \sum_{k=1}^{p} (y_{ik} - y_{jk})^2 \quad (i = 2, k; j = 1, i - 1)$$

for every pair of cluster centres (y_i and y_j) where p is the number of dimensions in the feature space and n_i is the number of pixels assigned to cluster i. If clusters $i = r$

and $j = s$ give the lowest value of this quantity then the centroids of clusters r and s are combined by a weighted-average procedure, the weights being the numbers of pixels in clusters r and s. If the number of clusters is still greater than or equal to a user-supplied minimum value k_{min} then the relabelling procedure is employed to reallocate the pixels to the reduced number of centres and the overall root-mean-square deviation is computed. The procedure is repeated for every integral value of k (the number of clusters) between k_{max} and k_{min} or until the analyst terminates the procedure after visually inspecting the classified image. As with the ISODATA algorithm, empty clusters can be thrown away at each iteration. Mather (1976) provides a Fortran program to implement this procedure.

The result of an unsupervised classification is a set of labelled pixels, the labels being the numerical identifiers of the classes. The label values run from 1 to the number of classes (k) picked out by the procedure. The class numbered zero (or $k + 1$) can be used as the label for uncategorised pixels. The image made up of the labels of the pixels can be displayed by assigning a colour or a grey tone to each label. In the MIPS display program, described in Appendix A, these colours are defined in a lookup table (LUT). You can create a lookup table manually using program `make_lut`, which is described in Appendix B. This lookup table can be read by the MIPS display program using the `Utilities|Read-LUT` menu item. From a study of the geographical location of the pixels in each class, an attempt is normally made to relate the spectral classes (groups of similar pixels) to corresponding information classes (categories of ground cover). Alternatively, a method of hierarchical classification can be used to produce a linkage tree or dendrogram from the centroids of the unsupervised classes, and this linkage tree can be used to determine which spectral classes might best be combined. The relationship between spectral classes and information classes is likely to be tenuous unless external information can be used for, as noted earlier, unsupervised techniques of classification are used when little or no detailed information exists concerning the distribution of ground cover types. An initial unsupervised classification can, however, be used as a preliminary step in refining knowledge of the spectral classes present in the image so that a subsequent supervised classification can be carried out more efficiently. The classes identified by the unsupervised analysis could, for example, form the basis for the selection of training samples for use in a supervised technique of classification (section 8.4). General references covering the material presented above are Anderberg (1973) and Hartigan (1975).

8.4   SUPERVISED CLASSIFICATION

Supervised classification methods are based on external knowledge of the area shown in the image. Unlike some of the unsupervised methods discussed in section 8.3, supervised methods require some input from the user before the chosen algorithm is applied. This input may be derived from fieldwork, air photo analysis, reports, or the study of appropriate maps of the relevant area. In the main, supervised methods are implemented using either statistical or neural algorithms. Statistical algorithms use parameters derived from sample data in the form of training classes, such as the minimum and maximum values on the features, or the mean values of the individual clusters, or the mean and variance–covariance matrices for each of the classes. Neural methods do not rely on statistical information derived from the sample data but are trained on the sample data directly. This is an important characteristic of neural methods of pattern recognition, for these methods make no assumptions concerning the frequency distribution of the data. In contrast, statistical methods such as the maximum-likelihood procedure are based on the assumption that the frequency distribution for each class is multivariate-normal in form. Thus, statistical methods are said to be *parametric* (because they use statistical parameters derived from training data) whereas neural methods are *non-parametric*. The importance of this statement lies in the fact that additional non-remotely-sensed data such as slope angle or soil type can more easily be incorporated into a classification using a non-parametric method.

Since all methods of supervised classification use training data samples it is logical to consider the characterisation of training data in the next section.

8.4.1   Training samples

Supervised classification methods require prior knowledge of the number and, in the case of statistical classifiers, certain aspects of the statistical nature of the information classes with which the pixels making up an image are to be identified. The statistical characteristics of the classes that are to be estimated from the training sample pixels depend on which method of supervised classification is used. The parallelepiped method requires estimates of the extreme values on each feature for each class, while the k-means or centroid method needs estimates of the multivariate means of the classes. The most elaborate statistical method discussed in this book, the maximum likelihood algorithm, requires estimates of the mean vector and variance–covariance matrix of each class. Neural classifiers operate directly on the training data, but are strongly influenced by misidentification of training samples as well as by the size of the training data sets. Misidentification of an individual training sample pixel may not have much influence on a statistical classifier, but the impact on a neural classifier could be considerable. The material contained in this section must be interpreted in the light of whichever of these methods is used.

It is of crucial importance to ensure that the *a priori* knowledge of the number and statistical characteristics of the classes is reliable. The accuracy of a supervised classification analysis will depend upon two factors: (i) the representativeness of the estimates of the number and statistical nature of the spectral classes present in the image data, and (ii) the degree of departure from the assumptions upon which the classification technique is based. These assumptions vary from one technique to another; in general, the most sophisticated techniques have the most stringent assumptions. These assumptions will be mentioned in the following subsections. In this section we concentrate on the estimation of statistical properties, in particular the mean and variance of each spectral band and the covariances of all pairs of spectral bands. These methods can be used to locate aberrant pixels that can then be eliminated or down-weighted.

The validity of statistical estimates depends upon two factors – the size and the representativeness of the sample. Sample size is not simply a matter of 'the bigger the better', for cost is, or should be, an important consideration. Sample size is related to the number of variables (spectral bands in this case) whose statistical properties are to be estimated, and the number of those statistical properties. In the case of a single variable and the estimation of a single property (such as the mean or the variance) a sample size of 30 is usually held to be sufficient. For the multivariate case the size should be at least $30p$ pixels per class where p is the number of features (spectral bands), and preferably more. Dobbertin and Biging (1996) show that classification accuracy improves as sample size increases. However, neural-based classifiers appear to work better than statistical classifiers for small training sets, though better results were achieved when training set size was proportional to class size and variability (Blamire, 1996; Foody *et al.*, 1995; Hepner *et al.*, 1990).

Training samples are normally located by fieldwork or from air photograph or map interpretation, and their positions on the image found either by visual inspection or by carrying out a geometric correction on the image to be classified. It is not necessary to carry out the procedure of geometric transformation on the full image set to be classified, unless the resulting classified image is to be input to a GIS. All that is required is the set of transform equations that will convert a map coordinate pair to the corresponding image column and row coordinates (section 4.3.1). Using these equations, the location on the image of a training sample whose map coordinates are known is a relatively simple matter. Geometric correction is best carried out on a single-band classified image (in which the pixel 'values' are the labels of the classes to which the pixels have been allocated) rather than on the p images to be classified. Not only is this less demanding of computer resources, but it ensures that the radiometric (pixel) values are not distorted by any resampling procedure (Khan *et al.*, 1995). If external data are used in the classification (section 8.7.2) then geometric correction of image and non-image data to a common reference system is a necessary prerequisite.

The minimum sample size specified in the preceding paragraphs is valid only if the individual members of the training sample are independent, as would be the case if balls were drawn randomly from a bag. Generally, adjacent pixels are not independent – if you were told that pixel a was identified as 'forest' you might be reasonably confident that its neighbour, pixel b, would also be a member of the class 'forest'. If a and b were statistically independent there would be an equal chance that b was a member of any other of the candidate classes, irrespective of the class to which a was allocated. The correlation between nearby points in an image is called *spatial autocorrelation*.

It follows that the number n of pixels in a sample is an overestimate of the number of fully independent pieces of information in the sample if the pixels making up the training sample are autocorrelated. The consequence is that the use of the standard statistical formulae to estimate the means and variances of each feature and the correlations among the features will give biased results. Correlations between spectral bands derived from spatially autocorrelated training data will, in fact, be underestimated and the accuracy of the classification will be reduced as a result. Campbell (1981) found that variance–covariance matrices (the unstandardised analogue of the correlation matrix) were considerably greater when computed from randomly selected pixels within a class rather than from contiguous blocks of pixels from the same class (Table 8.1). Figure 8.8 shows the ellipsoids defined by the mean vectors and variance–covariance matrices for Landsat MSS bands 6 and 7 of Campbell's (1981) contiguous and random data; the locations of the centres of the two ellipses are

Table 8.1 Variance–covariance matrices for four Landsat MSS bands obtained from a random sample (upper figure) and a contiguous sample (in parentheses) drawn from the same data. Source: table 7 of Campbell (1981).

1.09			
(0.40)			
1.21	3.50		
(0.21)	(1.01)		
−1.00	−1.65	23.15	
(−0.78)	(−0.19)	(14.00)	
−0.51	−1.85	12.73	11.58
(−0.43)	(−1.10)	(9.80)	(8.92)

not too far apart but their orientation, size and shape differ somewhat. Campbell (1981) suggests taking random pixels from within a training area rather than using contiguous blocks, while Labovitz and Matsuoko (1984) prefer a systematic sampling scheme with the

spacing between the samples being determined by the degree of positive spatial autocorrelation in the data. Dobbertin and Biging (1996) report that classification accuracy is reduced when images show a high level of spatial autocorrelation. Derived features such as texture might be expected to display a higher degree of spatial autocorrelation than the individual pixel values in the raw images, because such measures are often calculated from overlapping windows. Better results were obtained from randomly selected training pixels than from contiguous blocks of training pixels, a conclusion also reached by Gong *et al.* (1996) and Wilson (1992). The variances of the training samples were also higher when individual random training pixels were used rather than contiguous pixel blocks. The method of automatically collecting training samples, described by Bolstad and Lillesand (1992), may well generate training data that are highly autocorrelated.

The degree of autocorrelation will depend upon (i) the natural association between adjacent pixels, (ii) the pixel

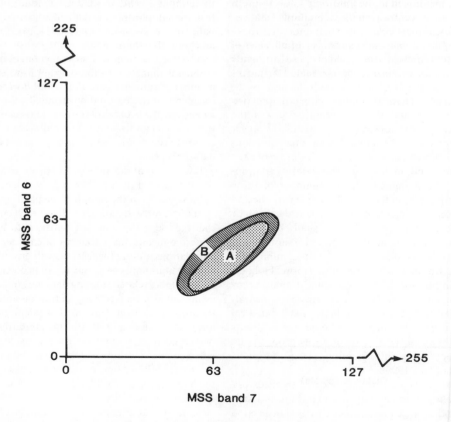

Figure 8.8 Ellipsoids derived from the variance–covariance matrices for training sets based on contiguous (A) and random (B) samples. Derived from Campbell (1981).

dimensions and (iii) the effects of any data pre-processing. The degree of autocorrelation can be calculated by taking sequences of pixels that are spaced 1, 2, 3, . . . units apart and plotting the correlations between a set of pixels and its first, second, third and subsequent nearest neighbours in the form of a correlogram. A diagram of the kind shown in Figure 8.9 might result, and the autocorrelation distance (in terms of number of pixels) can be read directly from it. As pixel size increases so the autocorrelation distance will diminish. The problem of spatially autocorrelated samples is considered in more detail in the papers cited above, and in Basu and Odell (1974), Craig (1979) and Labovitz *et al.* (1982). The definitive reference is still Cliff and Ord (1973). Geostatistical methods are based upon the spatial autocorrelation property, and are mentioned briefly in the introduction to section 8.5.

Another source of error encountered in the extraction of training samples is the presence in the sample of atypical values. For instance, one or more vectors of pixel measurements in a given training sample may be contaminated in some way, hence the sample mean and variance–covariance matrix for that class will be in error. Campbell (1980) considers ways in which these atypical values can be detected and proposes estimators of the mean and variance–covariance matrix that are robust (that is, they are not unduly influenced by the atypical values). These estimators give full weight to observations that are assumed to come from the main body of the data but reduce the weight given to observations identified as aberrant. A measure called the *Mahalanobis distance D* is used to identify deviant members of the sample. Its square is defined by:

$$D^2 = (\mathbf{x}_m - \bar{\mathbf{x}})'\mathbf{S}^{-1}(\mathbf{x}_m - \bar{\mathbf{x}})$$

where m is the index counting the elements of the sample, x_m is the mth sample value (pixel vector). The sample mean vector is $\bar{\mathbf{x}}$ and $\mathbf{S}$ is the sample variance–covariance matrix. The transpose of vector $\mathbf{x}$ is written as $\mathbf{x}'$. The Mahalanobis distance, or some function of that distance, can be plotted against the normal probabilities and outlying elements of the sample can be visually identified (Healy, 1968; Sparks, 1985). Robust estimates (that is, estimates that are less affected by outliers) of the mean and variance–covariance matrix are computed using weights that are functions of the Mahalanobis distance. The effect is to downgrade those pixel values with high Mahalanobis distances, which are associated with pixels that are relatively far from (dissimilar to) the mean of the training class taking into account the shape of the probability distribution of training-class members. For uncontaminated data the robust estimates are close to those obtained from the usual estimators. The procedure for obtaining the weights is described and illustrated by Campbell (1980). It is summarised here for completeness:

$$\bar{x}_k = \frac{\sum_{i=1}^{n} w_i x_{ki}}{\sum_{i=1}^{n} w_i} \qquad (k = 1, 2, \ldots, p)$$

$$s_{jk} = \sum_{i=1}^{n} w_i^2 (x_{ji} - \bar{x}_j)(x_{ki} - \bar{x}_k) \qquad \begin{array}{l} (j = 1, 2, \ldots, p) \\ (k = j, j+1, \ldots, p) \end{array}$$

where n = number of pixels in the training sample, p = number of features, w_i = weight for pixel i, x_{ki} = value for pixel i on feature k, $\bar{x}_j$ = mean of jth feature for this class and s_{jk} = element j, k of the variance–covariance matrix for this class. The weights are found from:

$$w_i = F(d_i)/d_i$$

given

$$F(d_i) = \begin{cases} d_i & d_i \leq d_0 \\ d_0 \exp[-0.5(d_i - d_0)^2/b_2^2] & \text{otherwise} \end{cases}$$

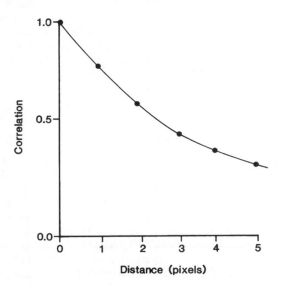

Figure 8.9 Illustrating the concept of autocorrelation. The diagram shows the correlation between the pixels in an image and their nth nearest neighbours in the x direction ($n = 1, 2, 3, 4, 5$).

where d_i = Mahalanobis distance of pixel i for this class, $d_0 = \sqrt{p} + b_1/\sqrt{2}$, $b_1 = 2$ and $b_2 = 1.25$. The weights w_i are initially computed from the Mahalanobis distances, which are themselves computed from $\bar{x}_j$ and S_j derived from the above formulae with unit weights. The Mahalanobis distances and the weights are recalculated iteratively until successive weight vectors converge within an acceptable limit, when any aberrant pixel vectors should have been given very low weights, and will therefore contribute only negligibly to the final (robust) estimates of $\bar{x}_j$ and S_j that are required in the maximum likelihood classification scheme.

The reason for going to such apparently great lengths to obtain robust estimates of the mean and variance–covariance matrix for each of the training samples for use in maximum likelihood classification is due to the fact that the probabilities of class membership of the individual pixels are dependent on these statistics. The performance of both statistical and neural classifiers depends to a considerable extent on the reliability and representativeness of the sample. It is easy to use an image processing system to extract 'training samples' from an image, but it is a lot more difficult to ensure that these training samples are not contaminated either by spatial autocorrelation effects or by the inclusion in the training sample of pixels that are not 'pure' but 'mixed' and therefore atypical of the class that they are supposed to represent.

The use of unsupervised classification techniques applied to the training classes has already been described as a method of ensuring that the classes have been well chosen to represent a single spectral class (that is, one with a single mode or peak in their frequency distributions). An alternative way to provide a visual check of the distribution of the training sample values is to employ an ordination method. Ordination is the expression of a multivariate data set in terms of a few dimensions (preferably two) with the minimum loss of information. The non-linear mapping procedure of Sammon (1969) projects the data in a p-dimensional space on to an m-dimensional subspace (m being less than p) whilst minimising the error introduced into the inter-point Euclidean distances. That is to say, the m-dimensional representation of the distances between data points is the best possible for that number of dimensions in terms of the maintenance of inter-point Euclidean relationships. If $m = 2$ or 3 the results can be presented graphically and the presence of multiple modes or outlying members of the sample can be picked out by eye. Figure 8.10 shows a training sample projected on to two dimensions using the non-linear mapping program described in Appendix B. The training data

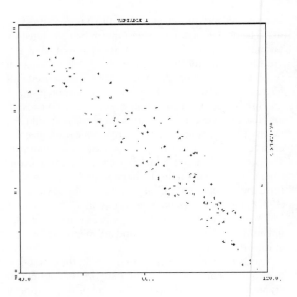

Figure 8.10 Non-linear mapping of a training sample. The raw data are measured on six spectral bands. Non-linear mapping has been used to project the sample data on to two dimensions. The program `nlmap` (Appendix B) allows the user to 'query' the identity of any aberrant or outlying pixels. Original in colour.

coordinates were collected using the `classify` option of the MIPS display program (Appendix A) and the pixel data were extracted from the log file produced by the program `ml_train` described in Appendix B. These data were then edited using Windows Notepad. The 'point and click' facility of program `nlmap` allows the identification of extreme points, perhaps representing aberrant pixels.

8.4.2   Statistical classifiers

Three algorithms are described in this section. All require that the number of categories (classes) be specified in advance, and that certain statistical characteristics of each class are known. The first method is called the parallelepiped or box classifier. A parallelepiped is simply a geometrical shape consisting of a body whose opposite sides are straight and parallel. A parallelogram is a two-dimensional parallelepiped. To define such a body, all that is required is an estimate for each class of the lowest and highest pixel values in each band or feature used in the analysis. Pixels are labelled by determining the identifier of the box into which they fall (see Figure 8.11 and section 8.4.2.1). The second method,

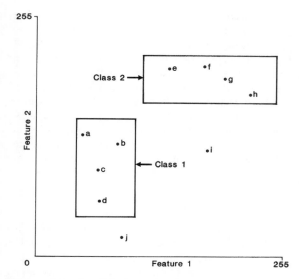

Figure 8.11   Parallelepiped classifier in two dimensions. Points *a*, *b*, *c* and *d* lie in the region bounded by parallelepiped 1 and are therefore assigned to class 1. Points *e*, *f*, *g* and *h* are similarly labelled '2'. Points *i* and *j* are unclassified.

which is analogous to the *k*-means unsupervised technique, uses information about the location of each class in the *p*-dimensional Cartesian space defined by the *p* bands (features) to be used as the basis of the classification. The location of each class in the *p*-space is given by the class mean or centroid (see Figure 8.12 and section 8.4.2.2). The third method also uses the mean as a measure of the location of the centre of each class in the *p*-space and, in addition, makes use of a measure summarising the disposition or spread of values around the mean along each of the *p* axes of the feature space. This third method is that of maximum likelihood (see Figure 8.14 and section 8.4.2.3). All three methods require estimates of certain statistical characteristics of the classes to which the pixels are to be allocated. These estimates are derived from samples of pixels, called training samples, which are extracted from the image to be classified (section 8.4.1).

8.4.2.1   *Parallelepiped classifier*

Of the statistical supervised classification methods described in this chapter, the parallelepiped classifier requires the least information from the user. For each of the *k* classes specified, the user provides an estimate of the minimum and maximum pixel values on each of the *p* bands or features. Alternatively, a range, expressed in terms of a given number of standard deviation units on

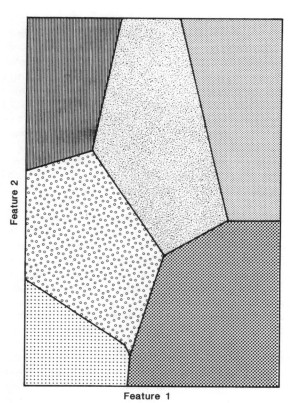

Figure 8.12   Two-dimensional feature space partitioned according to distance from the centroid of the nearest group. There are six classes.

either side of the mean of each feature, can be used. These extreme values allow the estimation of the position of the boundaries of the parallelepipeds, which define regions of the *p*-dimensional feature space that are identified with particular land cover types (or information classes). Regions of the *p*-space lying outside the boundaries of the set of parallelepipeds form a *terra incognita* and pixels lying in these regions are usually assigned the label zero. The decision rule employed in the parallelepiped classifier is simple. Each pixel to be classified is taken in turn and its values on the *p* features are checked to see whether they lie inside any of the parallelepipeds. Two extreme cases might occur. In the first, the point in *p*-space representing a particular pixel does not lie inside any of the regions defined by the parallelepipeds. Such pixels are of an unknown type. In the second extreme case the point lies inside just one of the parallelepipeds, and the corresponding pixel is therefore labelled as a member of the class represented

by that parallelepiped. However, there is the possibility
that a point may lie inside two or more overlapping
parallelepipeds, and the decision then becomes more
complicated. The easiest way around the problem is to
allocate the pixel to the first (or some other arbitrarily
selected) parallelepiped inside whose boundaries it falls.
The order of evaluation of the parallelepipeds then becomes of crucial importance, and there is often no sensible rule that can be employed to determine the best
order.

The method can therefore be described as 'quick and
dirty'. If the data are well structured (that is, there is no
overlap between the classes) then the quick-and-dirty
method might generate only a very few conflicts but,
unfortunately, many image data sets are not well structured. A more complicated rule for the resolution of
conflicts might be to calculate the Euclidean distance
between the doubtful pixel and the centre point of each
parallelepiped and use a 'minimum distance' rule to
decide on the best classification. In effect, a boundary
is drawn in the area of overlap between the parallelepipeds concerned. This boundary is equidistant from
the centre points of the parallelepipeds, and pixels can
be allocated on the basis of their position relative to the
boundary line. On the other hand, a combination of the
parallelepiped and some other, more powerful, decision
rule could be used. If a pixel falls inside one single
parallelepiped then it is allocated to the class that is
represented by that parallelepiped. If the pixel falls inside two or more parallelepipeds, or is outside all of the
parallelepiped areas, then a more sophisticated decision
rule could be invoked to resolve the conflict.

Figure 8.11 shows a geometric representation of a
simple case illustrating the parallelepiped classifier in
action. Points *a*, *b*, *c* and *d* are allocated to class 1 and
points *e*, *f*, *g* and *h* are allocated to class 2. Points *i* and *j*
are not identified and are labelled as 'unknown'. The
technique is easy to program and is relatively fast in
operation. Since, however, the technique makes use only
of the minimum and maximum values of each feature
for each training set, it should be realised that (i) these
values may be unrepresentative of the actual spectral
classes that they allegedly represent, and (ii) no information is garnered from those pixels in the training set
other than the largest and the smallest in value on each
band. Furthermore it is assumed that the shape of the
region in *p*-space occupied by a particular spectral class
can be enclosed by a box. This is not necessarily so.
Consequently the parallelepiped method should be considered as a cheap and rapid, but not particularly accurate, method of associating image pixels with information classes.

The MIPS display program (Appendix A) provides
an interactive (supervised) parallelepiped classification
procedure, which is based on training data collected
from a three-band (RGB) image display. The program
operates on the three displayed bands only. However,
the training data file can be saved and used in program
`piped`, which is described in Appendix B. This program accepts multiple image bands, or other features.
Both programs use the minimum and maximum values
present in the training data for a given class in order to
define the limits in feature space of the parallelepiped for
that class.

8.4.2.2   Centroid (k-means) classifier

The centroid or *k*-means method does make use of all
the data in each training class, for it is based upon the
'nearest-centre' decision rule that is described in section
8.3. The centroid (mean centre) of each training class is
computed – it is simply the vector comprising the mean
of each of the *p* features used in the analysis, perhaps
weighted to diminish the influence of extreme values as
discussed in section 8.4.1. The Euclidean distance from
each unknown pixel is then calculated for each centre
in turn and the pixel is given the label of the centre to
which its Euclidean distance is smallest. In effect, the
p-space is divided up into regions by a set of rectilinear
boundary lines, each boundary being equidistant from
two or more centres (Figure 8.12). Every pixel is classified by this method, for each point in *p*-space must be
closer to one of the *k* centres than to the rest, excluding
the case in which a pixel is equidistant from two or
more centres. A modification to the 'closest-distance'
rule could be adopted to prevent freak or outlying pixel
values from being attached to one or other of the
classes. This modification could take the form of a distance threshold, which could vary for each class depending upon the expected degree of compactness of
that class. Compactness might be estimated from the
standard deviation for each feature of the pixels making up the training sample for a given class. Any pixel
further away than the threshold distance is left unclassified. This modified rule is actually changing the geometry of the decision boundaries from that shown in
Figure 8.12 to that shown in Figure 8.13. In the latter,
the *p*-space is subdivided into *k* hyperspherical regions
each centred on a class mean. In the same way that the
parallelepiped method gets into difficulties with overlapping boxes and has to adopt a nearest-centre rule to
break the deadlock, so the *k*-means method can be
adapted to utilise additional information in order to
make it intuitively more efficient. The alteration to the

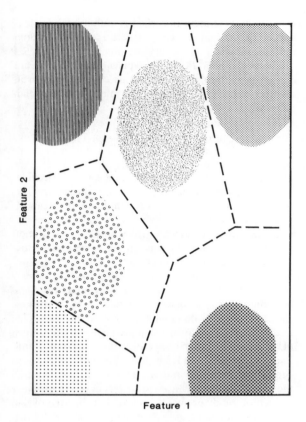

Feature 1

Figure 8.13 Group centroids are the same as in Figure 8.12 but a threshold has been used to limit the extent of each group. Blank areas represent regions of feature space that are not associated with a spectral class. Pixels located in these regions are not labelled.

decision rule involving a threshold distance is effectively acknowledging that the shape of the region in p-space that is occupied by pixels belonging to a particular class is important. The third classification technique described in this section (the maximum-likelihood method) begins with this assumption and uses a rather more refined method of describing the shapes of the regions in p-space that are occupied by the members of each class.

A k-means procedure is implemented in the MIPS display program (Appendix A) using training data collected interactively from the display. The resulting classification is based only on three features – the three images displayed on the screen in red, green and blue. It is possible to save the training data collected using the MIPS display program, however, and use them with program kmeans, which is described in Appendix B.

The program applies the k-means method to up to 12 features.

8.4.2.3 Maximum likelihood method

The geometrical shape of a cloud of points representing a set of image pixels belonging to a class or category of interest can often be described by an ellipsoid. This knowledge is used in Chapter 6 in the discussion of the principal components technique. In that chapter it is shown that the orientation and the relative dimensions of the enclosing ellipsoid (strictly speaking, a hyper-ellipsoid if p is greater than 3) depend on the degree of covariance among the p features defining the pattern space. Two examples of two-dimensional ellipses are shown in Figure 8.14. A shape such as that of ellipse A (oriented with the longer axis sloping upwards to the right) implies high positive covariance between the two features. If the longer axis sloped upwards to the left the direction of covariance would be negative. The more nearly circular shape of ellipse B implies zero covariance between the features represented by x and y. The lengths of the axes of the ellipses projected on to the x- or y-axes are proportional to the variances of the two variables. The location, shape and size of the ellipse therefore reflect the means, variances and covariances of the two features, and the idea can easily be extended to three or more dimensions. The ellipses in Figure 8.14 do not enclose all the points that fall into a particular class;

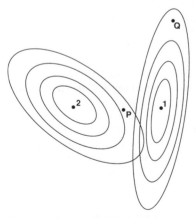

Figure 8.14 Showing the equi-probability contours for two bivariate-normal distributions with means located at points 1 and 2. Point P is closer to the mean centre of distribution 1 than it is to the centre of distribution 2 yet, because of the shapes of the two ellipses, P is more likely to be a member of class 2. Similarly, point P is closer to the centre of distribution 1 than is point Q, yet Q is more likely to be a member of class 1.

indeed, we could think of a family of concentric ellipses centred on the *p*-variate mean of a class. A small ellipse centred on this mean point might enclose only a few per cent of the pixels that are members of the class, and progressively larger ellipses will enclose an increasingly larger proportion of the class members. These concentric ellipses represent contours of probability of membership of the class, with the probability of membership declining away from the mean centre. Thus, membership probability declines more rapidly along the direction of the shorter axis than along the longer axis. Distance from the centre of the training data is not now the only criterion for deciding whether a point belongs to one class or another, for the shape of the probability contours depends on the relative dimensions of the axes of the ellipse as well as on its orientation. In Figure 8.14 point P is closer than point Q to the centre of class 1 yet, because of the shape of the probability contours, point Q is seen to be more likely to be a member of class 1 while point P is more likely to be a member of class 2.

If equi-probability contours can be defined for all *k* classes of interest then the probability that a pixel shown by a point in the *p*-dimensional feature belongs to class *i* ($i = 1, 2, \ldots, k$) can be measured for each class in turn, and that pixel assigned to the class for which the probability of membership is highest. The resulting classification might be expected to be more accurate than those produced by either the parallelepiped or the *k*-means classifiers because the training sample data are being used to provide estimates of the shapes of the distribution of the membership of each class in the *p*-dimensional feature space as well as of the location of the centre point of each class. The coordinates of the centre point of each class are the mean values on each of the *p* features, while the shape of the frequency distribution of the class membership is defined by the covariances among the *p* features for that particular class, as we saw earlier.

It is important to realise that the maximum likelihood method is based on the assumption that the frequency distribution of the class membership can be approximated by the multivariate-normal probability distribution. This might appear to be an undue restriction for, as an eminent statistician once remarked, there is no such thing as a normal distribution. In practice, however, it is generally accepted that the assumption of normality holds reasonably well, and that the procedure described above is not too sensitive to small departures from the assumption provided that the actual frequency distribution of each class is unimodal (i.e. has one peak frequency). A clustering procedure (unsupervised classification) could be used to check the training sample data

for each class to see if that class is multimodal, for the clustering method is really a technique for finding multiple modes.

The probability $P(\mathbf{x})$ that a pixel vector $\mathbf{x}$ of *p* elements (a pattern defined in terms of *p* features) is a member of class *k* is given by the multivariate-normal density:

$$P(\mathbf{x}) = 2\pi^{-0.5p} |\mathbf{S}_i|^{-0.5} \exp[-0.5(\mathbf{y}'\mathbf{S}_i^{-1}\mathbf{y})]$$

where $|\ \ |$ denotes the determinant of the specified matrix, $\mathbf{S}_i$ is the sample variance–covariance matrix for class *i*, $\mathbf{y} = (\mathbf{x} - \bar{\mathbf{x}}_i)$ and $\bar{\mathbf{x}}_i$ is the multivariate mean of class *i*. Note that the term $(\mathbf{y}'\mathbf{S}^{-1}\mathbf{y})$ is the Mahalanobis distance, used in section 8.4.1 to measure the distance of an observation from the class mean, corrected for the variance and covariance of class *i*.

Understanding of the relationship between equi-probability ellipses, the algebraic formula for class probability, and the placing of decision boundaries in feature space will be enhanced by a simple example. Figure 8.15 shows the bivariate frequency distributions of two samples, drawn respectively from (a) bands 1 and 2 and (b) bands 5 and 7 of the Littleport TM images shown in Figures 1.8 and 1.9. The contours delimiting the equi-probability ellipses are projected on to the base of each of the diagrams. It is clear that the area of the two-dimensional feature space that is occupied by the band 5–band 7 combination (Figure 8.15(b)) is greater than that occupied by the band 1–band 2 combination (Figure 8.15(a)). The orientation of the probability ellipses is similar, and the two ellipses are located at approximately the same point in the feature space. These observations can be related to the elements of the variance–covariance matrices on which the two plots are based. Figure 8.15 uses the following matrices to calculate the bivariate probabilities:

$$\bar{\mathbf{x}}_{12} = \begin{bmatrix} 65.812 \\ 28.033 \end{bmatrix} \qquad \mathbf{S}_{12} = \begin{bmatrix} 78.669 & 45.904 \\ 45.904 & 31.945 \end{bmatrix}$$

$$\mathbf{S}_{12}^{-1} = \begin{bmatrix} 0.079 & -0.113 \\ -0.113 & 0.194 \end{bmatrix} \qquad |\mathbf{S}_{12}| = 405.904$$

$$\bar{\mathbf{x}}_{57} = \begin{bmatrix} 64.258 \\ 23.895 \end{bmatrix} \qquad \mathbf{S}_{57} = \begin{bmatrix} 329.336 & 181.563 \\ 181.563 & 128.299 \end{bmatrix}$$

$$\mathbf{S}_{57}^{-1} = \begin{bmatrix} 0.014 & -0.020 \\ -0.020 & 0.035 \end{bmatrix} \qquad |\mathbf{S}_{57}| = 9288.356$$

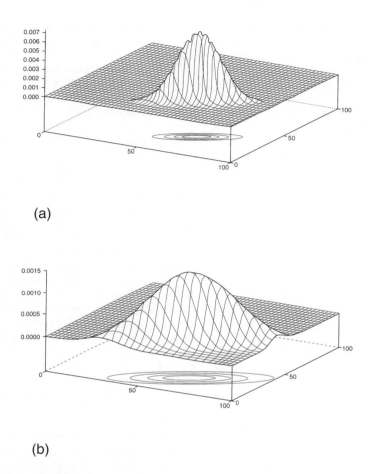

(a)

(b)

Figure 8.15   Bivariate frequency distribution for the Landsat Thematic Mapper images of the Littleport area: (a) bands 1 and 2, (b) bands 5 and 7. The mean vectors for both data sets are approximately the same, and both show a positive covariance. The variances of bands 5 and 7 are substantially larger than the variances of bands 1 and 2 (see text).

The variances are the diagonal elements of the matrix **S**, and it is clear that the variances of bands 5 and 7 (329.336 and 128.299) are much larger than the variances of bands 1 and 2 (78.669 and 31.945). Thus, the 'spread' of the two ellipses in the x and y directions is substantially different. The covariances of bands 5 and 7 (181.563) and bands 1 and 2 (45.904) are both positive, so the ellipses are oriented upwards towards the $+x$ and $+y$ axes (if the plot were a two-dimensional one, we could say that the ellipses sloped upwards to the right, indicating a positive correlation between x and y). The covariance of bands 5 and 7 is larger than that of bands 1 and 2, so the degree of scatter is less for bands 5 and 7, relative to the magnitude of the variances. This example illustrates the fact that the mean controls the location of the ellipse in feature space, while the variance–covariance matrix controls the 'spread' and orientation of the ellipse. It is not possible to illustrate these principles in higher-dimensional spaces, though if it were then the same conclusions would be drawn.

The function $P(\mathbf{x})$ can be used to evaluate the probability that an unknown pattern $\mathbf{x}$ is a member of class i ($i = 1, 2, \ldots, k$). The maximum value in this set can be chosen and $\mathbf{x}$ allocated to the corresponding class. However, the cost of carrying out these computations can be reduced by simplifying the expression. Savings can be made by first noting that we are only interested in the rank order of the values of $P(\mathbf{x})$. Since the logarithm to base e of a function has the same rank order as the function, the evaluation of the exponential term can be avoided by evaluating

$$\ln[P(\mathbf{x})] = -0.5p\ln(2\pi) - 0.5\ln|\mathbf{S}| - 0.5(\mathbf{y}'\mathbf{S}_i^{-1}\mathbf{y})$$

The rank order is unaffected if the expression is multiplied by -2 and the constant term $p\ln(2\pi)$ is dropped. The smallest value for all k classes is now chosen, rather than the largest. These modifications reduce the expression to

$$-2\ln[P(\mathbf{x})] = \ln|\mathbf{S}| + \mathbf{y}'\mathbf{S}_i^{-1}\mathbf{y}$$

Further savings can be made if the inverse and determinant of each $\mathbf{S}_i$ (the variance–covariance matrix for class i) are computed in advance and read from a file when required, rather than calculated when required. The computations then reduce to the derivation of the Mahalanobis distance, the addition of the logarithm of the determinant of the estimated variance–covariance matrix for each of the k classes in turn, and the selection of the minimum value from among the results. Note that, because we have multiplied the original expression by -2 we minimise $-2\ln[P(\mathbf{x})]$ so as to achieve the same result as maximising $P(\mathbf{x})$.

The maximum likelihood equations given above are based upon the presumption that each of the k classes is equally likely. This may be the safest assumption if we have little knowledge of the extent of each land cover type in the area covered by the image. Sometimes the proportion of the area covered by each class can be estimated from reports, land use maps, aerial photographs or previous classified images. An unsupervised classification of the area would also provide some guide to the areal extent of each cover type. A significant advantage of the maximum-likelihood approach to image classification is that this prior knowledge can be taken into account. *A priori* knowledge of the proportion of the area to be classified that is covered by each class can be expressed as a vector of *prior probabilities*. The probabilities are proportional to the area covered by each class, and can be thought of as weights. A high prior probability for class i in comparison with class j means that any pixel selected at random is more likely to be placed in class i than class j because class i is given greater weight. These weights are incorporated into the maximum likelihood algorithm by subtracting twice the logarithm of the prior probability for class i from the log likelihood of the class as given by the equation above. Strahler (1980) provides further details and shows how different sets of prior probabilities can be used in cases

where the image area can be stratified into regions according to an external variable such as eleva-tion (for instance, regions described as high, intermediate or low elevation might have separate sets of prior probabilities as described in section 8.7.2). Maselli *et al.* (1995b) discuss a non-parametric method of estimating prior probabilities for incorporation into a maximum likelihood classifier.

In the same way that the parallelepiped classifier (section 8.4.2.1) allows for the occurrence of pixels that are unlike any of the training patterns by consigning such pixels to a 'reject' class, so the probability of class membership can be used in the maximum-likelihood classifier to permit the rejection of pixel vectors for which the probability of membership of any of the k classes is considered to be too low. The Mahalanobis distance is distributed as chi-square, and the probability of obtaining a Mahalanobis distance as high as that observed for a given pixel can be found from tables, with degrees of freedom equal to p, the number of feature vectors used (Moik, 1980). For $p = 4$ the tabled chi-square values are 0.3 (99%), 1.06 (90%), 3.65 (50%), 7.78 (10%) and 15.08 (1%). The figures in brackets are probabilities, expressed in per cent form. They can be interpreted as follows: a Mahalanobis distance as high as (or higher than) 15.08 would, on average, be met in only 1% of cases in a long sequence of observations of four-band pixel vectors drawn from a multivariate-normal population whose true mean and variance–covariance matrix are estimated by the mean and variance–covariance matrix on which the calculation of the Mahalanobis distance is based. A Mahalanobis distance of 1.06 or more would be observed in 90% of all such observations. It is self-evident that 100% of all observations drawn from the given population will have Mahalanobis distances of 0 or more. A suitable threshold probability (which need not be the same for each class) can be specified. Once the pixel has been tentatively allocated to a class using the maximum likelihood decision rule, the value of the Mahalanobis distance (which is used in the maximum likelihood calculation) can be tested against a threshold chi-square value. If this chi-square value is exceeded then the corresponding pixel is placed in the 'reject' class, conventionally labelled '0'. The use of the threshold probability helps to weed out atypical pixel vectors. It can also serve another function – to indicate the existence of spectral classes that may not have been recognised by the analyst and which, in the absence of the probability threshold, would have been allocated to the most similar (but incorrect) spectral class.

A program, ml.exe, is provided on the accompanying CD and is described in Appendix B. Program `ml.exe` requires training statistics output by the program `ml_train`, which uses training data pixel coordinates derived from the classification module in the MIPS display program (Appendix A). Program `ml.exe` caters for class prior probabilities and also allows the user to select a chi-square reject probability for each class as described above.

8.4.3 Neural classifiers

The best image interpretation system that we possess is the combination of our eyes and our brain. Signals received by two sensors (our eyes) are converted into electrical impulses and transmitted to the brain, which interprets them in real time, producing labelled (in the sense of recognised) three-dimensional images of our field of view (section 5.2). Operationally speaking, the brain is composed of a very large number of simple processing units called *neurons*. Greenfield (1997, p. 79) estimates the number of neurons as being of the order of a hundred billion, a number that is of the same order of magnitude as the number of trees in the Amazon rainforest. Each neuron is connected to perhaps 10 000 other neurons (Beale and Jackson, 1990; Dayhoff, 1990). Many of these neurons are dedicated to image processing, which takes place in a parallel fashion (Greenfield, 1997, p. 50). The brain's neurons are connected together in complex ways so that each neuron receives as input the results produced by other neurons, and it in turn outputs its signals to other neurons. It is not possible at the moment to specify how the brain actually works, or even whether the connectionist model really represents what is going on in the brain. It has been suggested that if there were fewer neurons in the brain then we might have a chance of understanding how they interact but, unfortunately, if our brains possessed fewer neurons we would probably be too stupid to understand the implications of the present discussion.

The brain can thus be described in terms of sets of neural networks that perform specific functions such as vision or hearing. Artificial neural networks (ANN) attempt, in a very simple way, to simulate the workings of the brain by composing sets of linked processing units (by analogy with the neurons of the brain) and using these to solve problems. Each neuron is a simple computer, which receives weighted inputs from other neurons, sums these weighted inputs, performs a simple calculation on this sum such as thresholding, and then sends this output to other neurons (Figure 8.16).

The two functions of the artificial neuron are to sum the weighted inputs and to apply a thresholding func-

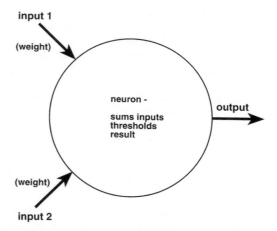

Figure 8.16 A basic neuron receives two weighted inputs, sums them, applies a threshold and outputs the result.

tion to this sum. The summation procedure can be expressed by:

$$S = \sum_{i=1}^{n} w_i x_i$$

where S represents the sum of the n weighted inputs, w_i is the weight associated with the ith input and x_i is the value of the ith input (which is an output from some other neuron). The thresholding procedure, at its simplest, is a comparison between S and some pre-set value, say T. If S is greater than T then the neuron responds by sending an output to other neurons to which it is connected further 'down the line'. The term *feed-forward* is used to describe this kind of neural network model because information progresses from the initial inputs to the final outputs.

A simple ANN such as the one presented above lacks a vital component – the ability to learn. Some training is necessary before the connected set of neurons can perform a useful task. Learning is accomplished by providing training samples and comparing the actual output of the ANN with the expected output. If there is a difference between the two then the weights associated with the connections between the neurons forming the ANN are adjusted so as to improve the chances of a correct decision and diminish the chances of the wrong choice being made, and the training step is repeated. The weights are initially set to random values. This 'supervised learning' procedure is followed until the ANN gets the correct answer. To a parent, this is perhaps reminiscent of teaching a child to read; repeated correction of mistakes in identifying the letters of the alphabet and the sounds associated with them eventually results in the development of an ability to read. The method is called *Hebbian learning* after its developer, D.O. Hebb.

This simple model is called the *single-layer perceptron* and it can solve only those classification problems in which the classes can be separated by a straight line (in other words, the decision boundary between classes is a straight line as in the simple example given in section 8.2 and shown in Figure 8.1). Such problems are relatively trivial and the inability of the perceptron to solve more difficult problems led to a lack of interest in ANN on the part of computer scientists until the 1980s when a more complex model, the *multi-layer perceptron*, was proposed. First, this model uses a more complex thresholding function rather than a step function, in which the output from the neuron is 1 if the threshold is exceeded and 0 otherwise. A sigmoid function is often used and the output from the neuron is a value somewhere between 0 and 1. Secondly, the neurons forming the ANN are arranged in layers as shown in Figure 8.17. There is a layer of input neurons that provide the link between the ANN and the input data, and a layer of output neurons that provide information on the category to which the input pixel vector belongs (for example, if output neuron number 1 has a value near to 1 and the remaining output neurons have values near 0, then the input pixel will be allocated to class 1).

The multi-layer perceptron shown in Figure 8.17 could, for example, be used to classify an image obtained from the SPOT HRV sensor. The three neurons on the input layer represent the values for the three HRV spectral bands for a specific pixel. These inputs might be normalised, for example to the 0–1 range, as this procedure has been shown to improve the network's performance. The four neurons in the output layer provide outputs (or *activations*), which form the basis for the decision concerning the pixel's class membership, and the two neurons in the centre (hidden) layer perform summation and thresholding, possibly using a sigmoidal thresholding function. All the internal links (input to hidden and hidden to output) have associated weights. The input neurons do not perform any summation or thresholding – they simply pass on the pixel's value to the hidden layer neurons (so that input layer neuron 1 transmits the pixel value in band HRV-1, neuron 2 in the input layer passes the pixel value for band HRV-2 to the hidden layer neurons, and the third input neuron provides the pixel value for band HRV-3). The multi-layer perceptron is trained using a slightly

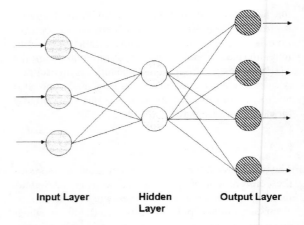

Figure 8.17 Multi-layer perceptron. The input layer (left side) connects to the hidden layer (centre), which in turn connects to the output layer (right).

more complex learning rule than the one described above; the new rule is called the *generalised delta function rule* or the *back-propagation rule*. Hence the network is a feed-forward multi-layer perceptron using the back-propagation learning rule, which is described by Paola and Schowengerdt (1995a).

Assume that a certain training data pixel vector, called a *training pattern*, is fed into the network and it is known that this training pattern is a member of class i. The output from the network consists of one value for each neuron in the output layer. If there are k possible classes then the expected output vector **o** should have elements equal to zero except for the ith element, which should be equal to one. The actual output vector, **a**, differs from **o** by an amount called the error, e. Thus,

$$e = \tfrac{1}{2} \sum_{j=1}^{k} (o_j - a_j)^2$$

Multiplication by $\tfrac{1}{2}$ is performed for arcane reasons of mathematics. The error is used to adjust the weights by a procedure which (in effect) maps the isolines or contours of the distribution of the error against the values of the weights, and then uses these isolines to determine the direction to move in order to find the minimum point in this map. The idea is known as the method of *steepest descent*. It requires (i) the direction of the steepest downhill slope and (ii) the length of the step that should be taken. Imagine that you have reached the summit of a hill when a thick mist descends. To get down safely you need to move downhill, but you may not be confident enough to take long strides downhill for fear of falling over a cliff. Instead, you may prefer to take short steps when the slope is steep and longer steps on gentle slopes. The steepest-descent method uses the same approach. The gradient is measured by the first derivative of the error in terms of the weight; this gives both the magnitude and direction of the gradient. The step length is fixed and, in the terminology of ANNs, it is called the *learning rate*. A step is taken from the current position in the direction of maximum gradient and new values for the weights are determined. The error is propagated backwards through the net from the output layer to the input data, hence the term back-propagation.

In your descent down a fog-bound hill you may come to an enclosed hollow. No matter which way you move you will go uphill. You might think that you were at the bottom of the hill. In mathematical terms this is equivalent to finding a local minimum of the function relating weights to error, rather than a global minimum. The method of steepest descent thus may not give you the right answer; it may converge at a point that is far removed from the real (global) minimum. Another problem may arise if you take steps of a fixed length. During the descent you may encounter a small valley. As you reach the bottom of this valley you take a step across the valley floor and immediately start to go uphill. So you step back, then forward, and so on in an endless dance rhythm. But you never get across the valley and continue your downhill march. This problem is similar to that of the local minimum and both may result in oscillations. Some ANN software allows you to do the equivalent of taking a long jump across the surface when such oscillations occur, and continuing the downhill search from the landing point. There are other, more powerful, methods of locating the minimum of a function that can be used in ANN training, but these are beyond the scope of this book.

The *advantages* of the feed-forward multi-layer ANN using the back-propagation training method are as follows.

(i) It can accept all kinds of numerical inputs whether or not these conform to a statistical distribution or not. So non-remotely-sensed data can be added as additional inputs, and the user need not be concerned about multivariate-normal distributions or multimodality. This feature is useful when remote-sensing data are used within a GIS, for different types of spatial data can be easily registered and used together in order to improve classifier performance.

(ii) ANNs can generalise. That is, they can recognise inputs that are similar to those which have been used to train them. They can generalise more successfully when the unknown patterns are intermediate between two known patterns, but they are less good at extending their ability to new patterns that exist beyond the realms of the training patterns. Bigger networks (with more neurons) tend to have a poorer generalisation capability than small networks.

(iii) Because ANNs consist of a number of layers of neurons, connected by weighted links, they are tolerant to noise present in the training patterns. The overall result may not be significantly affected by the loss of one or two neurons as a result of noisy training data.

Disadvantages associated with the use of ANN in pattern recognition are as follows.

(i) Problems occur in designing the network. Usually one or two hidden layers suffice for most problems, but how many hidden layers should be used in a given case? How many neurons are required on each hidden layer? (A generally used but purely empirical rule is that the number of hidden layer neurons should equal twice the number of input neurons.) Are all of the inter-neuron connections required? What is the best value for the learning rate parameter? Many authors do not say why they select a particular network architecture. For example, Kanellopoulos *et al.* (1992) use a four-layer network (one input, one output and two hidden layers). The two hidden layers contain, respectively, 18 and 54 neurons. Foody (1995), on the other hand, uses a three-layer architecture with the hidden layer containing three neurons. Ardö *et al.* (1997) state that

'... no significant difference was found between networks with different numbers of hidden nodes, or between networks with different numbers of hidden layers'

a conclusion also reached by Gong *et al.* (1996) and Paola and Schowengerdt (1997), though it has already been noted that network size and generalisation capability appear to be inversely related. This factor is the motivation for network pruning (see below, point (v)).

(ii) Training times may be long, possibly of the order of several hours. In comparison, the maximum likelihood statistical classifier requires collection of training data and calculation of the mean vector and variance–covariance matrix for each training class. The calculations are then straightforward rather than iterative. The maximum likelihood algorithm can classify a 512×512 image on a 486-based PC in less than a minute, while an ANN may require several hours of training in order to achieve the same level of accuracy.

(iii) The steepest-descent algorithm may reach a local

rather than a global minimum, or may oscillate.

(iv) Results achieved by an ANN depend on the initial values given to the inter-neuron weights, which are usually set to small, random values. Differences in the initial weights may cause the network to converge on a different (local) minimum and thus different classification accuracies can be expected. Ardö *et al.* (1997), Blamire (1996) and Skidmore *et al.* (1997) note differences in classification accuracy resulting from different initial (but still randomised) choice of weights (around 6% in Blamire's experiments, but up to 11% in the results reported by Ardö *et al.* (1997)). Skidmore *et al.* (1997, p. 511) remark that

'. . . the oft-quoted advantages of neural networks . . . were negated by the variable and unpredictable results generated'.

Paola and Schowengerdt (1997) also find that where the number of neurons in the hidden layer is low then the effects of changes in the initial weights may be considerable.

(v) The generalising ability of ANNs is dependent in a complex fashion on the numbers of neurons included in the hidden layers and on the number of iterations achieved during training. 'Over-training' may result in the ANN becoming too closely adjusted to the characteristics of the training data and losing its ability to identify patterns that are not present in the training data. Pruning methods, which aim to remove inter-neuron links without reducing the classifier's performance, have not been widely used in remote-sensing image classification, but appear to have some potential in producing smaller networks that can generalise better and run more quickly.

Further details of applications of ANNs in pattern recognition are described by Aleksander and Morton (1990), Bishoff *et al.* (1992), Cappellini *et al.* (1995) and Gopal and Woodcock (1996). Austin *et al.* (1997) and Kanellopoulos *et al.* (1997) provide an excellent summary of research problems in the use of ANNs in remote sensing. The topic of pruning ANNs in order to improve their generalisation capability is discussed by Le Cun *et al.* (1990), Sietsma and Dow (1988) and Tidemann and Nielsen (1997). Jarvis and Stuart (1996) summarise the factors that affect the sensitivity of neural nets for classifying remotely-sensed data. A good general textbook is Bishop (1995). Hepner *et al.* (1990) compare the performance of an ANN classifier with that of maximum likelihood (ML), and find that (with a small training data set) the ANN gives superior results, a finding that is repeated by Foody *et al.* (1995). Kanellopoulos *et al.* (1992) compare the ML classifier's

performance with that of a neural network with two hidden layers (with 18 and 54 neurons, respectively), and report that the classification accuracy (section 8.10) rises from 51% (ML) to 81% (ANN), though this improvement is very much greater than that reported by other researchers. Paola and Schowengerdt (1995b) report on a detailed comparison of the performance of a standard ANN and the ML statistical technique for classifying urban areas. Their paper includes a careful analysis of decision boundary positions in feature space. Although the ANN slightly out-performed the ML classifier in terms of percentage correctly classified test pixels, only 62% or so of the pixels in the two classified images were in agreement, thus emphasising the point that measures of classification accuracy based upon error matrices (section 8.10) do not take the spatial distribution of the classified pixels into account.

The feed-forward multi-layer perceptron is not the only form of artificial neural net that has been used in remote sensing nor, indeed, is it necessarily true that ANNs are always used in supervised mode. Chiuderi and Cappellini (1996) describe a quite different network architecture, the Kohonen self-organising map (SOM). This is an unsupervised form of ANN. It is first trained to learn to distinguish between patterns in the input data ('clusters') rather than to allocate pixels to predefined categories. The clusters identified by the SOM are grouped on the basis of their mutual similarities, and then identified by reference to training data. The architecture differs from the conventional perceptron in that there are only two layers. The first is the input layer, which – as in the case of the perceptron – has one input neuron per feature. The input neurons are connected to all of the neurons in the output layer, and input pixels are allocated to a neighbourhood in the output layer, which is arranged in the form of a grid. Chiuderi and Cappellini (1996) use a 6×6 output layer and report a classification accuracy in excess of 85% in an application to land cover classification using airborne Thematic Mapper data. Schaale and Furrer (1995) successfully use a SOM network also to classify land cover, while Hung (1993) gives a description of the learning mechanism used in SOM. Carpenter *et al.* (1997) report on the use of another type of neural network, the ART network, to classify vegetation. The use of neural networks to classify multitemporal data is considered by German and Gahegan (1996). A special issue of the *International Journal of Remote Sensing* (volume 18, number 4, 1997) is devoted to 'Neural Networks in Remote Sensing'.

8.5 FUZZY CLASSIFICATION AND LINEAR SPECTRAL UNMIXING

The techniques described in the first part of Chapter 8 are concerned with 'hard' pixel labelling. All of the different classification schemes require that each individual pixel is given a single, unambiguous, label. This objective is a justifiable one whenever regions of relatively homogeneous land cover occur in the image area. These regions should be large relative to the instantaneous field of view of the sensor, and may consist of fields of agricultural crops or deep, clear water bodies that are tens of pixels in each dimension. In other instances, though, the instantaneous field of view of the sensor may be too large for it to be safely assumed that a single pixel contains just a single land cover type. In many cases, a 1 km × 1 km pixel of an AVHRR or ATSR image is unlikely to contain just one single cover type. In areas covered by semi-natural vegetation, natural variability will be such as to ensure that, even in a 20 or 30 m square pixel, there will be a range of different cover types such as herbs, bare soil, bushes, trees and water. The question of scale is one that bedevils all spatial analyses. The resolution of the sensor is not the only factor that relates to homogeneity, for much depends on what is being sought. If generalised classes such as wheat, barley or rice are the targets then a resolution of 30 m rather than 1 km may be appropriate. A 30 m resolution would be quite inappropriate, however, if the investigator wished to classify individual plants. Fuzziness and hardness, heterogeneity and homogeneity are properties of the landscape at a particular geographical scale of observation that is related to the aims of the investigator. Questions of scale are considered by de Cola (1994), Levin (1991) and Ustin *et al.* (1996). The use of geostatistics in estimating spatial scales of variation is summarised by Curran and Atkinson (1998). Other references relevant to the use of geostatistics are Hyppänen (1996), Jupp *et al.* (1988, 1989), van Gardingen *et al.* (1997), Woodcock *et al.* (1988a, b) and Woodcock and Strahler (1987).

To the investigator whose concern is to label each pixel unambiguously, the presence of large heterogeneous pixels or smaller pixels containing several cover types is a problem, since they do not fall clearly within one or other of the available classes. If a conventional 'hard' classifier is used, the result will be low classification accuracy. 'Mixed pixels' represent a significant problem in the description of the Earth's terrestrial surface where that surface is imaged by an instrument with a large (1 km or more) instantaneous field of view, when natural variability occurs over a small area, or where the scale of variability of the target of interest is less than the size of the observation unit, the pixel.

Several alternatives to the standard 'hard' classifier have been proposed. The method of mixture modelling starts from the explicit assumption that the characteristics of the observed pixels constitute mixtures of the characteristics of a small number of basic cover types, or end-members. Alternatively, the investigator can use a 'soft' or 'fuzzy' classifier, which does not reach a definite conclusion in favour of one class or another. Instead, these soft classifiers present the user with a measure of the degree (termed membership grade) to which the given pixel belongs to some or all of the candidate classes, and leaves to the investigator the decision as to the category into which the pixel should be placed. In this section, the use of linear mixture modelling is described, and the use of the maximum-likelihood and artificial neural net classifiers to provide 'soft' output is considered.

8.5.1 The linear mixture model

If it can be assumed that a single photon impinging upon a target on the Earth's surface is reflected into the field of view of the sensor without interacting with any other ground-surface object, then the totality of photons reflected from a single pixel area on the ground and intercepted by a sensor are describable in terms of a simple linear model, as follows:

$$r_i = \sum_{j=1}^{n} a_{ij} f_j + e_i$$

in which r_i is the reflectance of a given pixel in the ith of m spectral bands. The number of mixture components is n, f_j is the jth fractional component (proportion of end-member j) in the make-up of r_i, and a_{ij} is the reflectance of end-member j in spectral band i. The term e_i expresses the difference between the observed pixel reflectance r_i and the reflectance for that pixel computed from the model. In order for the components of $\mathbf{r}$ ($= r_i$) to be computable, the number of end-members n must be less than the number of spectral bands, m. This model is simply expressing the fact that if there are n land cover types present in the area on the land surface that is covered by a single pixel, and if each photon reflected from the pixel area interacts with only one of these n cover types, then the integrated signal received at the sensor in a given band (r_i) will be the linear sum of the n individual interactions.

A simple example shows how the process works.

Assume that we have a pixel of which 60% is covered by material with a spectral reflectance curve given by the lower curve in Figure 8.18 and 40% is covered by material with a spectral reflectance curve like the upper curve in Figure 8.18. The values 0.6 and 0.4 are the proportions of these two end-members contributing to the pixel reflectance. The values of these mixture components are shown in the first two columns of Table 8.2, labelled C1 and C2. A 60:40 ratio mixture of the two mixture components is shown as the middle (dashed) curve in Figure 8.18 and in column M of Table 8.2. The rows of the table (b1, b2 and b3) represent three spectral bands in the green, red and near-infrared respectively, so C1 may be turbid water and C2 may be vigorous vegetation of some kind. The data in Table 8.2 therefore show two end-members, with contributions shown by the proportions in columns C1 and C2, from which a mixed pixel vector, M, is derived. We will now try to recover the values of the mixture proportions f_1 and f_2 from these data, knowing in advance that the correct answer is 0.6 and 0.4.

First, define a matrix **A** with three rows (the spectral bands) and two columns (the end-member proportions, represented by columns C1 and C2 in Table 8.2). Vector **b** holds the measurements for each spectral band of the mixed pixel (column M of Table 8.2). Finally **f** is an unknown vector, which will contain the proportions f_1 and f_2 as the elements of its two rows. The relationship between **A** and **b** is of the form **Af** = **b**, which is equivalent to the following set of simultaneous linear equations:

$$52.4 = 46f_1 + 62f_2$$
$$35.4 = 31f_1 + 42f_2$$
$$71.2 = 12f_1 + 160f_2$$

Because we have made up the columns of **A** and **b** we know that f_1 and f_2 must equal 0.6 and 0.4 respectively, but readers should check that this is in fact the case by, for example, entering the values of **A** and **b** into a spreadsheet and using the multiple regression option, remembering to set the intercept to 'zero' rather than 'computed'. This should give values of f_1 and f_2 equal to 0.6 and 0.4 respectively. If linear mixture modelling were being applied to an image then the values f_1 and f_2 would be scaled on to the range 0–255 and written to file as the two output fraction images, thus generating an output fraction image for each mixture component. This example assumes that the spectral reflectance curve that is derived as a mixture of two other spectral reflectance curves is unique and is different from other spectral reflectance curves. In reality, this may not be the case, as Price (1994) shows. For instance, the spectral reflectance curve for corn is intermediate between the spectral reflectance curves of soybeans and winter wheat. Hence, the procedure outlined in the example above would be incapable of distinguishing a pixel that was split between two agricultural crops (soybeans and winter wheat) and a pure pixel covered by corn. Applications of spectral unmixing with multispectral data use general classes such as 'green vegetation' and 'soil', so this confusion may not be too serious. However, hyperspectral data may be used to identify specific mineral/rock types, and Price's point is relevant to such studies. Sohn and McCoy (1997) also consider problems in end-member selection.

In a real application, rather than a fictitious example, it is necessary to go through a few more logical steps before determining the values of the fractional (endmember) components, otherwise serious errors could result. First of all, in our example we knew that the mixed pixel values are composed of two components, C1 and C2 in Table 8.2. In reality, the number of endmembers is not known. Secondly, the values of the fractions (the proportions of the mixture components

Table 8.2 Columns C1 and C2 show the reflectance spectra for two pure types. Column M shows a 60:40 ratio mixture of C1 and C2. See text for discussion.

	C1	C2	M
b1	46	62	52.4
b2	31	42	35.4
b3	12	160	71.2

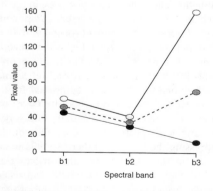

Figure 8.18 Top and bottom curves represent end-member spectra. The centre curve (dashed line) is the reflectance spectrum of an observed pixel, and is formed by taking 60% of the value of the lower curve and 40% of the value of the upper curve.

used to derive column M from columns C1 and C2) were known in advance to satisfy two logical requirements, namely:

$$0.0 \leq f_i \leq 1.0$$

$$\sum_{j=1}^{n} f_j \leq 1.0$$

These constraints specify that the individual fractions f_i must take values between 0% and 100%, and that the fractions for any given mixed pixel must sum to 100% or less. These two statements are a logical part of the specification of the mixture model which, in effect, assumes that the reflectance of a mixed pixel is composed of a linear weighted sum of a set of end-member reflectances, with no individual proportion exceeding the range 0–1 and the sum of the proportions being 1.0 at most. This statement specifies the linear mixture model.

There are several possible ways of proceeding from this point. One approach, called 'unconstrained', is to solve the mixture model equation without considering the constraints at all. This could result in values of f_i that lie outside the 0–1 range (such illogical values are called under-shoots or over-shoots, depending whether they are less than 0 or greater than 1). However, the value of the unconstrained approach is that it allows the user to estimate how well the linear mixture model describes the data. The following criteria may be used to evaluate the goodness of fit of the model:

1. *The size of the residual terms e_i in the mixture equation.* There is one residual value in each spectral band, representing the difference between the observed pixel value and the value computed from the linear mixture model equation. In the simple example given above, the three residuals are all zero. It is normal to take the square root of the sum of squares of all the residuals for a given pixel divided by the number of spectral bands, m, to give the root-mean-square (RMS) error for that pixel:

$$\text{RMS} = \sqrt{\frac{\sum_{b=1}^{m} e_b^2}{m}}$$

The RMS error is calculated for all image pixels and scaled to the 0–255 range in order to create an RMS error image. The larger the RMS error, the worse the fit of the model. Since the residuals are assumed to be random, then any spatial patterns that are visible in

the RMS images can be taken as evidence that the model has not fully accounted for the systematic variation in the image data, which implies that potential end-members have been omitted from the model, or that the selected end-members are deficient.

2. *The number of pixels that have proportions f_i that lie outside the logical range of* 0–1. These undershoots (f_i less than 0.0) and overshoots (f_i greater than 1.0) indicate that the model does not fit. If there are only a few per cent of pixels showing under- and overshoots then the result can be accepted, but if large numbers (say greater than 5%) of pixels under- or overshoot then the model does not fit well.

Under-shoots and over-shoots can be coded by a method that maps the legitimate range of the fractions f_i (i.e. 0–1) to the range 100–200 by multiplying each f_i by 100 and adding 100 to the result, which is constrained to lie in the range 0–255. This ensures that under-shoots and overshoots are mapped on to the range 0–99 and 201–255 respectively. Any very large under- and overshoots map to 0 and 255, respectively. Each fraction image can then be inspected by, for example, applying a pseudocolour transform to the range 100–200. All of the legitimate fractions will then appear in colour and the out-of-range fractions will appear as shades of grey. Alternatively, the ranges 0–99 and 201–255 can be pseudocoloured (section 5.4).

If unconstrained mixture modelling is used then the number of under- and over-shoots, and the overall RMS error, should be tabulated in order to allow others to evaluate the fit of the model. Some software packages simply map the range of the fractions for a given mixture component, including under- and over-shoots, on to a 0–255 scale without reporting the presence of these under- and overshoots. Whilst this approach may produce an interpretable image for a naive user, it fails to recognise a basic scientific principle: that sufficient information be provided to permit others independently to replicate your experiments and test your results.

The model may not fit well for one or more of several reasons. First, the pixels representing mixture components may be badly chosen. They should represent 'pure' pixels, composed of a single land cover type, that determine the characteristics of all other image pixels. It is easy to think of simple situations where mixture component selection is straightforward; for example, a flat, shadow-free landscape composed of dense forest, bare soil and deep, clear water. These three types can be thought of as the vertices of a triangle within which all image pixels are located, by analogy with the triangular sand–silt–clay diagram used in sedimentology. If the forest end-member is badly chosen and, in reality, represents a ground area that is only 80% forest, then any pixel with a forest cover of more than 80% will be an over-shoot. There may be cases in which no single pixel is completely forest-covered, so the forest end-member will not represent the pure case.

Secondly, a mixture component may have been omitted. This is a problem where the landscape is complex and the number of image bands is small because, as noted earlier, the number of mixture components cannot exceed the number of bands. Thirdly, the data distribution may not be amenable to description in terms of the given number of mixture components. Consider the triangle example again. The data distribution may be circular or elliptical, and thus all of the pixels in the image may not fit within the confines of the triangle defined by three mixture components. Fourthly, the assumption that each photon reaching the sensor has interacted with only a single object on the ground may not be satisfied, and a non-linear mixing model may be required (Ray and Murray, 1996). The topic of choice of mixture components is considered further in the following paragraphs.

The program `mixmod`, which is contained on the CD-ROM supplied with this book, implements the unconstrained linear mixture model, using the singular-value decomposition (SVD) method of solving for the f_i (Boardman, 1989). The SVD method is described below, and details of the operation `mixmod` are given in Appendix B. Output from the program is a log file containing details of the fit of the model, plus fraction images and the RMS error image.

Some software uses a simple procedure to make the model appear to fit. Negative fractions (under-shoots) are set to zero, and the remaining fractions are scaled so that they lie in the range 0–1 and add to 1. This might seem like cheating, and probably is; nevertheless, it is offered as an option in the `mixmod` program. An alternative is to use an algorithm that allows the solution of the linear mixture model equation subject to the constraints. This is equivalent to searching for a solution within a specified range, and is no different from the use of the square-root key on a calculator. Generally speaking, when we enter a number and press the square-root key we want to determine the positive square root, and so we ignore the fact that a negative square root also exists. If anyone were asked 'What is the square root of four?' he or she would be unlikely to answer 'Minus two', though this is a correct answer. Similarly, the solution of a quadratic equation may result in an answer that lies in the domain of complex numbers.

That solution, in some instances, would be unacceptable and the alternative solution is therefore selected.

Lawson and Hansen (1995) provide a Fortran-90 subroutine, BVLS, which solves the equations $\mathbf{Af} = \mathbf{b}$ subject to constraints on the elements of $\mathbf{f}$. This routine, which is available on the World Wide Web via the Netlib library, does not allow for a constraint on the sum of the f_i. An alternative is a Fortran-90 routine from the IMSL mathematical library provided with Microsoft Powerstation Fortran v4.0 Professional Edition. This routine, LCLSQ, allows the individual f_i to be constrained and it also allows the sum of the f_i to be specified. The programs `constmx1` and `constmx` (Appendix B) use routines BVLS and LCLSQ respectively to provide a means of constraining the mixture model. Note that both routines frequently return a value '0.0' in situations in which the unconstrained model would have computed an under-shoot.

The solution of the mixture model equation in the unconstrained mixture model requires some thought. The standard solution of the matrix equation $\mathbf{Af} = \mathbf{b}$ generally involves the calculation of the inverse of $\mathbf{A}$ (represented as $\mathbf{A}^{-1}$) from which $\mathbf{f}$ is found from the expression $\mathbf{f} = \mathbf{A}^{-1}\mathbf{b}$. A number of pitfalls are likely to be encountered in evaluating this innocent-looking expression. If $\mathbf{A}$ is an orthogonal matrix (that is, its columns are uncorrelated) then the derivation of the inverse of $\mathbf{A}$ is easy (it is the transpose of $\mathbf{A}$). If the columns of $\mathbf{A}$, which contain the reflectances of the end-members, are correlated, then error enters the calculations, and the greater the degree of interdependence between the columns, the more likely it is that error may become significant. When the columns of $\mathbf{A}$ are highly correlated (linearly dependent) then matrix $\mathbf{A}$ is said to be *near-singular*. If one column of $\mathbf{A}$ can be calculated from the values in the other columns then $\mathbf{A}$ is *singular*, which means that it does not possess an inverse (just as the value 0.0 does not possess a reciprocal). Consequently, the least-squares equations cannot be solved. However, it is the problem of near-singularity of $\mathbf{A}$ that should concern users of linear mixture modelling because it is likely that the end-member spectra are similar. The solution (the elements of vector $\mathbf{f}$) may, in such cases, be significantly in error, especially in the case of multispectral (as opposed to hyperspectral) data when the least-squares equations are being solved with relatively small numbers of observations.

Boardman (1989) suggests that a procedure based on the singular-value decomposition (SVD) handles the problem of near-singularity of $\mathbf{A}$ more effectively. The inverse of matrix $\mathbf{A}$ (i.e. $\mathbf{A}^{-1}$) is found using the SVD:

$$\mathbf{A} = \mathbf{UWV}'$$

where **U** is an $m \times n$ column-orthogonal matrix, **W** is an $n \times n$ matrix of singular values and **V** is an $n \times n$ matrix of orthogonal columns. **V'** is the matrix transpose of **V**. The inverse of **A** is found from the following expression:

$$\mathbf{A}^{-1} = \mathbf{VW}^{-1}\mathbf{U'}$$

The singular values, contained in the diagonal elements of **W**, give an indication of the dimensionality of the space containing the spectral end-members. They can be thought of as analogous to the principal components of a correlation or covariance matrix. If the spectral end-members are completely independent then the information in each dimension of the space defined by the spectral bands is equal and the singular values $\mathbf{W}_{ii}$ are all equal. Where one spectral end-member is a linear combination of the remaining end-members then one of the diagonal elements of **W** will be zero. Usually neither of these extreme cases is met with in practice, and the singular values of **A** (like the eigenvalues of a correlation matrix, used in principal components analysis) take different magnitudes. An estimate of the true number of end-members can therefore be obtained by observing the magnitudes of the singular values, and eliminating any singular values that are close to zero. If this is done then the inverse of **A** can still be found, whereas if matrix inversion methods are used then the numerical procedure either fails or produces a result that is incorrect.

If the landscape is composed of a continuously varying mixture of idealised or pure types, it might appear to be illogical to search within that landscape for instances of these pure types. Hence, in a number of studies, laboratory spectra have been used to characterise the mixture components. Adams *et al.* (1993) term these spectra *reference end-members*. They are most frequently used in geological studies of arid and semi-arid areas. Since laboratory spectra are recorded in reflectance or radiance units, the image data must be calibrated and atmospherically corrected before use, as described in Chapter 4.

In other cases, empirical methods are used to determine which of the pixels present in the image set can be considered to represent the mixture components. Murphy and Wadge (1994) use principal components analysis (PCA, section 6.4) to identify candidate image end-members. A graphical display of the first two principal components of the image set is inspected visually and

'. . . the pixels representing the end member for each cover type should be located at the vertices of the

polygon that bounds the data space of the principal components which contain information.'

Murphy and Wadge (1994, p. 73)

This approach makes the assumption that the m-dimensional space defined by the spectral bands can be collapsed on to two dimensions without significant loss of proximity information; in other words, it is assumed that pixels that lie close together in the PC1–PC2 plot are actually close together in the m-dimensional space, and vice versa. Principal components analysis is discussed in section 6.4, where it is shown that the technique is based on partitioning the total variance of the image data set in such a way that the first principal component accounts for the maximum variance of a linear combination of the spectral bands, the second principal component accounts for a maximum of the remaining variance and so on, with the restriction that the principal components are orthogonal. Since PCA has the aim of identifying dimensions of variance (in that the principal components are ordered on the basis of their variance), the first and second principal components will, inevitably, contain much of the information present in the image data set. However, significant variability may remain in the lower-order principal components. Furthermore, if the PCA is based on covariances, then the spectral bands may contribute unequally to the total variance analysed. Tompkins *et al.* (1997) present an elaborate methodology for end-member selection; their procedure estimates both the end-member proportions and the end-member reflectances simultaneously.

A possible alternative is based on the work of Imbrie (1963), who uses a method called Q-mode factor analysis to identify end-members in a study of the mineralogy of geological samples. The topic of factor analysis is too extensive to be considered here; interested readers are referred to Imbrie (1963), Harman (1967) and Mather (1976). In summary, Q-mode factor analysis considers the similarity relationships – measured by the cosine-theta ($\cos \theta$) coefficient of similarity, which is described below – between k pixels measured over m images. In contrast, R-mode analysis considers the relationships between the m spectral bands over the k pixels. Principal components analysis is an R-mode operation. The eigenvalues of the $k \times k$ inter-pixel similarity matrix (the *cosine-theta* matrix) are studied in order to determine the number of end-members. The scaled eigenvectors of the cosine-theta matrix (equivalent to principal component loadings, and representing the strength of the relationship between the corresponding factor and the individual pixels) are then rotated using the varimax procedure (Harman, 1967; Mather, 1976), which attempts to maximise the number of loadings that are either near to zero or close to ± 1 while keeping the factors orthogonal. The highest absolute loadings on each Q-mode varimax factor are selected as end-members. In Imbrie's (1963) study, the varimax rotation stage is followed by an oblique vector rotation in which each varimax axis is moved independently until the loading on one of the end-members is equal to unity. Each Q-mode factor then has one loading of exactly 1.0 on one of the end-members. The rest of the loadings on that oblique factor should range from zero to one, and represent proportions of the associated end-member. Imbrie (1963) gives examples which demonstrate that this is, indeed, the case.

In practice, image sizes are such that the computation of the eigenvalues and eigenvectors of a $k \times k$ similarity matrix is far too time-consuming to be a realistic proposition. For a 512×512 image the value of k will be $68\,719\,476\,736$ ($= 262\,144^2$). The memory required to store this similarity matrix is 275 Gb (since 4 bytes are used to store each element of the similarity matrix) and the time required to compute an eigensolution would be considerable, assuming that it were feasible. The problem becomes considerably more severe as the image size increases. A sampling procedure can be adopted and 600 or so pixels chosen randomly from the image area of interest. Although the resulting similarity matrix has only around 360 000 elements, this number is still far too large to be dealt with using standard numerical procedures. Wilkinson and Reinsch (1971) present algorithms to compute a specified number of eigenvalues of a matrix whose values exceed a threshold of 0.0001. The scaled eigenvectors corresponding to these eigenvalues are rotated using a varimax rotation followed by an oblique vector rotation (Imbrie, 1963). Unfortunately, the procedure does not appear to work well on real data. The number of significant under- and overshoots in the oblique vector solution is considerable, and the pixels selected as end-members do not represent all of the target areas in the image. The procedure is included in the MIPS programs on the accompanying CD-ROM as `qmode.exe` (Appendix B). Readers may wish to experiment in order to compare the results achieved by this method with those derived from PCA.

A third empirical method for selecting mixture components is Sammon's (1969) non-linear mapping. The method uses a measure of goodness of fit between the inter-pixel distances measured in the original m-dimensional space and those measured in the reduced-dimensionality space, usually of two or three dimensions (see

Figure 8.10). The two- or three-dimensional representation is updated so as to reduce the error in the inter-pixel distances measured in the two- or three-dimensional space compared to the equivalent distance measured in the full space of m spectral bands. A minimum of the error function is sought using the method of steepest descent, similar to that used in the back-propagating artificial neural net. A program, `nlmap`, is included on the CD-ROM and is described in Appendix B. As in the case of Q-mode factor analysis, discussed above, a sample of k pixels is selected randomly from the image data set using program `sel_pix` (Appendix B). Once the iterative procedure has converged, it is possible to display plots of the dimensions a pair at a time using program `plot_pix`, and to identify the extreme points using the mouse. Study of the plots of pairs of dimensions provides some interesting insights into the nature of the materials in the image. Non-linear mapping analysis appears to be a better way of analysing image information content than either the principal components approach (because the analysis is based on inter-pixel distances) or Q-mode factor analysis. Bateson and Curtiss (1996) discuss another multidimensional visualisation method of projecting points (representing pixels) on to two dimensions. Their method is based on a procedure termed parallel coordinate representation (Wegman, 1990), which allows the derivation of 'synthetic' end-member spectra that do not coincide with image spectra. The method is described in some detail by Bateson and Curtiss (1996).

A simple alternative to linear spectral unmixing is provided by the method of spectral angle mapping, which is based on the well-known coefficient of proportional similarity, or cosine-theta ($\cos \theta$). This coefficient measures the difference in the shapes of the spectral curves (Imbrie, 1963; Weinand, 1974). In the context of image processing, a small number of pixels are selected as the reference set and the remaining image pixels are compared to these reference pixels. The cosine-theta coefficient takes the value 0.0 when the two pixel vectors (reference and test) are completely dissimilar and the value 1.0 when the two pixel vectors coincide. If the reference pixel vector is represented by r_i and any other image pixel vector is written as p_i then the value of cosine-theta is:

$$\cos \theta = \frac{\sum\limits_{i=1}^{N} r_i p_i}{\left(\sum\limits_{i=1}^{N} r_i^2 \right)^{0.5} \left(\sum\limits_{i=1}^{N} p_i^2 \right)^{0.5}}$$

given N bands of image data. The value of $\cos \theta$ can be converted to an angle in the range 0–$90°$ or the range of cosine-theta (0–1) can be directly translated to the 0–255 scale for display purposes. One image of cosine-theta values is produced per reference vector $\mathbf{r}$. There are no statistical assumptions or other limitations on the number or nature of the reference vectors used in these calculations. Kruse *et al.* (1993), Ben-Dor and Kruse (1995) and van der Meer (1996c) provide examples of the use of the method, in comparison with the linear spectral unmixing technique. Program `spec_ang` is provided on the accompanying CD-ROM, and a description is given in Appendix B.

Applications of linear mixture modelling in remote sensing are numerous, though some authors do not give enough details for their readers to discover whether or not the linear mixture model is a good fit. If the data do not fit the model then, at best, the status of the results is questionable. Good surveys of the method are provided by Adams *et al.* (1989, 1993), Ichoku and Karnieli (1996), Settle and Drake (1993) and Settle and Campbell (1998). A selection of typical applications is: Cross *et al.* (1991), who use AVHRR data to provide sub-pixel estimates of tropical forest cover; and Gamon *et al.* (1993), who use mixture modelling to relate AVIRIS (imaging spectrometer) data to ground measurements of vegetation distribution. Hill (1993–4) describes the technique in the context of land degradation studies in the European Mediterranean basin, while Shipman and Adams (1987) assess the detectability of minerals on desert alluvial fans. Smith *et al.* (1990) provide a critical appraisal of the use of mixture modelling in determining vegetation abundance in semi-arid areas. Other useful references are Bateson and Curtiss (1996), Bryant (1996), García-Haro *et al.* (1996), Hill and Horstert (1996), Kerdiles and Grondona (1995), Roberts *et al.* (1993), Shimabukuro and Smith (1991), Thomas *et al.* (1996), Ustin *et al.* (1991, 1996) and Wu and Schowengerdt (1993). Roberts *et al.* (1993) consider an interesting variant on the standard mixture modelling technique, which, in a sense, attempts a global fit in that each pixel in the image is assumed to be the sum or mixture of the same set of end-members. Roberts *et al.* (1993) suggest that the number and nature of the end-members may vary over the image. This opens the possibility of using an approach similar to stepwise multiple regression (Grossman *et al.*, 1996) in which the best k from a pool of m possible end-members are selected for each pixel in turn. Foody *et al.* (1996, 1997) describe an approach to mixture modelling using artificial neural networks (section 8.4.3), while Bernard *et al.* (1996) consider the classification of mixtures using neural networks. The use of the ART network in fuzzy classification is discussed by Carpenter *et al.* (1992) and Mannan *et al.* (1998).

8.5.2 Fuzzy classifiers

The distinction between 'hard' and 'soft' classifiers is discussed in the opening paragraphs of the introduction to section 8.5. The 'soft' or fuzzy classifier does not assign each image pixel to a single class in an unambiguous fashion. Instead, each pixel is given a 'membership grade' for each class. Membership grades range in value from 0 to 1, and provide a measure of the degree to which the pixel belongs to or resembles the specified class, just as the fractions or proportions used in linear mixture modelling (section 8.5.1) represent the composition of the pixel in terms of a set of end-members. It might appear that membership grade is equivalent to probability, and the use of probabilities in the maximum-likelihood classification rule might lend support to this view. Bezdek (1993) differentiates between precise and fuzzy data, vague rules and imprecise information. Crisp sets contain objects that satisfy unambiguous membership requirements. Bedzek points out that $H = \{r \in \Re \mid 6 \leq 8\}$ precisely represents the crisp or hard set of numbers H from 6 to 8. Either a number is a member of the set H or it is not. If a set F is defined by a rule such as 'numbers that are close to 7' then a given number such as 7.2 does not have a precise membership grade for F, whereas its membership grade for H is 1.0. Bezdek (1993, p. 1) also notes that

'. . . the modeler must decide, based on the potential applications and properties desired for F, what m_F should be.'

The membership function m_F for set F can take any value between 0 and 1, so that numbers far away from 7 still have a membership grade. This, as Bezdek notes, is simultaneously both a strength and a weakness. Bezdek (1994) points out that fuzzy memberships represent similarities of objects to imprecisely defined properties, while probabilities convey information about relative frequencies.

Fuzzy information may not be of much value when decisions need to be taken. For example, a jury must decide, often on the basis of fuzzy information, whether or not the defendant is guilty. A fuzzy process may thus have a crisp outcome. The process of arriving at a crisp or hard conclusion from a fuzzy process is called *defuzzification*. In some cases, and the mixture model discussed in the previous section is an example, we may wish to retain some degree of flexibility in presenting our results. The membership grades themselves may be of interest as they may relate to the proportions of the pixel area that are represented by each of the end-members. In semi-natural landscapes one may accept that different land cover types merge in a transition zone. For instance, a forested area may merge gradually into grassland, and grassland in turn may merge imperceptibly into desert. In such cases, trying to draw hard and fast boundaries around land cover types may be a meaningless operation, akin to putting

> '. . . boundaries that do not exist around areas that do not matter.'
>
> Kimble (1951, p. 159)

On the other hand, woodland patches, lakes, and agricultural fields have sharp boundaries, though of course the meaning of 'sharp' depends upon the spatial scale of observation, as noted previously. A lake may shrink or dry up during a drought, or the transition between land cover types may be insignificant at the scale of the study. At a generalised scale it may be perfectly acceptable to draw sharp boundaries between land cover types. Wang (1990) discusses these points in the context of the maximum likelihood decision rule, and proposes a fuzzy version of maximum likelihood classification using weighted means and variance–covariance matrices similar to those described above (section 8.4.1) in the context of deriving robust estimates from training samples. Gong *et al.* (1996) use the idea of membership functions to estimate classification uncertainty. If the highest membership function value for a given pixel is considerably greater than the runner-up, then the classification output is reasonably certain, but where two or more membership function values are close together then the output is less certain.

One of the most widely used unsupervised fuzzy classifiers is the fuzzy *c*-means clustering algorithm. Bezdek *et al.* (1984) describe a Fortran program. Clustering is based on the distance (dissimilarity) between a set of cluster centres and each pixel. Either the Euclidean or the Mahalanobis distance can be used. These distances are weighted by a factor *m* that the user must select. A value of *m* equal to 1 indicates that cluster membership is 'hard' while all membership function values approach equality as *m* gets very large. Bezdek *et al.* (1984) suggest that *m* should lie in the range 1–30, though other users appear to choose a value of *m* of less than 1.5. Applications of this method are reported by Bastin (1997), Cannon *et al.* (1986), Du and Lee (1996), Foody (1996) and Key *et al.* (1989). Other uses of fuzzy classification procedures are reported by Blonda and Pasquariello (1991) and Maselli *et al.* (1995a, 1996).

The basic ideas of the fuzzy *c*-means classification algorithm can be expressed as follows. **U** is the *membership grade matrix* with *n* columns (one per pixel) and *p* rows (one per cluster):

$$\mathbf{U} = \begin{bmatrix} u_{11} & \cdots & u_{1n} \\ \vdots & & \vdots \\ u_{p1} & \cdots & u_{pn} \end{bmatrix}$$

The sum of the membership grades for a given class (row of **U**) must be non-zero. The sum of the membership grades for a given pixel (column of **U**) must add to one, and the individual elements of the matrix, u_{ij}, must lie in the range 0–1 inclusive. The number of clusters *p* is specified by the user, and the initial locations of the cluster centres are either generated randomly or supplied by the user. The Euclidean distance from pixel *i* to cluster centre *j* is calculated as usual, that is,

$$d_{ij} = \sqrt{\sum_{l=1}^{k} (x_{il} - c_{jl})^2}$$

and the centres *c* are computed from

$$c_{jl} = \frac{\sum_{i=1}^{n} u_{ji}^m x_{il}}{\sum_{i=1}^{n} u_{ji}^m}$$

This is simply a weighted average (with the elements *u* being the weights) of all pixels with respect to centre *j* $(1 \leq j \leq p)$. The term x_{il} is the measurement of the *i*th pixel $(1 \leq i \leq n)$ on the *l*th spectral band or feature. The exponent *m* is discussed above.

Each of the membership grade values u_{ij} is updated according to its Euclidean distance from each of the cluster centres:

$$u_{ij} = \frac{1}{\sum_{c=1}^{p} \left(\dfrac{d_{ij}}{d_{cj}} \right)^{2/(m-1)}}$$

where $1 \leq i \leq p$ and $1 \leq j \leq n$ (Bezdek *et al.*, 1984). The procedure converges when the elements of **U** differ by no more than a small amount between iterations.

The columns of **U** represent the membership grades for the pixels on the fuzzy clusters (rows of **U**). A process of defuzzification can be used to determine the cluster membership for pixel *i* by choosing the element in column *i* of **U** that contains the largest value. Alternatively, and perhaps more informatively, a set of classified images could be produced, one per class, with the class membership grades scaled from 0–1 to 0–255. If this is done, then the pixels with membership grades close to 1.0 would appear white, and pixels whose membership grade was close to zero would appear black. Colour composites of the classified images, taken three at a time, would be interpretable in terms of class membership. Thus, if class 1 were shown in red and class 2 in green, then a yellow pixel would be equally likely to belong to class 1 as to class 2.

The activations of the output neurons of an artificial neural network, or the class membership probabilities derived from a standard maximum likelihood classifier, can also be used as approximations of the values of the membership function. In an ideal world, the output from a neural network for a given pixel will be represented by a single activation value of 1.0 in the output layer of the network, with all other activation levels being set to 0.0. Similarly, an ideal classification using the maximum likelihood method would be achieved when the class membership probability for class *i* is 1.0 and the remaining class membership probabilities are zero. These ideal states are never achieved in practice. The use of a 'winner takes all' rule is normally used, so that the pixel to be classified is allocated to the class associated with the highest activation in the output layer of the network, or the highest class membership probability. It is, however, possible to generate one output image per class, rather than a single image representing all the classes of a 'hard' representation. The first image will show the activation for output neuron number one (or the class membership probability for that class, if the maximum likelihood method is used) scaled to a 0–255 range. Pixels with membership function values close to 1.0 are bright, while pixels with a low membership function value are dark. This approach allows the evaluation of the degree to which each pixel belongs to a given class. Such information will be useful in determining the level of confidence that the user can place in the results of his or her classification procedure. Further reading on the use of neural networks in fuzzy classification and spectral unmixing is provided by Buckley and Hayashi (1994), Foody *et al.* (1997) and Bernard *et al.* (1996).

8.6 OTHER APPROACHES TO IMAGE CLASSIFICATION

It is sometimes the case that different patterns can be distinguished on the basis of one, or a few, features. The one (or few) features may not be the same for each pair of patterns. On a casual basis, one might separate a vulture from a dog on the basis that one has wings and the other does not, while a dog and a sheep would be distinguished by the fact that the dog barks and the sheep baas. The feature 'possession of wings' is thus quite irrelevant when the decision to be made is 'Is this a dog or a sheep?' The decision-tree classifier is an attempt to use this 'stratified' or 'layered' approach to the problem of discriminating between spectral classes. The design of decision trees is a research topic of considerable interest (Wang 1986a, b; Lee and Richards, 1985); the latter authors found that the computer time for the decision-tree approach increased only linearly with the number of features while the time for the maximum likelihood classification increased as the square of the number of features, while classification accuracy for both approaches was not dissimilar.

Hansen *et al.* (1996) discuss a method based on the use of a 'classification tree', which, in principle, is similar to the decision-tree approach mentioned in the preceding paragraph. Rule-based classifications are also related to the decision-tree approach, and represent an attempt to apply artificial intelligence methods to the problem of determining the class to which a pixel belongs (Srinivasana and Richards, 1990). The number of rules can become large, and require advanced search algorithms to perform an exhaustive analysis. The *genetic algorithm*, which uses a randomised approach based on the processes of mutation and crossover, has become popular in recent years as a robust search procedure. Seftor and Larch (1995) illustrate its use in optimising a rule-based classifier, and Clark and Cañas (1995) compare the performance of a neural network and a genetic algorithm for matching reflectance spectra. The genetic algorithm is also described by Zhou and Civco (1996) as a tool for spatial decision-making. Stolz and Mauser (1996) combine fuzzy logic (based on maximum-likelihood probabilities) and a rule base to classify agricultural land use in a mountainous area of southern Germany and report a substantial improvement in accuracy.

If a single image set is to be classified on the basis of spectral data alone then the 'reliability' of the respective features is not usually taken into consideration. All bands are considered to be equally reliable, or important. Where multi-source data are used, the question of reliability needs to be considered. For example, one might use data derived from maps, such as elevation, and may even use derivatives of these features, such as slope and aspect, as well as data sets acquired by different sensor systems (section 8.7). The Dempster–Shafer theory of evidence has been used by some researchers to develop the method of *evidential reasoning*, which is a formal procedure that weights individual data sources according to their reliability or importance. Duguay and Peddle (1996), Kim and Swain (1995), Lee *et al.* (1987), Peddle (1993, 1995a, b), Peddle *et al.* (1994), Srinivasana and Richards (1990) and Wilkinson and Mégier (1990) provide a survey of the uses of the evidential reasoning approach in land cover classification, and compare its effectiveness with that of other classifiers. They note that, in comparison to the artificial neural network, the evidential reasoning approach is faster, is also non-parametric, is independent of scales of measurement, and can handle uncertainty about label assignments. Another approach to multi-source classification is based on Bayesian theory (Benediktsson *et al.*, 1990).

The image classification techniques discussed thus far are based on the labelling of individual pixels in the expectation that groups of neighbouring pixels will form regions or patches with some spatial coherence. This method is usually called the *per-pixel* approach to classification. An alternative approach is to use a process termed *segmentation*, which – as its name implies – involves the search for homogeneous areas in an image set and the identification of these homogeneous areas with information classes. A hierarchical approach to segmentation, using the concept of geographical scale, is used by Woodcock and Harward (1992) to delineate forest stands. This paper also contains an interesting discussion of the pros and cons of segmentation versus the per-pixel approach. Shandley *et al.* (1996) test the Woodcock–Harward image segmentation algorithm in a study of chaparral and woodland vegetation in southern California. Mason *et al.* (1988) present a method of image segmentation that uses digital map data within a GIS.

Another alternative to the per-pixel approach requires *a priori* information about the boundaries of objects in the image, for example, agricultural fields. If the boundaries of these fields are digitised and registered to the image, then some property or properties of the pixels lying within the boundaries of the field can be used to characterise that field. For instance, the means and standard deviations in the six non-thermal TM bands of pixels lying within agricultural fields could be used as features defining the spectral reflectance properties of the fields. The fields, rather than the pixels, are then classified, hence this method is termed a *per-field* approach. If field boundaries digitised from a suitable map are not available, it may also be possible to use edge-detection and line-following techniques to derive boundary information from the image (section 7.4), though it is unlikely that a full set of boundaries will be extractable. Normally, the use of map and image data would take place within a geographical information system (GIS), which provides facilities for manipulating digitised boundary lines (for example, checking the set of lines to eliminate duplicated boundaries, ensuring that lines 'snap on' to nodes, and identifying illogical lines that end unexpectedly). One useful feature of most GIS is their ability to create buffer zones on either side of a boundary line. If a per-field approach is to be used then it would be sensible to create a buffer around the boundaries of the objects to be classified in order to remove pixels that are likely to be mixed. Such pixels may represent tractor-turning zones (headlands) as well as field boundary vegetation. The per-field approach is often used with SAR imagery because the individual pixels contain speckle noise, which leads to an unsatisfactory per-pixel classification. Averaging of the pixels within a defined geographical area such as a field generally gives better results (Schotten *et al.*, 1995; Wooding *et al.*, 1993). Lobo *et al.* (1996) discuss the per-pixel and per-field approaches in the context of Mediterranean agriculture.

8.7   INCORPORATION OF NON-SPECTRAL FEATURES

Two types of feature in addition to spectral values can be included in a classification procedure. The first kind are measures of the texture of the neighbourhood of a pixel, while the second kind represent external (i.e. non-remotely-sensed) information such as terrain elevation values or information derived from soil or geology maps. Use of textural information has been limited in passive remote sensing, largely because of two difficulties. The first is the operational definition of texture in terms of its derivation from the image data, and the second is the computational cost of carrying out the texture calculations relative to the increase in classification accuracy, if any. External data have not been widely used either, though digital cartographic data have become much more readily available in recent years. A brief review of both these topics is provided in this section.

8.7.1   Texture

Getting a good definition of texture is almost as difficult as measuring it. While the grey level of a single pixel in a greyscale image can be said to represent 'tone', the texture of the neighbourhood in which that pixel lies is a more elusive property, for several reasons. At a simple level, texture can be thought of as the variability in tone within a neighbourhood, or the pattern of spatial relationships among the grey levels of neighbouring pixels, and is usually described in terms such as 'rough' or 'smooth'. Variability is a variable property, however; it is not necessarily random – indeed, it may be structured with respect to direction as, for instance, a drainage pattern on an area underlain by dipping beds of sandstone. The observation of texture depends on two factors. One is the scale of the variation that we are willing to call 'texture' – it might be local or regional. The second is the scale of observation. Micro-scale textures that might be detected by the panchromatic band of the SPOT HRV would not be detected by the NOAA AVHRR because of the different spatial resolutions of the two sensor systems (10 m and 1.1 km respectively). We must also be careful to distinguish between the real-world texture present, for example, in a field of potatoes (which are generally planted in parallel rows), and the texture that is measurable from an image of the same area at a given spatial resolution.

The fact that texture is difficult to define is no reason to ignore it. It has been found to be an important contributor to the ability to discriminate between targets of interest where the spatial resolution of the image is sufficient to make the concept a meaningful and useful one, for example in manual photo interpretation. On the other hand, an investigation of a number of texture measurements for satellite remotely-sensed data led Irons and Petersen (1981) to conclude that none of these measures was useful in thematic mapping. This divergence of opinion demonstrates the difficulty in defining the property known as texture, the dependence of this property on the scale of the image and the variation in scale of 'texture'.

The earliest application of texture measurements to digital remotely-sensed image data was published by Haralick *et al.* (1973). These authors proposed what has become known as the grey-level co-occurrence matrix (GLCM), which represents the distance and angular spatial relationships over an image sub-region of specified size. Each element of the GLCM is a measure of the probability of occurrence of two grey-scale values separated by a given distance in a given direction. The concept is more easily appreciated via a simple numerical example. Table 8.3(a) shows a small segment of a digital image quantised to four grey levels (0–3). The number of adjacent pixels with grey levels i and j is counted and placed in element (i, j) of the GLCM **P**. Four definitions of adjacency are used: horizontal ($0°$), vertical ($90°$), diagonal (bottom left to top right, i.e. $45°$)

Table 8.3 Example data and derived grey-level co-occurrence matrices: (a) test data set: (b)–(e): GLCMs for angles of $0°$, $45°$, $90°$ and $135°$, respectively.

(a) Data set.

0	0	0	2	1
0	1	1	2	2
0	1	2	2	3
1	1	2	3	3

(b) Angle $0°$.

	0	1	2	3
0	4	2	1	0
1	2	4	4	0
2	1	4	4	2
3	0	0	2	2

(c) Angle $45°$.

	0	1	2	3
0	2	2	0	0
1	1	4	3	0
2	0	3	6	0
3	0	0	0	2

(d) Angle $90°$.

	0	1	2	3
0	4	3	0	0
1	3	4	2	0
2	0	2	6	2
3	0	0	2	2

(e) Angle $135°$.

	0	1	2	3
0	0	4	1	0
1	4	0	3	0
2	1	3	2	3
3	0	0	3	0

and diagonal (top left to bottom right, i.e. $135°$). The distance used in these calculations is one pixel. Thus, four GLCM are calculated, denoted $\mathbf{P}_0$, $\mathbf{P}_{90}$, $\mathbf{P}_{45}$ and $\mathbf{P}_{135}$ respectively. For example, the element $\mathbf{P}_0(0, 0)$ is the number of times a pixel with grey-scale value 0 is horizontally adjacent to a pixel that also has the grey-scale value 0, scanning from left to right as well as right to left. Element $\mathbf{P}_0(1, 0)$ is the number of pixels with value 1 that are followed by pixels with value 0, while $\mathbf{P}_0(0, 1)$ is the number of pixels with value 0 that are followed by pixels with value 1, again looking in both the left–right and right–left directions. The four GLCM are shown in Tables 8.3(b)–(e). The program glcm.exe that is described in Appendix B is designed to read in a small digital image in ASCII format and to calculate the GLCM for the four directions as described above, with a distance of one pixel. Readers wishing to discover how the GLCM is generated are encouraged to run the program using invented data and to check the results using pencil-and-paper methods.

Haralick *et al.* (1973) originally proposed 32 textural features to be derived from each of the four GLCM. Few instances of the use of all these features can be cited; Jensen and Toll (1982) use only one, derived from the Landsat-1 to -3 MSS band 5 image. The first two of the Haralick measures will be described here to illustrate the general approach. The first measure (f_1) is termed the *angular second moment*, and is a measure of homogeneity. It effectively measures the number of transitions from one grey level to another and is high for few transitions. Thus, low values indicate heterogeneity. The second Haralick texture feature, *contrast* (f_2), gives non-linearly increasing weight to transitions from low to high grey-scale values. The weight is the square of the difference in grey level. Its value is a function of the number of high–low or low–high transitions in grey level. The two features are formally defined by:

$$f_1 = \sum_{i=1}^{N} \sum_{j=1}^{N} \left[\frac{\mathbf{P}(i,j)}{R} \right]^2$$

$$f_2 = \sum_{n=0}^{N-1} \sum_{i=1}^{N} \sum_{j=1}^{N} \frac{\mathbf{P}(i,j)}{R} \qquad |i-j| = n$$

where N is the number of grey levels, $\mathbf{P}(i,j)$ is an element of one of the four GLCM listed above, and R is the number of pairs of pixels used in the computation of the corresponding $\mathbf{P}$. For the horizontal and vertical directions R is equal to $2N^2$, while in the diagonal direction R equals $2(N-1)^2$. Haralick *et al.* (1973) and Haralick and Shanmugam (1974) give examples of the images and corresponding values of f_1 and f_2. A grassland area gave low (0.064 to 0.128) values of f_1 indicating low homogeneity and high contrast whereas a predominantly water area had values of f_1 ranging from 0.0741 to 0.1016 and of f_2 between 2.153 and 3.129 (higher homogeneity, lower contrast). These values are averages of the values of f_1 and f_2 for all four angular GLCM. The values of f_1 for the example data in Table 8.3 are (for angles of $0°$, $45°$, $90°$ and $135°$) 0.074, 0.247, 0.104 and 0.216, while the values of f_2 for the same data are 0.688, 0.444, 0.438 and 1.555.

Rather than compute the values of these texture features for windows surrounding a central pixel, Haralick and Shanmugam (1974) derive them for 64×64 pixel sub-images. This shortcut is unnecessary nowadays, as sufficient computer power is available to compute local texture measures for individual pixels. They also use 16 rather than 256 quantisation levels in order to reduce the size of the matrices $\mathbf{P}$. If all 256 quantisation levels of the Landsat TM were to be used, for example, then $\mathbf{P}$ would become very large. The reduction in the number of levels from 256 to 16 or 32 might be seen as an unacceptable price to pay, though if the levels are chosen after a histogram equalisation enhancement (section 5.3.2) to ensure equal probability for each level then a reduction from 256 to 64 grey levels will give acceptable results (Tso, 1997).

Improvements in computing power since the 1980s have led to an increased interest in the use of texture measures. Paola and Schowengerdt (1997) incorporate texture features into a neural network-based classification by including as network inputs the grey-scale values of the eight neighbours of the pixel to be classified. The central pixel is thus classified on the basis of its spectral reflectance properties plus the spectral reflectance properties of the neighbourhood. Although the number of input features is considerably greater than would be the case if the central pixel alone were to be input, Paola and Schowengerdt report that the extra size of the network is compensated by faster convergence during training. Mather *et al.* (1998) use a number of texture measures in a study of lithological mapping using SIR-C SAR and Landsat TM data. Four approaches to the measurement of texture are compared in this study. The first is based on the GLCM, which is discussed above.

A second approach uses filters in the frequency domain to measure the proportion of high-frequency information (section 7.5) in each of the moving windows for which texture is measured (see also Riou and Seyler, 1997). The third approach is based on the calculation of the fractal dimension of the region surrounding the pixel of interest, and the fourth method uses a model of the spatial autocorrelation properties of the same region as texture features. The conclusions drawn from this study are: (i) Classification accuracy, in terms of lithological mapping, is poor when Landsat TM spectral information is used alone. Values of accuracy (section 8.10) in the range 47–57% were achieved, with the feed-forward neural network performing best. Classification accuracy rose by 12% at best when SAR-based texture measures were added. (ii) The two best-performing texture descriptors were produced by the GLCM and autocorrelation modelling approaches, which raised classification accuracy to 69.5% and 68.8% respectively. Results from the GLCM approach using 256 and 64 grey levels are compared by Tso (1997), who shows that accuracy is not significantly affected by the reduction in the number of grey levels, though the computational requirements are significantly reduced. The size of the moving windows was estimated by calculating a geostatistical function, the semi-variogram, for each lithological unit. Geostatistical methods are summarised above (section 8.5, introduction). Further reading on alternative approaches to and applications of texture measures in remote-sensing image classification is provided by Bruzzone *et al.* (1997), Carlson and Ebell (1995), de Jong and Burrough (1995), Dikshit (1996), Fioravanti (1994), Franklin and Peddle (1987), Frankot and Chellappa (1987), Keller *et al.* (1989), Lark (1996), Marceau *et al.* (1990), Sarkar and Chaudhuri (1994), Schistad and Jain (1992), Soares *et al.* (1997), Stromberg and Farr (1986), Tso (1997) and Wezka *et al.* (1976).

With few exceptions, texture measures have not been found to be cost-effective in terms of the improvement in classification accuracy resulting from their use. Two reasons could be proposed to account for this: (i) the difficulty of establishing the relationship between land-surface texture and scale in terms of the spatial resolution of the image and the scale of the textural feature on the ground relative to pixel size, and (ii) the cost of calculating the texture feature. Nevertheless, the value of texture as one of the fundamental pattern elements used in manual photo interpretation guarantees a continuing interest in the topic.

8.7.2 Use of external data

The term external (or ancillary) is used to describe any data other than the original image data or measures derived from these data. Examples include elevation and soil-type information or the results of a classification of another image of the same spatial area. Some such data are not measured on a continuous (ratio or interval) scale and it is therefore difficult to justify their inclusion as additional feature vectors. Soil type or previous classification results are examples, both being categorical variables. Where a continuous variable, such as elevation, is used, difficulties are encountered in deriving training samples. Some classes (such as water) may have little relationship with height and the incorporation of elevation information in the training class may well reduce rather than enhance the efficiency of the classifier in recognising those categories.

An external variable may be used to stratify the image data into a number of categories. If the external variable is land elevation then, for example, the image may be stratified in terms of land below 500 m, between 500 and 800 m and above 800 m. For each stratum of the data an estimate of the frequency of occurrence of each class must be provided by the user. This estimate might be derived from field observation, from sampling of a previously classified image or from sample estimates obtained from air photographs or maps. The relative frequencies of each class are then used as estimates of the prior probabilities of a pixel belonging to each of the k classes and the maximum likelihood algorithm used to take account of these prior probabilities (section 8.4.2.3). The category or level of the external variable is used to point to a set of prior probabilities, which are then used in the estimation of probabilities of class membership.

This would assist in the distinction between classes that are spectrally similar but have different relationships with the external variable. Strahler *et al.* (1978) used elevation and aspect as external variables; both were separated into three categories and used as pointers to sets of prior probabilities. They found that the elevation information contributed considerably to the improvement in the accuracy of forest cover classification. Whereas the spectral features alone produced a classification with an accuracy estimated as 57%, the addition of terrain information and the introduction of prior probability estimates raised this accuracy level to a more acceptable 71%. The use of elevation and aspect to point to a set of prior probabilities raised the accuracy further to 77%. Strahler (1980) provides an excellent review of the use of external categorical variables and associated sets of prior probabilities in maximum likelihood classification. He concludes that the method 'can be a powerful and effective aid to improving classification accuracy'. Another accessible reference is Hutchinson (1982), while Maselli *et al.* (1995b) discuss integration of ancillary data using a non-parametric method of estimating prior probabilities.

Elevation data sets are now available for many parts of the world, generally at a scale of 1 : 50 000 or coarser. Digital elevation models (DEM) can also be derived from stereo SPOT images as well as from interferometric data from SAR sensors. The widespread availability of GIS means that many users of remotely-sensed data can now derive DEM by digitising published maps, or by using photogrammetric software to generate a DEM from stereoscopic images such as SPOT HRV or IRS-1 LISS. Care should be taken to ensure that the scale of the DEM matches that of the image. GIS technology allows the user to alter the scale of a data set, and if this operation is performed thoughtlessly then error is inevitable. Users of DEM derived from digitised contours should refer to one of the many GIS textbooks now available (for example, Bonham-Carter, 1994) to ensure that correct procedures are followed.

Rather than use an external variable to stratify the image data for improved classifier performance, users may prefer to use what has become known as the *stacked vector* approach, in which each feature (spectral, textural, external) is presented to the classifier as an independent input. Where a statistical classifier, such as maximum likelihood, is used then this approach may well not be satisfactory. Some external variables, such as elevation, may be measured on a continuous scale but may not be normally distributed or even unimodal for a given class. Other variables, such as lithology or soil type, may be represented by a categorical label that the maximum likelihood classifier cannot handle. The value of artificial neural network classifiers is that they are non-parametric, meaning that the frequency distribution and scale of measurement of the individual input feature are not restricted. Thus, the ANN-based classifier can accept all kinds of input features without any assumption concerning the normality or otherwise of the associated frequency distribution and without consideration of whether the feature is measured on a continuous, ordinal or categorical scale. One problem with an indiscriminate approach, however, is that all features may not have equal influence on the outcome of the classification process. If one is trying to distinguish between vultures and dogs, then 'possession of wings' is a more significant discriminating feature than 'colour of eyes', though the latter may have some value. Evidential reasoning (section 8.6) offers a more satisfactory approach.

8.8 CONTEXTUAL INFORMATION

Geographical phenomena generally display order or structure, as shown by the observation that landscapes are not generally randomly organised. Thus, trees grow together in forests and groups of buildings form towns. The relationship between one element of a landscape and the whole defines the context of that element. So too the relationship between one pixel and the pixels in the remainder of the image is the context of that pixel. Contextual information is often taken into account after a preliminary classification has been produced, though at the research level investigations are proceeding into algorithms that can incorporate both contextual and spectral information simultaneously (Kittler and Föglein, 1984). The simplest methods are those which are applied following the classification of the pixels in an image using one of the methods described in section 8.4. These methods are similar in operation to the spatial filtering techniques described in Chapter 7 for they use a moving window algorithm.

The first of these methods is called a 'majority filter'. It is a logical rather than numerical filter since a classified image consists of labels rather than quantised counts. The simplest form of the majority filter involves the use of a filter window, usually measuring 3 rows × 3 columns, centred on the pixel of interest. The number of pixels allocated to each of the k classes is counted. If the centre pixel is not a member of the majority class (containing five or more pixels within the window) it is given the label of the majority class. A threshold other than five (the absolute majority) can be applied – for example, if the centre pixel has fewer than n neighbours (in the

window) that are not of the same class then relabel that pixel as a member of the majority class. The effect of this algorithm is to smooth the classified image by weeding out isolated pixels that were initially given labels that were dissimilar to the labels assigned to the surrounding pixels. These initial dissimilar labels might be thought of as noise or they may be realistic. If the latter is the case then the effect of the majority filter is to treat them as detail of no interest at the scale of the study, just as contours on a 1:50 000 map are generalised (smoothed) in comparison with those on a map of the same area at a 1:25 000 scale. A modification of the algorithm just described is to disallow changes in pixel labelling if the centre pixel in the window is adjacent to a pixel with an identical label. In this context adjacent can mean having a common boundary (i.e. to the left or right, above or below) or having a common corner. The former definition allows four pixels to be adjacent to the centre pixel, the latter eight. Thomas (1980) reports a method based on what he terms a 'proximity function'. Again, it uses a moving 3 × 3 window with the pixels numbered as shown in Figure 8.19. The function is defined by

$$F_j = \sum_i \frac{q_i q_5}{d_{i5}^2} \qquad (j = 1, \ldots, k; i = 2, 4, 6, 8)$$

In this expression the q_i are weights. If pixel i has been placed in class j then $q_i = 2$, otherwise $q_i = 0$ for i = 2, 4, 6 and 8. If pixel 5 has been placed in the jth class then $q_5 = 2$, otherwise $q_5 = 1$. The d_{i5} are distances (in metres) from the centre pixel to its neighbours. The function is evaluated for all classes j and the centre pixel is reallocated if the maximum value of F_j exceeds a threshold. Thomas (1980) suggests a value of 12×10^{-4} m for this threshold, and illustrates the use of the method with reference to classified Landsat MSS imagery. Like the majority filter, its effect is to remove isolated pixels and relabel them with the most frequently occurring label considering the vertical and horizon-

1	2	3
4	5	6
7	8	9

Figure 8.19 Pixel labelling scheme for Thomas's (1980) filter window. See text for explanation.

tal neighbours. It might also reallocate a previously unclassified pixel that had been placed in the 'reject' class by the classification algorithm. Another example of a post-classification context algorithm is given by Wharton (1980).

Harris (1981, 1985) describes a method of post-classification processing that uses a probabilistic relaxation model. An estimate of the probability that a given pixel will be labelled l_i $(i = 1, 2, \ldots, k)$ is required. Examination of the pixels surrounding the pixel under consideration is then undertaken to attempt to reduce the uncertainty in the pixel labelling by ensuring that pixel labels are locally consistent. The procedure is both iterative and rather complicated. The results reported by Harris (1985) showed the ability of the technique to clean up a classified image by eliminating improbable occurrences (such as isolated urban pixels in a desert area) while at the same time avoiding smoothing out significant and probably correct classifications. However, the computer time requirements are severe. Further discussion of the probabilistic relaxation model is given in Rosenfeld (1976), Peleg (1980), Kittler (1983) and Kontoes and Rokos (1996), while an alternative approach to the problem of specifying an efficient spectral–spatial classifier is discussed by Landgrebe (1980). An excellent general survey is by Gurney and Townshend (1983).

More recently, attention has been given to the use of geostatistical methods of characterising the spatial context of a pixel that is to be classified. Geostatistical methods are summarised in section 8.5. Image data are used to characterise the spectral properties of the candidate pixel, and geostatistical methods provide a summary of its spatial context, so that both are simultaneously considered in the decision-making process. See Lark (1998) and van der Meer (1994, 1996a, b) for a fuller exposition. Flygare (1997) gives a review of advanced statistical methods of characterising context. Wilson (1992) uses a modified maximum likelihood approach to include neighbourhood information by the use of a penalty function that increases the 'cost' of labelling a pixel as being different from its neighbours. Sharma and Sarkar (1998) review a number of approaches to the inclusion of contextual information in image classification.

8.9   FEATURE SELECTION

Developments in remote-sensing instruments over the past 10 years have resulted in image data of increasingly higher resolution becoming available in more spectral channels. Thus, the volume and dimensionality of data sets being used in image classification are exceeding the ability of both available software systems and computer hardware to deal with them. However, as shown in the discussion of the decision-tree approach to classification (section 8.6), it is possible to base a classification on the consideration of the values measured on one spectral band at a time. In this section the idea will be extended, so that we will ask: 'Can the dimensions of the data set be reduced (in order to save computer time) without losing too much of the information present in the data?' If a subset of the available spectral bands (and other features such as textural and ancillary data) will provide almost as good a classification as the full set, then there are very strong arguments for using the subset. We will consider what 'almost as good' means in this context in the following paragraphs.

Reduction of the dimensionality of a data set is the aim of principal components analysis, which is discussed in Chapter 6. An obvious way of performing the feature selection procedure would be to use the first m principal components in place of the original p features (m being smaller than p). This does not, however, provide a measure of the relative performance of the two classifications – one based on all p features, the other on m principal components. Methods of accuracy assessment (section 8.10) might be used on training and test sites to evaluate the performance directly. Information, in terms of principal components, is directly related to variance or scatter (Chapter 6) and is not necessarily a function of inter-class differences. Thus, the information contained in the last $(p - m)$ components might represent the vital piece of information needed to discriminate between class x and class y, as shown in the example of principal components analysis in Chapter 6. Principal components analysis might therefore be seen as a crude method of feature selection if it is employed thoughtlessly. It could be used in conjunction with a suitable method for determining which of the possible p components should be selected in order to maximise inter-class differences, as discussed below. Spanner *et al.* (1984) give an example.

Two methods of feature selection are discussed in this section. The first is based on the derivation of a measure of the difference between all pairs from the k groups. It is called *divergence*. The second is more empirical. It evaluates the performance of a classifier in terms of a set of test data for which the correct class assignments have been established by ground observations or by the study of air photographs or maps. The classifier is applied to subsets of the p features and classification accuracy measured for each subset using the techniques described in section 8.10. A subset is selected that gives a sufficiently high accuracy for a specific problem.

The technique based on the divergence measure requires that the measurements on the members of the k classes are distributed in multivariate-normal form. The effect of departures from this assumption is not known, but one can be certain that the results of the analysis would be less reliable as the departures from normality increased. If the departures are severe then the results could well be misleading. Hence, the divergence method is only to be recommended for use in conjunction with statistical (rather than neural) classifiers. The divergence measure J based on a subset m of the p features is computed for classes i and j as follows (Singh, 1984) with a zero value indicating that the classes are identical – the greater the value of $J(i, j)$ the greater is the class separability based on the m selected features:

$$J(i,j) = 0.5 \, \mathrm{tr}[(\mathbf{S}_i - \mathbf{S}_j)(\mathbf{S}_j^{-1} - \mathbf{S}_i^{-1})]$$
$$+ 0.5 \, \mathrm{tr}[(\mathbf{S}_i^{-1} + \mathbf{S}_j^{-1})(\bar{\mathbf{x}}_i - \bar{\mathbf{x}}_j)(\bar{\mathbf{x}}_i - \bar{\mathbf{x}}_j)']$$

The symbol tr( ) means the trace or the sum of the diagonal elements of the indicated matrix. $\mathbf{S}_i$ and $\mathbf{S}_j$ are the $m \times m$ sample variance–covariance matrices for classes i and j, computed for the m selected features, and $\bar{\mathbf{x}}_i$ and $\bar{\mathbf{x}}_j$ are the corresponding sample mean vectors. For $m = 1$ (a single feature) the divergence measure for classes i and j is:

$$J(i,j) = 0.5 \left(\frac{s_i^2}{s_j^2} + \frac{s_j^2}{s_i^2} - 2 \right)$$

where s_i^2 and s_j^2 are the variances of the single feature calculated separately for classes i and j. Since the divergence measure takes into account both the mean vectors and the variance–covariance matrices for the two classes being compared, it is clear that the inter-class difference is being assessed in terms of (i) the shape of the frequency distribution and (ii) the location of the centre of the distribution. The divergence will therefore be zero only when the variance–covariance matrices and the mean vectors of the two classes being compared are identical.

The distribution of $J(i, j)$ is not well known so a measure called the *transformed divergence* is used instead. This has the effect of reducing the range of the statistic, the effect increasing with the magnitude of the divergence. Thus, when averages are taken, the influence of one or more pairs of widely separated classes will be reduced. The transformed divergence is obtained from:

$$J_T(i,j) = c\{1 - \exp[-J(i,j)/8]\}$$

with c being a constant used to scale the values of J_T on to a desired range. Sometimes the value 2000 is used as a scaling factor, but 100 seems equally reasonable as the values of J_T can then be interpreted in the same way as percentages. A value of J_T of 80 or more indicates good separability of the corresponding classes i and j. The values of $J_T(i, j)$ are averaged for all possible mutually exclusive pairs of classes i and j and the average pairwise divergence is denoted by J_{Tav}.

$$J_{Tav} = \frac{2}{k(k-1)} \sum_{i=1}^{k-1} \sum_{j=1}^{i} J_T(i,j)$$

Study of the individual $J_T(i, j)$ might show that some pairs of classes are not statistically separable on the basis of any subset of the available features. The feature selection process might then also include a class amalgamation component. It might be worth following another line of thought. If the aim of feature selection is to produce the subset of m features that best combines classification accuracy and computational economy then, instead of considering the average separability of all pairs of classes, why not try to find that set of m features which maximises the minimum pair-wise divergence? In effect, this is trying to find the subset of m features that best performs the most difficult classification task. The minimum pair-wise divergence is:

$$J_{min}(i,j) = \min J(i,j) \qquad i < j$$

Given that the *raison d'être* of feature selection is the availability of several (more than four) features, the selection of combinations of m from p features is a problem. The number of subsets of size m that can be drawn from a set with p elements is

$$\binom{p}{m} = \frac{p!}{m!(p-m)!}$$

The symbol '!' indicates 'factorial'; for example, 3! is $3 \times 2 \times 1 = 6$. If p is large then the number of subsets soon becomes very considerable. Take the Daedalus airborne scanner as an example. This instrument generates 12 channels of spectral data. If we assume that no texture features or ancillary data are added, then the number of subsets of size $m = 4$ is 495. If subsets of size $m = 12$ are to be drawn from a data set with $p = 24$ features then the number of subsets is 2 704 156. Clearly any brute-force method involving the computation of the average pair-wise divergence for such a large number of subsets is out of the question. The problem of selection of more likely subsets is not dissimilar to the problem of determining the best subset of independent variables in multiple linear regression. Any of three approaches can be used – the forward selection, backward elimination and stepwise procedures.

The forward selection method starts with the best subset of size $m = 1$. Call this feature f_1. Now find the best subset of size $m = 2$ including f_1, that is, f_1 plus one other feature. The best subset at the end of the second cycle will be $\{f_1, f_2\}$. The procedure continues to determine subsets $\{f_1, f_2, f_3\}$ and so on, until all features are included. The user can then evaluate the list of features included and corresponding divergence value, and must weigh up the advantages of using fewer features against the cost of lower classification accuracy.

The backward elimination method works the opposite way round. Starting with the complete set $\{f_1, f_2, \ldots, f_p\}$, remove that feature which contributes least to the average pair-wise divergence. This is done by computing the average pair-wise divergence for all subsets of size $p - 1$. Repeat until $m = 1$.

Neither procedure is guaranteed to produce the optimal subset; indeed, both may produce differing results unless the data set is so clearly structured that no selection procedure is needed. Kumar (1979) describes an experiment in which the exhaustive search, forward and backward selection algorithms were employed. He found that the forward selection method produced results that were almost as good as exhaustive search and were better than those produced by the backward elimination method. Goodenough *et al.* (1978) report an implementation of a branch-and-bound algorithm based on a method published by Narendra and Fukunaga (1977). This is claimed to give a globally optimum solution. Their method is based on the assumption that, given a certain criterion B then, if a subset of size greater than m has a value of average pair-wise divergence that is less than B, all sub-subsets of this subset will have values that are also less than B. The advantage of the branch-and-bound approach is that many subsets are discarded without the need to evaluate them.

A measure called the *Bhattacharyya distance* is sometimes used in place of the divergence to measure the statistical separability of a pair of spectral classes. It measures the probability of correct classification rather than the statistical separability and is computed from the expression:

$$B_{12} = \frac{1}{8}(\bar{\mathbf{x}}_1 - \bar{\mathbf{x}}_2)'\frac{\mathbf{S}_1 + \mathbf{S}_2}{2}(\bar{\mathbf{x}}_1 + \bar{\mathbf{x}}_2) + \frac{1}{2}\ln\left[\frac{(\mathbf{S}_1 - \mathbf{S}_2)/2}{|\mathbf{S}_1|^{0.5}|\mathbf{S}_2|^{0.5}}\right]$$

(Haralick and Fu, 1983). The quantity B_{ij} is computed

for every pair of classes given m features. The sum of the B_{ij} for all $k(k-1)/2$ classes is obtained and is a measure of the overall separability of the k classes using m features. All possible combinations of m out of p features are used to decide the best combination. Again, selection algorithms such as those described above for the transformed divergence could be used to improve the efficiency of the method. Like the divergence measure the Bhattacharyya distance is based on the assumption of multivariate normality.

Swain and King (1973) report a comparative study of feature selection using divergence, transformed divergence and Bhattacharyya distance. Their results indicate that the transformed divergence and Bhattacharyya distance performed best, though it should be noticed that they used artificial data that were made to be normally distributed. Other studies of feature selection are provided by Decell and Guseman (1979), Kailath (1967), Kanal (1974), Muasher and Landgrebe (1984) and Ormsby (1992). Yool *et al.* (1986) compare the use of transformed divergence and empirical approaches to the assessment of classification accuracy (section 8.10). They found no clear agreement between the results from the two alternative approaches, and attributed the differences – which in some instances were considerable – to departures from normality and conclude that

> '. . . a divergence algorithm requiring normally-distributed data may not be a reliable indicator of performance.'
>
> Yool *et al.* (1986, p. 689)

However, if the empirical classification accuracy approach is used then a classification analysis must be carried out on test samples for each subset of m features.

The availability of high-dimensional multispectral image data is thus seen to be a mixed blessing. The additional spectral channels provide further information regarding the spectral response of the ground-cover targets, though their use requires additional computer time. Classification accuracy is dependent on feature-set size, yet no clear and recommendable algorithm is available to determine the subset that will produce the best compromise between accuracy and cost. Factors other than dimensionality will affect the choice of subset; the number of classes and their relative separability will have some influence on the number and choice of features needed to discriminate between them. Studies that have been carried out to date indicate that statistical methods (based on the assumption of normal distributions) should be used with caution.

Non-parametric feature selection methods do not rely on assumptions concerning the frequency distribution of the features. One such method, which has not been widely used, is proposed by Lee and Landgrebe (1993). Benediktsson and Sveinsson (1997) demonstrate its application.

8.10   CLASSIFICATION ACCURACY

The methods discussed in section 8.9 have as their aim the establishment of the degree of separability of the k spectral classes to which the image pixels are to be allocated (though the Bhattacharyya distance is more like a measure of the probability of misclassification). Once a classification exercise has been carried out there is a need to determine the degree of error in the end-product. These errors could be thought of as being due to incorrect labelling of the pixels. Conversely, the degree of accuracy could be sought. First of all, if a method allowing a 'reject' class has been used then the number of pixels assigned to this class (which is conventionally labelled '0') will be an indication of the overall representativeness of the training classes. If large numbers of pixels are labelled '0' then the representativeness of the training data sets is called into question – do they adequately sample the feature space? The most commonly used method of representing the degree of accuracy of a classification is to build a $k \times k$ *confusion matrix* (or *error matrix*). The elements of the rows i of this matrix give the number of pixels that the operator has identified as being members of class i that have been allocated to classes 1 to k by the classification procedure (see Table 8.4). Element i of row i (the ith diagonal element) contains the number of pixels identified by the operator as belonging to class i that have been correctly labelled by the classifier. The other elements of row i give the number and distribution of pixels that have been incorrectly labelled. The classification accuracy for class i is therefore the number of pixels in cell i divided by the total number of pixels identified by the operator from ground data as being class i pixels. The overall classification accuracy is the average of the individual class accuracies, which are usually expressed in percentage terms.

Some analysts use a statistical measure, the kappa coefficient (κ), to summarise the information provided by the contingency matrix (Bishop *et al.*, 1975). Kappa is computed from:

$$\kappa = \frac{N \sum\limits_{i=1}^{r} x_{ii} - \sum\limits_{i=1}^{r} x_{i+}x_{+i}}{N^2 - \sum\limits_{i=1}^{r} x_{i+}x_{+i}}$$

The x_{ii} are the diagonal entries of the confusion matrix. The notation x_{i+} and x_{+i} indicates, respectively, the sum of row i and the sum of column i of the confusion matrix. N is the number of elements in the confusion matrix. Row totals (x_{i+}) for the confusion matrix shown in Table 8.4 are listed in the column headed (i) and column totals are given in the last row. The sum of the diagonal elements (x_{ii}) is 350 ($\sum_{i=1}^{r} x_{ii}$ for $r = 6$), and the sum of the products of the row and column marginal totals ($\sum_{i=1}^{r} x_{i+}x_{+i}$) is 28 820. Thus the value of kappa is:

$$\kappa = \frac{410 \times 350 - 28\,820}{168\,100 - 28\,820} = \frac{114\,680}{139\,280} = 0.82$$

Table 8.4   Confusion or error matrix for six classes. The row labels (Ref.) are those given by an operator using ground reference data. The column labels (Class.) are those generated by the classification procedure. See text for explanation. The four right-hand columns are as follows: (i) number of pixels in class from ground reference data; (ii) estimated classification accuracy (per cent); (iii) class i pixels in reference data but not given label by classifier; and (iv) pixels given label i by classifier but not class i in reference data. The sum of the diagonal elements of the confusion matrix is 350, and the overall accuracy is therefore $(350/410) \times 100 = 85.4\%$.

Ref.	Class.						(i)	(ii)	(iii)	(iv)
	1	2	3	4	5	6				
1	50	3	0	0	2	5	60	83.3	10	21
2	4	62	3	0	0	1	70	88.5	8	10
3	4	4	70	0	8	3	89	81.4	19	6
4	0	0	0	64	0	0	64	100.0	0	3
5	3	0	2	0	71	1	77	92.2	6	10
6	10	3	1	3	0	33	50	66.0	17	10
Col. sums	71	72	76	67	81	43	410		60	60

A value of zero indicates no agreement, while a value of 1.0 shows perfect agreement between the classifier output and the reference data. Montserud and Leamans (1992) suggest that a value of kappa of 0.75 or greater shows a 'very good to excellent' classifier performance, while a value of less than 0.4 is 'poor'. However, these guidelines are only valid when the assumption that the data are randomly sampled from a multinomial distribution, with a large sample size, is met.

Values of kappa are often cited when classifications are compared. If these classifications refer to different procedures (such as maximum likelihood and artificial neural networks) applied to the same data set, then comparisons of kappa values are acceptable, though the percentage accuracy (overall and for each class) provides as much, if not more, information. If the two classifications have different numbers of categories then it is not clear whether a straightforward logical comparison is valid. It is hard to see what additional information is provided by kappa over and above that given by a straightforward calculation of percentage accuracy. See Congalton (1991), Kalkhan *et al.* (1997), Stehman (1997) and Zhuang *et al.* (1995).

The confusion matrix procedure stands or falls by the availability of a test sample of pixels for each of the *k* classes. The use of training-class pixels for this purpose is dubious and is not recommended – one cannot logically train and test a procedure using the same data set. A separate set of test pixels should be used for the calculation of classification accuracy. Users of the method should be cautious in interpreting the results if the ground data from which the test pixels were identified were not collected on the same date as the remotely-sensed image, for crops can be harvested or forests cleared. Other problems may arise as a result of differences in scale between test and training data and the image pixels being classified. So far as is possible, the test pixel labels should adequately represent reality.

The literal interpretation of accuracy measures derived from a confusion matrix can lead to error. Would the same level of accuracy have been achieved if a different test sample of pixels had been used? Figure 8.20 shows an extract from a hypothetical classified image and the corresponding ground reference data. If the section outlined in the solid line in Figure 8.20(a) had been selected as test data the user would infer that the classification accuracy was 100%, whereas if the area outlined by the dashed line had been selected then the accuracy would appear to be 75%. For a given spectral class there are a very large number of possible configurations of test data and each might give a different accuracy statistic. It is likely that the distribution of accuracy

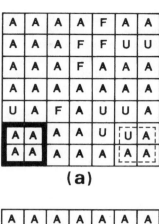

(a)

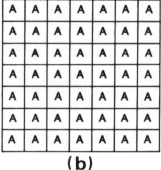

(b)

Figure 8.20   Cover type categories derived from (a) ground reference data and (b) automatic image classifier. The choice of sample locations (solid or dashed lines in (a)) will influence the outcome of accuracy assessment measures.

values could be summarised by a conventional probability distribution, for example the hypergeometric distribution, which describes a situation in which there are two outcomes to an experiment, labelled *P* (successful) and *Q* (failure), and where samples are drawn from a population of finite size. If the population being sampled is large the binomial distribution (which is easier to calculate) can be used in place of the hypergeometric distribution. These statistical distributions allow the evaluation of confidence limits, which can be interpreted as follows: If a very large number of samples of size *N* are taken and if the true proportion *P* of successful outcomes is P_T then 95% of all the sample values will lie between P_L and P_U (the lower and upper 95% confidence limits around P_T). The values of the upper and lower confidence limits depend on (i) the level of probability employed and (ii) the sample size *N*. The confidence limits get wider as the probability level increases towards 100% so that we can always say that the 100%

confidence limits range from minus infinity to plus infinity. Confidence limits also get wider as the sample size N becomes smaller, which is self-evident.

Jensen (1986, p. 228) provides a formula for the calculation of the lower confidence limit associated with a classification accuracy value obtained from a training sample of N pixels. The formula used to determine the required $r\%$ lower confidence limits given the values of P, Q and N is:

$$s = P - \left[z\sqrt{\frac{PQ}{N}} + \frac{50}{N} \right]$$

where z is the $(100 - r)/100$th point of the standard normal distribution. Thus, if r equals 95% then the z value required will be that having a probability of $(100 - 95)/100$ or 0.05 under the standard normal curve. The tabled z value for this point is $z = 1.645$. If r were 99% then z would be 2.05. To illustrate the procedure assume that, of 480 test pixels, 381 were correctly classified, giving an apparent classification accuracy (P) of 79.375%. Q is therefore $(100 - 79.375) = 20.625\%$. If the lower 95% confidence limit was required then z would equal 1.645 and

$$s = 79.375 - \left[1.645 \sqrt{\left(\frac{79.375 \times 20.625}{480} \right)} + \frac{50}{480} \right]$$

$$= 79.375 - [1.645 \times 1.847 + 0.104]$$

$$= 76.223\%$$

This result indicates that, in the long run, 95% of training samples with observed accuracies of 79.375% will have true accuracies of 76.223% or greater. As mentioned earlier, the size of the training sample influences the confidence level. If the training sample in the above example had been composed of 80 rather than 480 pixels then the lower 95% confidence level would be

$$s = 79.375 - \left[1.645 \sqrt{\left(\frac{79.375 \times 20.625}{80} \right)} + \frac{50}{80} \right]$$

$$= 71.308\%$$

This procedure can also be applied to individual classes in the same way as described above with the exception that P is the number of pixels correctly assigned to class j from a test sample of N_j pixels.

The confusion matrix can be used to assess the nature of erroneous labels besides allowing the calculation of classification accuracy. Errors of omission are committed when patterns that are really class i become labelled as members of some other class, whereas errors of commission occur when pixels that are really members of some other class become labelled as members of class i. Table 8.4 shows how these error rates are calculated. From these error rates the user may be able to identify the main sources of classification accuracy and alter his or her strategy appropriately. Congalton *et al.* (1983), Congalton (1991) and Story and Congalton (1986) give more advanced reviews of this topic.

How to calculate the accuracy of a fuzzy classification might appear to be a difficult topic; refer to Gopal and Woodcock (1994) and Foody and Arora (1996). Burrough and Frank (1996) consider the more general problem of fuzzy geographical boundaries. The question of estimating area from classified remotely-sensed images is discussed by Canters (1997) with reference to fuzzy methods. Dymond (1992) provides a formula to calculate the root-mean-square error of this area estimate for 'hard' classifications (see also Lawrence and Ripple, 1996). Czaplewski (1992) discusses the effect of misclassification on areal estimates derived from remotely-sensed data, and Fitzgerald and Lees (1994) examine classification accuracy of multisource remote sensing data.

The use of single summary statistics to describe the degree of association between the spatial distribution of class labels generated by a classification algorithm and the corresponding distribution of the true (but unknown) ground cover types is rather simplistic. First, these statistics tell us nothing about the spatial pattern of agreement or disagreement. An accuracy level of 50% for a particular class would be achieved if all the test pixels in the upper half of the image were correctly classified and all those in the lower half of the image were incorrectly classified, assuming an equal number of test pixels in both halves of the image. The same degree of accuracy would be computed if the pixels in agreement (and disagreement) were randomly distributed over the image area. Secondly, statements of 'overall accuracy' levels can hide a multitude of sins. For example, a small number of generalised classes will usually be identified more accurately than would a larger number of more specific classes, especially if one of the general classes is 'water'. Thirdly, a number of researchers appear to use the same pixels to train and to test a supervised classification. This practice is illogical and cannot provide much information other than a measure of the 'purity' of the training classes. More thought should perhaps be given to the use of measures of confidence in pixel labelling. It is more useful and interesting to state that the analyst assigns label x to a pixel, with the probability of correct labelling being y, especially if this information can be presented in quasi-map form. A possible measure might be the relationship between the first and second highest membership grades output by a fuzzy classifier. The use of ground data to test the output from a classifier is, of course, necessary. It is not always sufficient, however, as a description or summary of the value or validity of the classification output.

8.11   SUMMARY

Compared to other chapters of this book, this chapter shows the greatest increase in size relative to the 1987 edition. To some extent this is a reflection of the author's own interests. However, the developments in classification methodology over the past 10 years have been considerable, and the problem has been what to omit. The introduction of artificial neural net classifiers, fuzzy methods, new techniques for computing texture features, and new models of spatial context have all occurred during the past decade. This chapter has hardly scratched the surface, and readers are encouraged to follow up the references provided at various points. I have deliberately avoided providing potted summaries of each paper or book to which reference is made in order to encourage readers to spend some of their time in the library. However, 'learning by doing' is always to be encouraged. The CD supplied with this book contains some programs for image classification. These programs are intended to provide the reader with an easy way into image classification. More elaborate software is required if methods such as artificial neural networks, evidential reasoning and fuzzy classification procedures are to be used. It is important, however, to acquire familiarity with the established methods of image classification before becoming involved in advanced methods and applications.

Despite the efforts of geographers following in the footsteps of Alexander von Humboldt over the past 150 years, we are still a long way from being able to state with any acceptable degree of accuracy the proportion of the Earth's land surface that is occupied by different cover types. At a regional scale, there is a continuing need to observe deforestation and other types of land cover change, and to monitor the extent and productivity of agricultural crops. More reliable, automatic, methods of image classification are needed if answers to these problems are to be provided in an efficient manner. New sources of data, at both coarse and fine resolution, are

becoming available. The early years of the new millennium will see a very considerable increase in the volumes of Earth observation data being collected from space platforms, and much greater computer power (with intelligent software) will be needed if the maximum value is to be obtained from these data. An integrated approach to geographical data analysis is now being adopted, and this is having a significant effect on the way image classification is performed. The use of non-remotely-sensed data in the image classification process is providing the possibility of greater accuracy, while – in turn – the greater reliability of image-based products is improving the capabilities of environmental GIS, particularly with respect to studies of temporal change.

All of these factors will present challenges to the remote sensing and GIS communities, and the focus of research will move away from specialised algorithm development to the search for methods that satisfy user needs and are broader in scope than the statistically based methods of the 1980s, which are still widely used in commercial GIS and image processing packages. If progress is to be made then high-quality interdisciplinary work is needed, involving mathematicians, statisticians, computer scientists and engineers as well as Earth scientists and geographers. The future has never looked brighter for researchers in this fascinating and challenging area.

8.12   QUESTIONS

1. Explain the following terms: labelling, classification, clustering, unsupervised, supervised, pattern, feature, pattern recognition, Euclidean space, per-pixel, per-field, texture, context, divergence, decision rule, spatial autocorrelation, prior probability, neuron, feed-forward, multi-layer perceptron, steepest descent, geostatistics, variogram, image segmentation, GLCM, fractal dimension, kappa.

2. What is meant by the term 'feature space'? How can you measure similarities between points (representing objects to be classified) in a feature space of n dimensions, where $n > 3$?

3. Compare the operation of the k-means and also ISODATA unsupervised classifiers. Use the programs `k-means` and `isodata` (described in Appendix B) to carry out two unsupervised classifications of one of the test images on the CD. Summarise your experiences in note form.

4. The parallelepiped, supervised k-means and maximum likelihood classifiers are described as *parametric*. Explain. These three classifiers use, respectively, the extreme pixel values in each band, the mean pixel

value in each band, and the means and covariances of the bands. What does this imply in terms of the shape of the decision boundaries in feature space? Use diagrams to illustrate your answer.

5. Why should clustered distribution patterns of training-class data be avoided? What are the implications of *spatial autocorrelation* for training-class selection?

6. What are the advantages and disadvantages of the artificial neural network classifier in comparison with the maximum-likelihood classifier? What are the implications, if any, for image classification within a GIS?

7. What are the differences and similarities between *linear mixture modelling* and *spectral angle mapping*? Can you distinguish the aims of these methods from those of *fuzzy classifiers*?

8. Define *classification accuracy*. Can it be measured in a meaningful way?

Appendix A

MIPS Display Program

A.1 INTRODUCTION

A.1.1 The MIPS display program

This appendix is a guide to the MIPS display software, which is contained on the CD-ROM that is included with this book. Reference to the software and its applications is made at appropriate points in the text. Before you start to read the book, please install the software so that you can try out the ideas and techniques that you read about in the text. You should also install the programs described in Appendix B. The MIPS programs can be used to process the sample image data sets, details of which are provided in Appendix D.

MIPS provides the following facilities: read images in raw format, contrast stretch, histogram equalisation, pseudocolour transform, RGB to HSI transform, decorrelation stretch, density slice, NDVI calculation, image minus Laplacian filter, median filter, Sobel edge detector, Roberts gradient, user-defined filter, principal components analysis, and image classification using the parallelpiped or the k-means procedures. In addition, MIPS will plot image histograms, save files as Windows bitmaps, read and write lookup tables from disk, zoom, display cursor coordinates and associated pixel values, perform arithmetic on images, and threshold images. All of these functions are described in this appendix.

MIPS is written for PCs running the Microsoft Windows 95 operating system. It should be stressed that MIPS is not a professional image processing system. It was developed specifically as a teaching aid. There are limitations to its performance, including the lack of online help (which is deliberate – this appendix is intended to provide the user with guidance) and the limitation of the displayed image size to a maximum of 512×512. To complement the MIPS display program, 38 stand-alone programs are also provided on the CD, together with some test images, to which reference is made in other parts of this book. These stand-alone programs are documented in Appendix B.

The MIPS display program was developed using Salford Software's *Fortran 95* compiler and *Clearwin+* library. It remains the author's copyright, and must not be sold or traded except as part of this book. The author cannot answer queries about the performance or use of the program. No warranty or guarantee is provided.

A.1.2 MIPS image format

A MIPS image file set consists of:

1. an image dictionary file (usually with the filename extension `inf`, and often called an INF file in this book),
2. one or more image files, and
3. an image histogram file.

The image dictionary file can be created using program `make_inf`, which is described in Appendix B. This program also generates a file containing the image histograms. The image files are in raw format, with the first *npix* bytes representing the first scan line of an image with *npix* pixels per line. The second *npix* bytes of the file represent the second line of the image, and so on. Note that other 'raw' formats may store the image in 'upside down' order, with the last line first. In a multiband image set the image corresponding to each spectral band is stored in a single file, so that for a set of *nbands* images each of size *npix* × *nlines* there are *nbands* image files, each of size *npix* ∗ *nlines* bytes, an image dictionary file and an image histogram file. Figure A.1 shows a set of three image files (*nbands* = 3) each with *nlines* = NL scan lines containing NPIX pixels per scan line.

Before you run MIPS you must create the appropriate image dictionary file and histogram file using the `make_inf` program (section B.3.16). MIPS cannot read any other file type (Windows bitmap, GIF, TIFF, etc.).

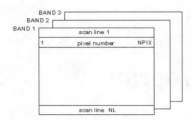

Figure A.1 Illustrating the dimensions of an image file set. Three images are shown (*nbands* = 3) each with NL scan lines of NPIX pixels.

You can use one of several widely available graphical display programs such as *xv* (for UNIX systems) or Paintshop Pro (for Windows) to read in file formats such as GIF and TIFF and export (reformat) them in 'raw' format. The MIPS stand-alone program `bmp2raw` (section B.3.1) will take a Windows bitmap file as input and will generate one raw image file or three raw image files depending on whether the bitmap is 8 or 24 bits in depth.

A.2   MINIMUM HARDWARE SPECIFICATION FOR MIPS

In order to operate properly, the software requires a PC with at least 16 Mb RAM, a graphics card capable of supporting an 800 by 600 pixel screen in 16-bit colour mode (i.e. a minimum of 32k colours), and a mouse. Five megabytes of disk space are needed to store the software, plus additional disk space to store images (the sample images can be read directly from the CD). A 133 MHz Pentium is recommended, though the software was developed and runs reasonably quickly on a 66 MHz 486/DX2 machine. By the time this book reaches the shelves the DX2 machine will be in a museum and a 133 MHz Pentium will be regarded as slow. You can check the properties of the Windows 95 display by selecting `My Computer`, then `Control Panel`, then `Display`. Inside `Display` choose `Settings`. The number of colours that you are able to select will depend on the type of display card in your computer. MIPS will run using 16 or 256 colours, but this is not recommended. Choose `High Colour-16 bits` (64k colours) or `True Colour-24 bits` (16 million colours), if your display card allows this. But make sure that the `Desktop Area` specification is at least 800 × 600 pixels.

A.3   SOFTWARE INSTALLATION

The MIPS software is provided on the CD that accompanies this book. A program, `Mips_Install`, to install the software described in this Appendix on to your hard

disk is described in Appendix C. `Mips_Install` also installs the stand-alone programs described in Appendix B and the sample images detailed in Appendix D.

You do not need to install the software on to your hard disk if your CD drive letter is `D:`. In other cases, you are recommended to install the MIPS display program, plus the stand-alone programs (Appendix B) and the image dictionary files (*INF files) for the sample images (Appendix D). The sample images can be kept on the CD.

Once the programs are installed, create a new desktop folder, and place short cuts to the programs you wish to use in this new folder. See the Microsoft Windows 95 documentation for instructions on creation of short cuts.

Note that the MIPS display program does not provide any facilities for printing images. If you wish to print an image, save it to the clipboard and then paste it into an image viewer or a word processor.

Ensure that the files `salflib.dll` and `imlib.dll` are always stored in the directory in which the MIPS software resides, otherwise Windows will tell you that it cannot access a missing DLL (dynamic link library) file.

These programs are supplied without warranty. They were developed using Microsoft Windows 95 and Salford Compilers Ltd.'s FTN95 compiler running on a Pentium PC with 32Mb memory.

See the WWW page http://22.geog. nottingham.ac.uk/~mather/booknews.html for details of any changes, comments or updates.

A.4   USING MIPS

MIPS is a Windows-based application for image display and processing. It is not a professional system and is intended to be used in teaching and in small projects. Hence it has certain restrictions, the principal one being that it can only display images of 512 × 512 pixels in size. These images are stored in raw format, as described in section A.1.2. MIPS cannot handle bitmap files, TIFF, GIF or any other image format, though a utility program is provided to convert bitmap images to raw form (see Appendix B). MIPS can, however, save images in bitmap form so that they can be imported by applications that provide facilities for adding text and printing, for example a word processor.

MIPS can only handle images that have a dynamic range of 8 bits per pixel (such as Landsat MSS and TM, and SPOT HRV). It cannot handle 10-bit images, or images stored in real (floating point) format. The images are stored in random access memory (RAM) as Fortran `character*1` arrays, and the `ichar()` function is used to convert from character to integer form. Some routines use Fortran virtual arrays, which grab and release memory as required. If no free memory is available, an error message will inform you that the program is being aborted. If this happens, ensure that all other Windows application are closed. Because Windows may develop 'memory leaks' if a number of applications have been run, you may find that rebooting the operating system will solve the problem.

A.4.1 Starting MIPS

Start MIPS by clicking on its icon if you created one by following the instructions above. Alternatives are switching to MS-DOS mode (from the `Start|Programs` menu item on the Windows 95 lower menu bar) and typing `C:\MIPS\mips` (though if you have placed `mips.exe` in a directory other than `C:\MIPS` then you will need to substitute the directory name for `C:\MIPS`) or using Windows Explorer to locate the file `mips.exe` and double-clicking on the filename.

A.4.2 Selecting an image to display

You will see an authorship notice that allows you to use MIPS for educational purposes. Click `OK` or press the keyboard `Enter` key. A second window to appear reminds you that the first two steps in using MIPS are:

1. Create a log file to hold a text listing of the processing steps that you followed during your use of MIPS. This log file is useful in that it contains many of the details that will help you understand the procedures that are described elsewhere in this book. You can delete the log file after you have inspected it (a good way to read it is to use Windows Notepad or the MS-DOS editor).
2. Select an image dictionary (INF) file which contains a description of the image set that you wish to use. The creation of dictionary files is dealt with in section B.3.16.

When you have read the information in the window, press `Enter`. A third window is now displayed and you are asked to enter the name of the directory that contains your images. This is called your working directory. The default is the current MS-DOS directory. Press `Enter` or click `OK` when you have typed in this information.

Now the full-screen MIPS Main Window appears with a menu bar at the top of the screen (Figure A.2). The menu items on the Main Window are: `Start`, `Enhance`, `Display`, `Filter`, `Transform`, `Plot`, `Classify`, `Utilities` and `End`. Following the instructions given in the second of the advisory windows that appear at program start-up, click on the menu item `Start`, and choose `Open_Log_File`. The familiar Windows file selection window appears, with your working directory selected, and the `Files of Type` box set to 'LOG files (*.log)'. You can double-click on the name of an existing text (log) file or type in a new one. I always give my log files the extension '.log' to avoid confusion with other files such as image files or image dictionary files; many students find it difficult to remember that the log file is a text file containing a record of their MIPS session, but this difficulty is avoided if the filename extension .log is always used.

Once the log file has been selected you must open a second file – an image dictionary file (an INF file). Click `Start` on the menu bar and choose `Open_Inf_File`. The same Windows file selection window appears, but this time the `Files of Type` box shows INF (*.inf) files'. A list of image dictionary files appears in the window. Choose one, either by highlighting and selecting `OK` or by double-clicking. Note that you can select a new INF file at any time, for example, to open a new image file set or to reread the image file you are working with in the event that you make an error and wish to start again.

MIPS is able to read files that are bigger than 512 lines × 512 pixels per line in size, but it cannot display such files. When MIPS detects that your image files are too big you are asked to specify (i) a start line, (ii) the number of lines (or pixels per line) to display, and (iii) a sampling rate. For example, if your images are 1024 × 1024 in size you could enter '1 512 2', meaning 'start at line (or pixel) number 1, select every second line (or pixel), and display 512 lines (or pixels)'. Alternatively you could type '384 512 1' which means 'use the 512 lines (or pixels) starting at line (pixel) number 384'. The dialogue appears once for each oversized dimension, so that if you have an image of size 1024 × 512 the dialogue will ask for a start line, number of lines and line sampling rate. If your image size is 512 × 1024 then the dialogue will refer to the number of pixels per line. If both of your image dimensions exceed 512 then the dialogue will be presented for both lines and pixels. If your specifications are incorrect the dialogue will continue to reappear until you make an acceptable response.

Once you have selected an image dictionary file you must decide whether you want to display a single-band image (mono or grey scale) or three bands of a multi-spectral image in red, green and blue (false colour). If your INF file references fewer than three image files

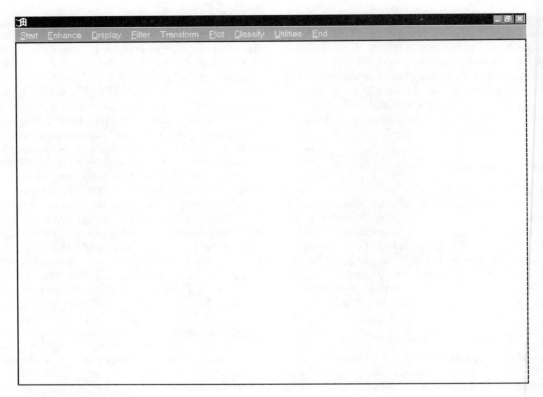

Figure A.2   MIPS Main Window.

then it is assumed that you will display a grey-scale image. If your INF file references three or more image files then the next window to appear (Figure A.3) asks you to choose between `Greyscale` and `False Colour`. Highlight your choice and press `OK`.

The next window is shown in Figure A.4. Use this window to choose the image band or bands to be displayed: one for grey scale and three for false colour. If

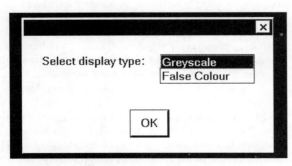

Figure A.3   Image type selection box.

you have chosen to display a `Greyscale` image then highlight your chosen image band and press `OK`. If you have chosen to display a `False Colour` image then you must highlight first the image band that is to be displayed in red, and press `OK`. The same window re-appears (except that it now has the title **GREEN**) and you then select the image band to be displayed in green. Finally select the image band to be displayed in blue when the selection window appears for the third time.

A progress bar now shows you that the selected image (`greyscale`) or images (`false colour`) are being read from the hard disk into the random access memory. Once the image has been selected and read into memory you are ready to proceed with image processing.

A.4.3   Displaying an image in MIPS

MIPS is able to display the following image types:

- A false-colour image with three components, displayed in red, green and blue respectively

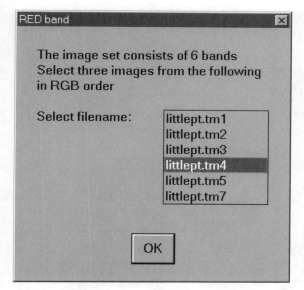

Figure A.4 Selection of image bands to form the red, green and blue components of a false-colour composite image. The same type of file selection checkbox is used in `Linear Function` and `PCA` functions.

- A mono or grey-scale image
- The individual **RGB** components of a false-colour image as separate grey-scale images

A.4.3.1 Displaying a false-colour or grey-scale image

It is possible to use a contrast enhancement procedure on the false-colour or grey-scale image stored in memory before it is displayed on the screen, but you will learn this as you become more familiar with MIPS. Initially it is suggested that you display the raw image, which will probably look rather dark because the lookup tables are set to a default state in which the image intensities 0–255 map to screen intensities 0–255, and it is likely that your raw image has an intensity range that is far less than 256 levels. To see your raw image, click the `Display` menu item on the Main Window menu bar then `Display_Image` in the submenu. An empty Image Display Window is generated. This Image Display Window sits on top of the MIPS Main Window and has three menu items, `Display`, `Cursor` and `Zoom`, which are explained below. When the Image Display Window appears press `Display` on its menu bar and your image will appear (Figure A.5). It is important to remember that you must exit from the Image Display Window by clicking the `OK` button at the

top right corner of the window. Do not use any other method of erasing the window or else the subsequent behaviour of MIPS will be entirely predictable – it will crash! You can, however, have several Image Display Windows open simultaneously. The number will depend on the size of the physical and virtual memory of your computer. If you open too many Image Display Windows then MIPS will get sluggish as virtual memory is used. You may find it useful to keep open the Image Display Window containing the raw image for a while so that you can compare the effects of different contrast enhancement operations, for example.

However, please note two warnings very carefully. First, you must always exit the Image Display Window only by pressing the `OK` button in that window. Secondly, if you display multiple image windows, as described in the previous paragraph, only one of them is 'live' and that is the one that you displayed last. Do not try to use the `Zoom`, `Display` or `Cursor` menu items except in the 'live' Image Display Window. Otherwise the system may crash.

(*a*) *The Cursor function.*   Once the image is displayed you can use the two other menu items that appear on the menu bar of the Image Display Window. Note that these functions (as well as the `Display` menu item) can only be used on the last display window that you opened (see above).

The `Cursor` menu item displays a pointer (normal cursor) and allows you to point at an individual pixel and left-click the mouse. The RGB values for that pixel, together with the cursor position, will be displayed in a new window (Figure A.6). Remember that the cursor position is with reference to the top left of the image, which has coordinates (0, 0), and that the x value is measured across the screen and the y value down the screen. The RGB values are those in the raw, not the enhanced, image. This function can be used to collect ground control points for geometric correction, or to inspect the range of RGB values present in the image.

You can repeat the selection of points without repeatedly choosing `Cursor` from the menu bar – in fact you can select as many points sequentially as you like. But remember: right-click the mouse to exit this procedure.

(*b*) *The Zoom function.*   The third item on the Image Display Window menu bar is `Zoom`. Like the `Display` and `Cursor` items, this function can only be used on the Image Display Window that was opened last (see above). When you select `Zoom` you are asked to left-click the mouse in the Image Display Window to indicate the top left corner of the area to be zoomed. Once

Figure A.5   The Image Display Window.

you have done this the zoom window appears. Click `Display` on the Zoom Window menu bar and the zoomed image appears (Figure A.7). The zoom factor of 2 is fixed. If you point to a top left corner that is below the centre of the main image or in the right-hand side of the main image then it is not possible to draw a 256 × 256 zoomed image and the undrawable part of the zoomed image is left black.

To exit the zoomed image press `Quit`. Do not use the `Select` menu option – it is used in the `Classify` procedure, which is described later. If you click on `Select` then you will find yourself in uncharted waters. You cannot use the Image Display Window's `Cursor` function from inside the zoom window.

Figure A.6   Cursor coordinates and colour values are displayed for each left mouse click.

A.4.3.2   Separate display of the RGB components of a false-colour image

The second item on the MIPS Main Window `Display` menu list is `Separate_RGB`. To use this function you must have specified a false-colour image using the MIPS Main Window menu item `Start| Open_InfFile`. You may also have contrast enhanced the image (section 5.3) and displayed it in an Image Display Window. Click `Display|Separate_RGB` then choose which of the components of the false-colour image (red, green or blue) you wish to display. A new display window opens and the selected component is displayed automatically, using whatever contrast enhancement (section 5.3) or filter (Chapter 7) you have applied. Click `Quit` – the only menu item associated with this window – to exit. You can use this function three times to see the three separate components of the false-colour image.

A.4.4   Contrast enhancement in MIPS

The second menu item from the left in the main window is `Enhance`. The drop-down list associated with `Enhance` contains the following options:

- Stretch
- Histogram equalisation
- Gaussian stretch
- Pseudocolour
- HSI to RGB
- Decorrelation stretch
- Density slice

Note that you do not have to display an image in the Image Display Window before you use contrast stretch functions, though you must have read an image into memory using `Start|Open_Inf_File`. Once you become familiar with MIPS you may find it convenient to look at the image histograms (using the `Plot` menu

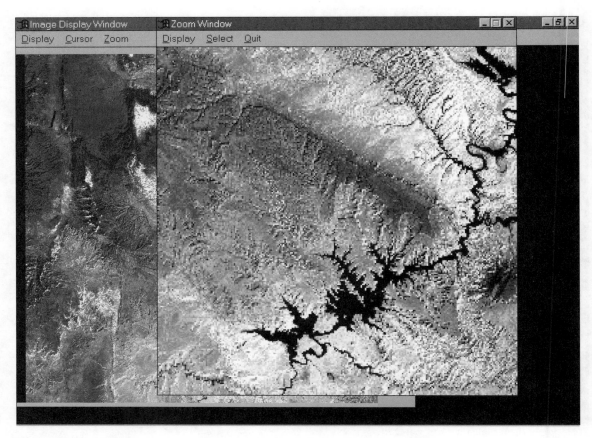

Figure A.7   Zoom Window.

item on the Main Window) and decide on a contrast stretch function (such as linear stretch or histogram equalisation) before displaying the enhanced image in the image display window. You can leave open the display window containing the unenhanced image if you wish, and put the enhanced image in a new display window so that you can compare the two.

A.4.4.1   Linear stretch

Before doing a linear stretch you should look at a plot of the image histogram, which is displayed when you click on the `Plot` menu item in the Main Window. `Plot` is described below (section A.4.7). For a grey-scale image you make the choices described in the next paragraph once only. For a false-colour image you make three (possibly different) choices for the images displayed in red, green, blue respectively.

Once you select `Stretch` you are asked whether you want to define the stretch limits manually or automati-

cally, or if you want to use the 5% and 95% points of the histogram. If you choose the automatic option then the histogram of the displayed channel (grey scale) or channels (false colour) is examined and the minimum and maximum pixel values in the relevant band are extracted. If you choose the 'manual' option you must give lower and upper pixel values in the range 0–255 for each of the displayed channels. Remember that for false-colour images the channel order is red, green then blue. If you choose the '5–95%' option then the lookup tables are set so that the pixel value at the 5% point in the histogram maps to an intensity of 0 and the pixel value at the 95% point in the histogram maps to 255. Values below the 5% point and above the 95% point map to 0 and 255 respectively.

All that the `Stretch` option does is to set the lookup tables. You have to choose `Display` from the menu in the MIPS Main Window in order to view the image after stretching, as described above. Remember to close each Image Display Window by clicking on the associated OK button. You can undo the effect of the contrast stretch by choosing Utilities|Reset_Luts on the MIPS Main Window menu bar.

If you want to see what effect the selected stretch has had on the lookup tables, use the Plot|Histograms menu item from the MIPS Main Window as described in section A.4.7.1. The linear stretch can be applied to grey-scale, pseudocolour and false-colour images.

A.4.4.2   Histogram equalisation

The histogram equalisation option is selected by choosing Enhance from the menu on the Main Window and clicking on Histogram Equalise. As in the case of the linear contrast stretch, all that the procedure does is to set the lookup tables based on the image histogram shape. You must choose the Display option from the MIPS Main Window in order to view the enhanced image. As noted in section A.4.3.1 you can have more than one Image Display Window open at any one time. Close each of them by clicking the associated OK button. Again you can use the Plot menu item from the MIPS Main Window to see the effect of histogram equalisation on the shape of the lookup tables.

You can apply the histogram equalisation stretch to the whole histogram or to a selected part of it. The first dialogue box to appear after you click on Enhance|Histogram Equalise asks you to make this choice. If you want to apply the stretch to a part of the histogram then click on Use specified range of histogram. If you want to perform equalisation using the whole histogram then click on Use full range of histogram. If you decide to base the stretch on a specified range of the histogram it would be a good idea to look at the histograms first, using the Plot|Histograms menu item, described below in section A.4.7.1. For each histogram (one for a grey-scale image, three for a false-colour image) give the lower and upper bounds of the part of the histogram that you want to specify. The LUT is set to zero outside this specified range. Histogram equalisation can be applied to grey-scale, pseudocolour and false-colour images.

A.4.4.3   Gaussian stretch

The Gaussian Stretch attempts to fit lookup table values to a normal (Gaussian) distribution, which has a mean of zero and a standard deviation of s. Values of s that are commonly used are in the range 2–4. The Gaussian Stretch can be applied using the full range of the image histogram or a selected range (see section A.4.4.2). For each band to be stretched you must

supply a value of s. A value of -1 (minus 1) means: do not apply a stretch to this band. The Gaussian Stretch can be applied to grey-scale, pseudocolour and false-colour images.

A.4.4.4   Pseudocolour enhancement

This enhancement operation can only be applied to grey-scale images. A warning message will be displayed if you attempt to convert a false-colour image to pseudocolour.

As in the case of the linear stretch procedure, you can select the limits of the pseudocolour mapping manually or have the program select them for you, using the histogram limits (minimum and maximum values). Before you use the manual option, have a look at the image histogram using the Plot|Histograms menu item on the MIPS Main Window menu. This will allow you to make a visual estimate of the appropriate values to use as maximum and minimum. The automatic option simply uses the minimum and maximum values in the histogram. You can reapply the Pseudocolour option if you want to change the limits of the enhancement in order to achieve a particular effect.

The program works by mapping the histogram minimum and maximum to a colour circle and interpolating intermediate values. The 49 elements of the table to which the interpolation procedure is applied are shown in Table A.1. Note that by using the Pseudocolour procedure the image type is changed irreversibly from grey-scale to pseudocolour, and the single-band (grey-scale) image now has three associated LUTs (red, green and blue).

A.4.4.5   HSI to RGB transform

The HSI to RGB (hue–saturation–intensity to red–green–blue) procedure is described in section 6.5. It can only be applied to a false-colour image. The following steps are involved:

1. transformation of the displayed (RGB) image to HSI,
2. linear stretch of intensity and saturation, if required, this stretch using the lower 5% and upper 95% points of the histograms of these two components as the stretch limits, and
3. transformation back from HSI to RGB representation.

Note that, since an RGB image is required, you cannot apply this enhancement to either a grey-scale image or a pseudocolour image.

Table A.1   Colour table used in the pseudocolour enhancement. The 256 grey levels in the image are mapped on to the 49 values shown in the table, so that a grey-scale value of 0 maps to row 1 of the table and a grey-scale value of 255 maps to row 49. Intermediate levels are then interpolated.

	R	G	B		R	G	B
1	1.000	0.000	0.000	26	0.000	0.613	1.000
2	1.000	0.064	0.000	27	0.000	0.548	1.000
3	1.000	0.193	0.000	28	0.000	0.484	1.000
4	1.000	0.258	0.000	29	0.000	0.419	1.000
5	1.000	0.323	0.000	30	0.000	0.355	1.000
6	1.000	0.387	0.000	31	0.000	0.290	1.000
7	1.000	0.452	0.000	32	0.000	0.226	1.000
8	1.000	0.516	0.000	33	0.000	0.161	1.000
9	1.000	0.548	0.000	34	0.000	0.000	1.000
10	1.000	0.613	0.000	35	0.516	0.000	1.000
11	1.000	0.677	0.000	36	0.581	0.000	1.000
12	1.000	0.742	0.000	37	0.645	0.000	1.000
13	1.000	0.806	0.000	38	0.710	0.000	1.000
14	1.000	0.871	0.000	39	0.774	0.000	1.000
15	1.000	0.936	0.000	40	0.839	0.000	1.000
16	1.000	1.000	0.000	41	0.903	0.000	1.000
17	0.903	1.000	0.000	42	0.968	0.000	1.000
18	0.806	1.000	0.000	43	1.000	0.000	0.935
19	0.677	1.000	0.000	44	1.000	0.000	0.839
20	0.516	1.000	0.000	45	1.000	0.000	0.742
21	0.000	1.000	0.452	46	1.000	0.000	0.613
22	0.000	1.000	0.710	47	1.000	0.000	0.484
23	0.000	1.000	1.000	48	1.000	0.000	0.097
24	0.000	0.871	1.000	49	1.000	0.000	0.000
25	0.000	0.677	1.000				

The messages displayed by MIPS after you have selected the HSI-to-RGB function follow the sequence listed above; an initial message informs you that the transform is in progress. Then you can select (by clicking the YES or NO buttons) whether you want to stretch the intensity and/or saturation components. Finally, you can choose to transform back to RGB. If you leave the procedure before transformation back to RGB then the channel assignments (initially red = 0, green = 1 and blue = 2) become hue = 0, saturation = 1 and intensity = 2.

The image resulting from this operation replaces the previously stored images in memory and the operation is not reversible. You should inspect the histograms and lookup tables using Plot|Histograms on the Main Window menu and use an enhancement function before displaying the image.

A.4.4.6   Decorrelation stretch

The ideas underlying the Decorrelation Stretch are provided in section 6.4.3. Like the RGB to HSI transform, the decorrelation stretch can only be performed on a false-colour image. To use this procedure, simply click on Enhance|Decorrelation Stretch in the MIPS Main Window. You are asked to select the basis of the stretch – the image correlation matrix or the image covariance matrix. The considerations on which this choice should be made are given in section 6.4.1. The use of the correlation matrix appears to give better results for images of semi-arid areas. The only further message is one that tells you that the transformation is in progress; wait until the message disappears before using Display from the MIPS Main Window menu to view the image, as detailed in section A.4.3.

The image resulting from this operation replaces the

previously stored images in memory and the operation is not reversible. Inspect the histograms and lookup tables using Plot|Histograms on the Main Window menu, and use an enhancement function before displaying the image. The method works well with some images, but not with others. If you wish to try some other enhancement procedure after applying a decorrelation stretch, then use Start|Read_Inf_File to read the original image files from disk.

The method used to calculate the decorrelation stretch differs somewhat from the more simplified procedure described in the text. The eigenvector matrix elements a_{ij} are multiplied by $\sqrt{\lambda_i}/s_j$ (for analysis based on the correlation matrix) or $\sqrt{\lambda_i}$ in the case of analysis based on the covariance matrix. The symbols λ_i and s_j represent the ith eigenvalue of matrix **R** (or **S**) and the variance of band j, respectively. Then scaling factors corresponding to 1.0, $0.5\sqrt{\lambda_i}/\sqrt{\lambda_2}$ and $0.2\sqrt{\lambda_i}/\sqrt{\lambda_3}$ are applied. The choice of scaling factors is one that has a significant effect on the appearance of the resulting decorrelation-stretched image. The factors used here seem to give acceptable results on most images. The method of scaling the output to the 0–255 range is that of resampling. The decorrelation stretch is first calculated for a sample of every fourth scan line and pixel, and the necessary factors are estimated from the sample. The procedure is then applied to the full image, and the result is scaled to 0–255 using the factors computed from the sample. Those images with bimodal histograms in any of the three bands are not generally amenable to the use of this technique. In most cases, the decorrelation stretch is followed by a linear contrast stretch or histogram equalisation, which is needed to compensate for the inadequacies involved in the method of converting the result of the procedure to the 0–255 scale.

Note that if any of the bands forming the RGB colour composite are identical then this procedure will exit with an error message 'One or more of the eigenvalues is equal to zero'. This should never happen.

A.4.4.7   Density slicing

The Density Slice option is applicable only to greyscale images. First, open the relevant INF file (Start|Open_Inf_File) and select a single band to display as a grey-scale image. You may wish to apply a linear stretch (5–95%) to the image before selecting the Density Slice option, which automatically displays the image in a Density Slice Window, using whatever contrast stretch you have applied. A colour palette and a grey-scale wedge are overlaid on the image. Select a colour from the wedge using the left mouse button, or click Edit to make up your own colour using the Windows palette. Then click on the grey-scale wedge (which corresponds to pixel values from 0 to 255, reading from the top to the bottom; the actual grey level reflects any contrast enhancements applied to the image) and the image is coloured up to that point with the chosen colour. At this point you can decide to continue with the same colour, select a new colour, extend the range of the current density slice to 255 ('fill to the top'), write the image to the clipboard and quit, or just quit. Note that you can reverse the direction of the operation; if you have filled 'too far' with a given colour, just click the 'use the same colour' button, and left-click the greyscale wedge above the lower boundary of the current colour, and the original grey-scale LUT values will replace the current colour.

To exit from the Density Slice procedure, do *not* click the Windows 'cross' symbol at the top right of the Density Slice Window, as this will crash the program. Click on the vertical colour bar, just as you would if you were starting another slice. You can then select either write image to clipboard and quit or just quit. In the former case the image in the Density Slice Window is written to the Windows clipboard, from which it can be retrieved by another Windows application. If you select just quit then the Density Slice Window is deleted.

Note that by using the Density Slice procedure the image type is changed from grey-scale to pseudocolour, and the single-band (grey-scale) image now has three associated LUTs (red, green and blue).

A.4.5   Image filtering options

The Filter menu item in the Main Window has six associated procedures:

- Image minus Laplacian
- Median
- Sobel
- Sigma
- Roberts gradient
- User defined

These procedures are described and illustrated in Chapter 7.

A.4.5.1   Image minus Laplacian operator

This operator is selected by clicking on the Filter|Image-Laplacian item in the Main Window. One (in the case of a grey-scale image) or three (in the case of a false-colour image) progress bars appear on the screen to show that calculations are taking place. The image in memory is altered by this operation, which cannot be reversed. You may need to reset the lookup tables by a contrast stretch (Enhance menu item in the MIPS Main Window) to take into account the fact that the image histogram(s) will have been altered by the procedure. Use the Display menu item on the Main Window (section A.4.3) to view the image.

The image resulting from this operation replaces the previously stored images in memory and the operation is not reversible. You should inspect the histograms and lookup tables using Plot|Histograms and change them using an enhancement function before displaying the image.

A.4.5.2 *Median filter*

Choose Median from the drop-down list associated with the Filter menu in the MIPS Main Window and a dialogue box will appear. Fill in the filter size in the *x* and *y* directions, remembering that these dimensions must be odd, positive integers. A maximum size of 9 × 9 has been built into MIPS. However, you can use the median filter option more than once to smooth an image repeatedly.

The median filtered image replaces the previously stored images in memory and the operation is not reversible. You should inspect the histograms and lookup tables using Plot and change them using an enhancement function before displaying the image.

A.4.5.3 *Sobel filter*

Choose Sobel from the drop-down list associated with the Filter menu in the MIPS Main Window. A 'busy' cursor is all that you will see as the operation is carried out. In the case of a false-colour image the edges highlighted by the Sobel filter are displayed in the colours corresponding to each image component, so that a yellow edge indicates strong red and strong green (but weak blue) edges.

The Sobel filtered image replaces the previously stored images in memory and the operation is not reversible. You should inspect the histograms and lookup tables using Plot and change them using an enhancement function before displaying the image. If you wish to see the strongest edges only, use the Enhance| Stretch function and set the lower limits manually to the threshold required. This will turn all values that are below the threshold to black. For example, if – after the Sobel filter has been applied – the Plot menu function

shows that the image histogram values range from 0 to 127 in each channel, then by using Enhance|Stretch to set the limits of the linear stretch to 95 and 127 respectively for the lower and upper limits you will suppress all the Sobel filter output values that are less than 95 to black (invisible) and will thus be able to view the strongest edges only.

A.4.5.4 *Sigma filter*

The sigma filter is normally applied to radar images in order to reduce speckle. The principles of the sigma filter are described in section 7.2.3. To select this function, choose the Filter menu item from the MIPS Main Window and click Sigma. You will be asked to supply values for (i) the window size, which must be specified in terms of two odd, positive integers in the range 3–11, (ii) *sigma* (pixels in the window will not be included in the calculation of the average if their value differs from the window central value by ± *sigma*) and (iii) *k* (if fewer than *k* pixels are included in the average then the window central value is left unchanged). The filter can operate any image type.

The sigma filtered image replaces the previously stored images in memory and the operation is not reversible. You should inspect the histograms and lookup tables using Plot and change them using an enhancement function before displaying the image.

A.4.5.5 *Roberts gradient*

The operation of this procedure is almost identical to the description in section A.4.5.3 (the Sobel filter) with the exception that the Roberts Gradient item is chosen from the drop-down list associated with the Filter menu item on the MIPS Main Window.

A.4.5.6 *User-defined filter*

This procedure allows you to define your own filter weights for a box filter with dimensions up to 7 × 7. Select the User Defined item from the drop-down list associated with the Filter menu item on the MIPS Main Window. An advisory message tells you that the maximum filter size is 7 × 7. Click OK. Now fill in the form which contains the filter weights and the divisor, as explained in section 7.2.1 (Figure A.8). Once you click OK one (for a grey-scale image) or three (for a false-colour image) progress boxes show that calculation is taking place.

The image resulting from this operation replaces the previously stored images in memory and the operation

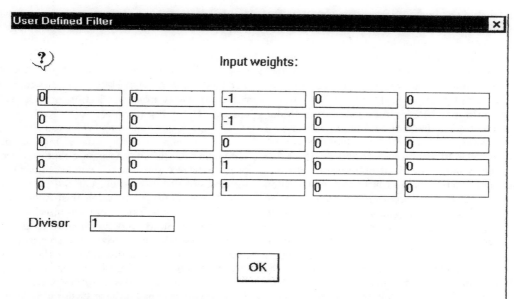

Figure A.8   Filter weight input form for User Defined Filter. Readers may like to determine the nature of the filtering operation that will result from the application of the weights shown in this figure.

is not reversible. You should inspect the histograms and lookup tables using `Plot` and change them using an enhancement function before displaying the image.

A.4.6   Transform menu

Three items are available from this menu: NDVI, linear functions, and PCA.

A.4.6.1   NDVI

The normalised difference vegetation index (NDVI) is described in Chapter 6. It involves the calculation of the ratio of the difference between the near-infrared and red bands and the sum of the same two bands. Hence you must select a multispectral image set in order to calculate the NDVI (see section A.4.2).

A warning appears on the screen after you select `Transform|NDVI` to tell you that the NDVI image will overwrite the image band currently stored in channel 0 (red). At this point you can opt to quit or to carry on. If you carry on then the image type will be changed to grey-scale. You are then asked to specify the locations of the infrared and red bands in the image you have chosen. The convention is that the longest waveband (e.g. near-infrared or band 3 in the case of a SPOT HRV image) is shown on the screen in red, and the two shorter wavebands (for example SPOT HRV bands 2 and 1,

corresponding to visible red and visible green) are shown in green and blue on the screen. Recall from section A.4.2 that the first band you select is displayed in red and is labelled 'channel 0'. The second and third bands that you select are displayed in bands 1 and 2. So, if you have used the conventional colour composite selection, the near-infrared and red bands will be displayed in channels 0 and 1, respectively, and it is these channel numbers that you must provide in the first selection box.

The program does an initial scan through the image, calculating the NDVI values for each pixel. In theory these values range from -1.0 to $+1.0$ and the output (NDVI) is scaled so that the NDVI value '-1.0' is represented by the pixel value '0' and the maximum possible NDVI value ($+1.0$) is represented by the pixel value '255'. Hence a pixel value of 127 represents an NDVI value of 0.0. In practice these limits are rarely reached and an image with greater detail and contrast can be obtained if the NDVI values are rescaled so that the minimum NDVI value in the image is given the pixel value 0, and the maximum NDVI value in the image is given the pixel value 255. The intermediate NDVI values are scaled to the pixel values 1–254. It is very important to remember that, if you select this option, the relationship between NDVI and pixel values cannot be determined by the use of the `Cursor` menu function in the Image Display Window. If you want to be able to select a pixel in the Image Display Window using the `Cursor` function and determine its NDVI value, do not select the `Rescale` option in `Transform|NDVI`. If you do not select this option then, once you display the NDVI image you will be able to compute the actual NDVI value from the pixel value by remembering that the range of NDVI values (-1.0 to $+1.0$) is linearly scaled on to the range 0–255. Just divide the pixel value given by the `Cursor` function by 128 and subtract 1.0 from the answer.

Once the NDVI function has terminated (the active cursor shape changes to an arrow) select a contrast enhancement function such as histogram equalisation (section A.4.4.2) before displaying the image (section A.4.3) otherwise the lookup tables will be unaltered from their previous state, which may be related to the false-colour image from which the NDVI was computed. Alternatively, click on `Utilities |Reset_LUTS` on the Main Window menu bar and the LUTS will return to their default state.

A.4.6.2   *Linear functions*

Unlike other MIPS display functions except PCA (section A.4.6.3) the `Linear Function` option operates on some or all of the image files contained in the current INF file (selected by the last `Start|Read_Inf` function). For example, if the INF file references six images you can choose one or more of the referenced images and calculate one or three linear functions of the form

$$o_{ij} = a_1 f_{ij1} + a_1 f_{ij2} + \ldots + a_k f_{ijk}$$

where o_{ij} is the (i, j)th pixel of the output image, $a_1, \ldots, a_k$ are the user-defined coefficients, and f_{ijk} is the (i, j)th pixel of input image k. The values of o_{ij} are scaled on to the range 0–255. If you choose to calculate one linear function it will be stored as a grey-scale image. You can choose to define three linear functions, in which case the first is stored as red, the second as green, and the third as blue.

When you choose this option an information window is displayed. You can either proceed or quit. The second input item requested is the number of output images to be generated. The choice is one or three. Then a check box showing the image filenames referenced by the current INF file is displayed. Click on the names of the files you wish to use. Your selection is verified next; you can accept your choice or try again. Finally, enter the coefficients a_{ij} for each file in turn and verify your choice. An hour-glass cursor appears while the procedure is executing. The final step (specification

and verification) is repeated twice if you request three output files.

A.4.6.3   *PCA*

The function `Transform|PCA` computes three principal components from a selection of files from the current INF file (the last one read using the `Start|Read_INF` function). When you select this option an information box appears giving the option of proceeding or quitting. Assuming that you decide to proceed the next step is to select at least three files from those referenced by the current INF file, using the checkboxes. You have to verify this choice, and have the option of re-inputting your selection. The final input box asks you to choose whether to base the analysis on the correlation matrix or the covariance matrix (see section 6.4.1 for a discussion of this choice). An information box tells you that the covariances are being calculated (this message appears irrespective of whether you have based the analysis on correlations or on covariances), then three information boxes indicate that component images (scores) are being computed for PC1, PC2 and PC3. These images are placed in the red, green and blue channels respectively. Remember to reset the LUTs before displaying the principal component colour composite, using `Display|Display Image`.

A.4.7   Plot function

The `Plot` function has two options – `Histograms` and `Cross-Sections`.

A.4.7.1   *Histograms function*

When you select this function the histogram or histograms for the single-band (grey-scale) or three-band (false-colour) image are drawn. The lookup table for each band is also plotted, and the maximum and minimum pixel values in the image are listed. The image (input) pixel values are plotted on the x-axis. The left-hand y-axis shows the number of pixel values achieving the given level and the right-hand y-axis is the output colour intensity. Thus, you use the left-hand y-axis to interpret the image histogram and the right-hand y-axis to interpret the lookup table (LUT), which is drawn as a black line. Note that for a pseudocolour image (generated using density slicing or the pseudocolour enhancement) the histogram of the original grey-scale image is drawn three times with the red, green and blue LUTS, respectively, superimposed.

On the right-hand side of the Plot Window there are two buttons. Click OK to exit the Plot|Histograms function. Click the Copy button to place the plot on the Clipboard. You can use the Paste option in your word processor to copy the histogram/LUT plot into your document, or you can paste it into a graphics manipulation program. Figure A.9 was 'captured' using this procedure.

A.4.7.2 *Cross-sections function*

The Cross-Sections function on the MIPS Main Window Plot menu provides the means to generate cross-sectional plots between specified points on a displayed image. You must have read an image using Start|Read_Inf before you attempt to draw a cross-section. When you select Plot|Cross-Sections a display window appears, and the image is drawn in this window, using any enhancement you have applied from the Enhancement menu. Move the mouse (cross-wire) pointer inside this display window and click and release the left mouse button to indicate the start point of the

cross-section. When you move the mouse a line is drawn over the image from this initial position, using XOR mode (the line colour changes in response to the colour of the underlying image). Click and release the right mouse button to indicate the end of the cross-section. An information box appears to tell you how many pixels (points) there are along the cross-section. Click OK when you have read and absorbed this information. A graphics window appears and the cross-section is drawn. For a false-colour image the red, green and blue components of the image (which you selected when the image was initially displayed using the Start| Read_Inf_File function) are shown in red, green and blue. If the image is grey-scale or pseudocolour a single grey line is used to plot the pixel values along the cross-section. Note that the unenhanced (raw) image values are used in the plot. The graph is written to the Clipboard. Press OK when you have read the message informing you of this action. You are then asked if you wish to draw another cross-section. Click Yes if you wish to do this, and proceed exactly as before. If you click No the Image Display Window is closed.

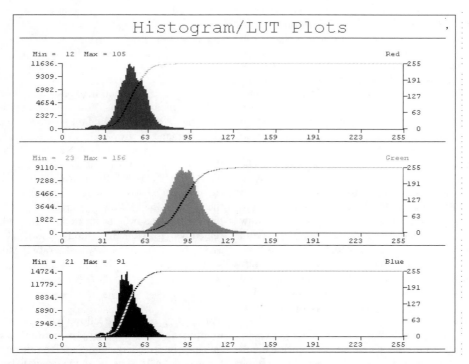

Figure A.9 Histograms and lookup tables (LUTS) captured from MIPS using the 'Save' button in the Plot|Histogram option. The right-hand *y*-axis is the output colour intensity (to interpret the LUT) and the left-hand *y*-axis is the number of pixel values (for interpretation of the histogram). The *x*-axis is the input image pixel value. The lookup table is shown by a black line. This set of histograms and LUTS was derived from a SPOT HRV image after contrast enhancement by histogram equalisation.

A.4.8 Classify function

This menu option has no sub-menu, and is only applicable to false-colour images. When you click `Classify` the first input window to appear asks whether you wish to save the training class area coordinates to a file that can be read by stand-alone program `ml_train` (section B.3.20). If you want to save the file, press the `YES` button then enter a filename in the input box which appears next. When you have done this, or if you pressed the `NO` button, a new graphics window entitled 'Training Class Selection' appears. This window has the following menu: `Display`, `Select`, `Zoom`, `Classify` and `Finish`.

The `Display` menu item simply displays the current image using the current lookup tables. This is the first step. The second step is to click the `Select` button. An input box appears, and choose one of `Start New Class`, `Add to Present` or `End`. You must select training data for at least one class. To do this, press the `Start New Class` button. Give the class name when asked. Now click the top left corner of the area on the image that you want to use as training data. Then click the bottom right corner of the training area. A box is drawn in the Training Data Selection Window to outline the training area, and the coordinates are given in an information box. Press `Select` again to add another training area to this class, or to start a new training class, or to end. Note that you must choose `Select` then click the `End` button before you do anything else except `Zoom`. Thus, you repeatedly use `Select` to add further training data until the training data collection phase is complete. Then you click `Select` again and click the `End` button.

While you are in `Select` mode you can use the `Zoom` menu item on the Training Data Selection Window. When you click `Zoom` you must point to the upper left corner of the area to be zoomed by left-clicking in the Training Class Selection Window. A zoom window appears with three menu items: `Display`, `Select` and `Quit`. Click `Display` to see the zoomed image. A fixed zoom factor of 2 is used, and zooming is performed by pixel replication, not interpolation. Click `Select` to continue to collect training class data, as described in the preceding paragraph. Click `Quit` to exit from the zoom procedure.

Once you have collected the training data (and have clicked the `End` button on the `Select` menu item's input box) you can click `Classify` on the Training Class Collection Window's menu. If you wish to perform a maximum-likelihood classification using the stand-alone program `ml` (section B.3.21) you can omit the `Classify` stage and simply press `Quit`. Assuming that you wish to do an online classification using the three image bands forming the current false-colour composite image, press `Classify` and you will be asked to choose between the `K-Means` and the `Parallelepiped` classifiers. Once this choice is made, a Classified Image Window appears. Press `Draw` and the cursor changes to the hour-glass shape as the image is classified using the three bands selected for the false-colour display. (Note – Appendix B contains details of stand-alone programs `ml`, `kmeans` and `piped` which classify an image set on the basis of up to 12 features.) Once the classified image is displayed an information window tells you to click `Key` on the annotation window. Click `OK` on this information window and the annotation window appears to the right of the classified image display. Choose the menu item `Key` as instructed and a key shows the relationship between the colours on the classified image display and the class number. Class 0 refers to pixels left unclassified by the parallelepiped classifier. Click `OK` to close all the open windows except the main window. The windows associated with the `Classify` option are illustrated in Figure A.10.

A.4.9 Utilities

A.4.9.1 Save image

The `Save Image` function saves the current image to a disk file in raw format. You are prompted for one or three filenames, depending on whether the current displayed image is grey-scale or false-colour, using the Windows File Selection box. No INF file is created for the saved file. You can either add the name of saved file(s) to an existing INF file using a text editor or you can create a new INF file to reference this image using the stand-alone MIPS program `make_inf`. If you add the saved filename(s) to an existing INF file you must remember to run the stand-alone MIPS program `make_his` to create a new histogram file for the image set referenced by the INF file.

A.4.9.2 Image arithmetic

This utility procedure allows you to add, subtract, multiply and divide images, or to add a constant to an image or intermediate image, and to rescale the result to the 0–255 scale. The final image may be created in several steps, involving the generation of an intermediate image. The available options are shown in Figure A.11.

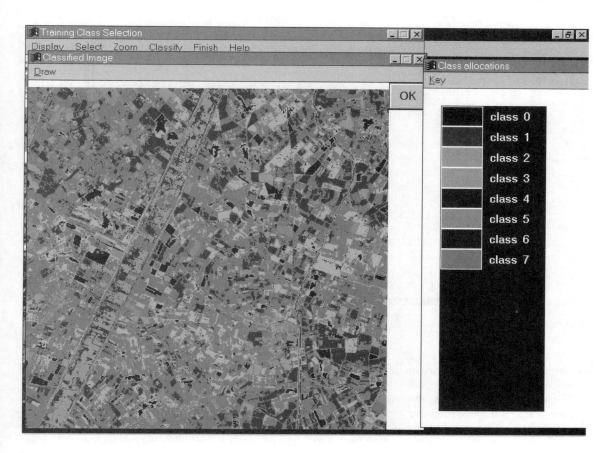

Figure A.10 Classified Image and Class Allocation Windows.

Operations of type 'add/subtract/multiply/divide image' involving a constant require (i) the choice of image, which is one of the images stored in the red (0), green (1) or blue (2) channels, and (ii) the specification of a constant in the range −255 to +255. The result of this operation (e.g. image + constant) is an intermediate image that is known as `image1`. If you perform any other operation involving a constant then this constant will be added to or subtracted from `image1`, or else `image1` will be multiplied or divided by the constant. If you select an operation such as `divide image by image` then the first image will be `image1` and the second will be the image stored in one of the red, green or blue channels.

If you choose image–image operations without specifying a constant, for example divide image by image, then the result of the first such operation will become the intermediate `image1`, so that if you decide to carry out another operation on the result, for example

'divide image by constant', then you will be asked only for the value of the constant and `image1` will be divided by this constant.

To exit from this procedure select `Scale to 0-255 and Finish`. Do not exit in any other way. The final image is rescaled and then stored as a grey-scale image. The original image or images selected in the last `Start |Read_Inf_file` procedure are overwritten.

If you try to divide an image by zero (for example, some pixels in the divisor image may be zero) then a divisor of 1 is used.

A.4.9.3 Threshold

The `Threshold` procedure allows the creation of a binary mask (in fact it is a 0/255 mask) from a selected image. This mask can be displayed and saved using the `SaveImage` function (section A.4.9.1) or applied to either one or three files selected from the current INF

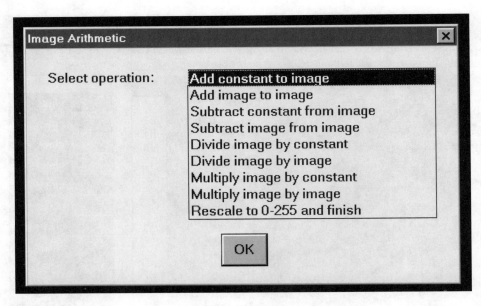

Figure A.11   Options available in Image Arithmetic function.

file to create a grey-scale or false-colour masked image, respectively.

The procedure is quite straightforward to use. Display the image from which you want to create the mask. This image can be displayed as a grey-scale image or as a false-colour image. If it is displayed as a false-colour image then you have to select one of the bands (1, 2, or 3 – or red, green, and blue) as the source image for the mask. Now enter the threshold value (it would be sensible to look at the image histogram using Plot|Histograms before deciding on the threshold value). The mask is created from the source image by setting pixels that have values less than or equal to this threshold to the value 0, and pixels that have values greater than the threshold to the value 255. At this point you can quit the procedure, leaving the 0/255 mask image in memory. You can view this mask using Display|Display Image.

Alternatively, you can apply the mask to one or three of the files referenced by the current INF file. If you decide to do this you must first select (using buttons) whether you are going to generate a grey-scale or a false-colour output image. If you select grey-scale output then you will choose one file from those referenced by the INF file. If you choose false-colour output then you will need to specify three files from the current INF files in the order red, green, blue.

You can decide to mask (set to zero) pixels in the output image that are equal to 0 in the mask file, or that are equal to 255 in the mask file. In other words, consider the mask file to consist of 0 (black) values and 255 (white) values. This mask file is overlaid on one (grey-scale) or three (false-colour) images, depending on your choice. You can use either the black area or the white area as a mask.

The result of the operation will become the current image. You will need to apply an enhancement procedure before viewing it.

A.4.9.4   Reset_LUTs

This procedure simply resets the LUTS so that input values of 0–255 produce corresponding colour intensities of 0–255.

A.4.9.5   Create_BMP

The current image is saved as an 8- or 24-bit bitmap depending on whether it is a grey-scale or false-colour image. You provide the name of the file into which the bitmap is to be saved. A file suffix of .bmp is normal for bitmap files.

A.4.9.6   Read_LUT_file

This procedure reads a text file containing a lookup

table. This lookup table file can be generated by procedure `Write_LUT_file` (section A.4.9.7) or by the MIPS stand-alone program `make_LUT` (section B.3.17). You provide the name of the LUT file using the Windows File Selection box.

A.4.9.7 *Write_LUT_file*

This procedure writes the current lookup tables to a text file, which is created using the Windows File Selection box. It is sometimes useful to save both the current image (for example, after filtering) using `Save_Image` (section A.4.9.1) and the associated lookup table, so that the result can quickly be re-created.

A.4.10 *Exit*

Click on the `Exit` item on the MIPS Main Window menu to close the log file and exit gracefully from the MIPS display program.

Appendix B

Stand-Alone Programs

B.1   INSTALLING THE PROGRAMS

B.1.1   Introduction

The CD-ROM accompanying this book contains the MIPS display program, which is described in Appendix A, plus a number of 'stand-alone' programs that are described in this appendix. Those programs which perform image processing use the MIPS image file format as described in Appendix A, section A.1.2. Salford Software Ltd's FTN95/WIN32 compiler version 1.12 was used to create these programs, except for CONSTMX, which uses a Microsoft ISML library routine that is available with Microsoft PowerStation Fortran version 4.0.

Like the MIPS display program described in Appendix A, these stand-alone programs require a PC running the Windows 95 operating system. A minimum of 16 Mb RAM is recommended (though the programs will run faster with 32 Mb). The programs use less than 4 Mb of disk space, though additional space will be needed to store images.

Note that this software remains the author's copyright, and is provided on condition that this copyright is acknowledged whenever results based upon any of the programs listed in this Appendix are published, and that the programs are not copied, sold, or traded separately from the book. The author cannot answer queries relating to the use of the programs or their behaviour, and they are supplied without any guarantee or warranty. No *Help* files are included on the CD, in order to encourage users to read this appendix and appropriate sections of the book.

B.1.2   Software installation

A program called Mips_Install is provided on the CD. This program copies the stand-alone programs from the CD to your hard disk. If your CD drive letter is D: you can run the stand-alone programs directly from the CD. However, if your CD drive letter is not D: then you must copy the MIPS display program (Appendix A), the stand-alone programs (Appendix B) and the image dictionary files that reference the sample images (Appendix D) to your hard disk. The sample images can be accessed from the CD drive irrespective of the CD drive letter.

Full details of MIPS_Install are given in Appendix C. After the programs are installed, you can create a desktop folder by right-clicking on the Windows desktop and selecting New then Folder. Open the new folder by double-clicking on its icon, then right-click inside the folder area. Select New and then Shortcut. Now use the Browse button to locate one of the stand-alone programs. Finally, change the icon to one of the Windows-provided (standard) icons. If you are not sure what to do, click the Start button on the Windows 95 lower tool bar, select Help, click Index and then type desktop. When the term Desktop is highlighted, look for icons, adding and click on this topic.

Note, however, that some of the stand-alone programs are DOS applications. This means that they do not run in a window but in what is called a 'DOS box'. These programs are identified as DOS applications in the descriptions below. To run a DOS application, click Start on the lower Windows 95 toolbar, then Programs then MS-DOS Prompt. A DOS box will appear. Pressing Alt and Return will change this DOS box from a window to full-screen mode (and *vice versa*). To run the DOS application, simply type its path and filename, for example: C:\myprograms\DOS_application. Further details are given in section B.2 and in the individual program descriptions.

B.2 RUNNING THE STAND-ALONE PROGRAMS

There are four ways of running the programs described in this appendix. I generally use the fourth method – start the program from MS-DOS – because a few of the programs are not genuine Windows applications. If they are started from a shortcut, or Explorer, or the Run command on the Start menu button, then they run in a 'DOS box'. Programs that are not Windows applications are described as such in the following text. The four methods of running the programs are as follows.

1. Use Windows Explorer to locate the file containing the program you wish to run. Double-click on the filename. This file should be in the directory that you selected when you copied the programs from the CD as described in Appendix C.
2. If you created a Desktop folder containing shortcuts to the stand-alone programs, as described above, then open the folder and double click on the icon corresponding to the program you wish to run.
3. Use the `Start|Run` option, and then type `disc:\path\progname` in the box provided, then click `OK`. Note: you must change the term `disc` to the appropriate drive letter, substitute the correct path for the term `path`, and change the term `progname` to the name of the program you wish to run, as listed in section B.3. For example, if the stand-alone programs are stored on drive `C:` and the path is `\programs\mips\programs` then, if I wish to run the geometric correction program (for example), I would type `C:\programs\mips\programs\geomcorr` in the filename box.
4. Open an MS-DOS window from the Windows `Start` button on the lower menu bar (it appears under `Programs`) and type the filename of the program to be executed. For example, to run program `constmx` from the CD, type the command `D:\constmx` at the DOS prompt (assuming that the drive letter for your CD is `D`).

These operations are described in the Windows 95 documentation.

B.3 ALPHABETICAL LIST OF STAND-ALONE PROGRAMS

The stand-alone programs are provided on the CD and described in the following sections.

B.3.1 BMP2RAW: convert Windows bitmap to raw image

All MIPS programs use 'raw' format images as described in section A.1.2. The `bmp2raw` utility is provided to enable you to convert 8- or 24-bit bitmaps to raw format suitable for use with MIPS. A Windows bitmap usually has the suffix `.bmp`. It can store either one 8-bit image or three (RGB) 8-bit images. The latter is called a 24-bit bitmap. An 8-bit bitmap contains a colour table (LUT) that defines each of the 256 palette (LUT) entries in terms of their RGB values in the range 0–255. Each palette entry may define a colour (different RGB values) or a grey level (identical RGB values). In the terminology of this book, the former is considered to be a pseudocolour image. The 24-bit bitmap normally does not store a colour table. There are many other file formats such as GIF, TIFF and JPEG. These can be converted to bitmap form using a graphics utility.

Program `bmp2raw` can handle image sizes up to 1024 × 1024 pixels. It cannot handle compressed bitmaps. You can save the colour table associated with the bitmap into a separate file and read it into the MIPS display program using `Utilities|Read_LUT`. Note that a 24-bit bitmap has either no stored colour table or a set of zero values.

Input to the program is the name of the bitmap file to be converted, the name of a file to hold the colour table (LUT), and the name or names of the image output files. There is one output file for an 8-bit bitmap and three output files for a 24-bit bitmap. The red component of the bitmap is placed in the first output file, the green component is placed in the second output file, and the blue component in the third. If you wish to view the resulting image files using the MIPS display program you will need to prepare an image dictionary (INF) file using the stand-alone program `make_inf` (section B.3.16). Use the `Utilities|Read_LUT_file` menu item in the MIPS display program (section A.4.9.6) to read in the colour table.

Bitmaps are peculiar in that there are a number of options available, some of which are not used and others of which appear to be ignored. This program is fairly simple and cannot deal with all the available options and idiosyncrasies, so in some cases you may get very peculiar results.

B.3.2 CHAVEZ: atmospheric haze correction

Program `chavez` is an implementation of the routine described by Chavez (1988) (see section 4.4) and is intended to demonstrate the principle of haze correction

Table B.1 The programs on the CD-ROM.

Section	Program name	Program function
B.3.1	BMP2RAW	Convert Windows bitmap to raw image
B.3.2	CHAVEZ	Chavez's atmospheric haze correction
B.3.3	CONSTMX	Fully constrained spectral unmixing
B.3.4	CONSTMX1	Constrained spectral unmixing
B.3.5	CUBCONV	Illustrating cubic convolution resampling
B.3.6	DIVERGEN	Divergence analysis
B.3.7	FOURIER	Two-dimensional Fourier transform
B.3.8	GEOMCORR	Geometric correction using least-squares polynomials
B.3.9	GLCM	Grey-level co-occurrence matrix
B.3.10	IMFACT	Noise-reduced principal components analysis
B.3.11	ISODATA	ISODATA unsupervised classification
B.3.12	JULDAY	Days elapsed since start of reference year
B.3.13	JULDAY__INV	Day and month from Julian day number
B.3.14	KMEANS	k-means classification
B.3.15	MAKE__HIS	Generate a histogram file
B.3.16	MAKE__INF	Generate a MIPS image dictionary (INF) file
B.3.17	MAKE__LUT	Manually generate a lookup table
B.3.18	MASK	Image masking
B.3.19	MIXMOD	Spectral unmixing
B.3.20	ML__TRAIN	Generate training statistics for maximum-likelihood classification
B.3.21	ML	Maximum-likelihood classification
B.3.22	NL__DESTR	Non-linear de-striping after Horn and Woodham
B.3.23	NLMAP	Non-linear mapping
B.3.24	NADIRCOR	Cross-scan illumination correction
B.3.25	PCA	Principal components analysis
B.3.26	PIPED	Parallelepiped classification
B.3.27	PLOT__PIX	Plot pixel spectra
B.3.28	QMODE	Q-mode factor analysis
B.3.29	READ__ESATM__CD	Read TM image from Eurimage CD
B.3.30	READ__SPOT__CD	Read a SPOT image from CD
B.3.31	SCATTER	Scatter plot
B.3.32	SEL__PIX	Select random sample of pixels
B.3.33	SPECINDX	Spectral indices
B.3.34	SPEC__ANG	Spectral angle mapping
B.3.35	SUB__EXTR	Extract sub-image from source image
B.3.36	TASSCAP	Tasselled cap transform
B.3.37	TEST__INF	Check consistency of INF file
B.3.38	UNPACK_GB	Unpacks a generic binary file

rather than to provide an operational routine. The procedure requires that the pixel values (sometimes referred to as DN, or digital number) are converted to radiances before the haze correction procedure is applied. This program uses the gains and offsets for the Landsat-4 TM as listed by Chavez. Five atmospheric models are available: very clear, clear, moderate, hazy and very hazy. The user must first provide a starting haze value in one of TM bands 1, 2 or 3, and then specify which of the five models is to be used. The program calculates haze corrections for the other five optical TM bands and lists them both as radiances and as DN.

B.3.3 CONSTMX: fully constrained spectral unmixing

Program `constmx` uses the Microsoft IMSL (Mathematical and Statistical Library) routine LCLSQ to solve the spectral unmixing equations subject to the following constraints:

- the sum of the proportions of the end-members at a given pixel is in the range 0–1, and
- the value of any individual proportion of an end-member at a given pixel is in the range 0–1.

Program `constmx` is not a true Windows application and is best run from MS-DOS. The program requests the names of two files. The first is a data file, which must be prepared beforehand using a standard text editor (if you use a word processor make sure that you save the data file in text mode). The layout of this data file is described below. The second is a log file to which the program directs text output. You can inspect this file afterwards using the MS-DOS `type` command or any word processor.

Program `constmx` can handle images with any number of scan lines, but image width is restricted to 2048 pixels per scan line. There is a maximum of 12 input image bands and 12 end-members. If you wish to view the resulting end-member image files using the MIPS display program you will need to prepare an image dictionary (INF) file using the stand-alone program `make_inf` (section B.3.16).

The data file holds several items. Each item may be one or more lines of text. The first element of each item must start on a new line. The format of the data file for `constmx` is as follows:

Record   Contents

1.   Number of bands of imagery (NBANDS).
        Number of end-members required (NEND).

2.   The end-member library matrix presented as NBANDS lines of NEND values.
        Thus, each end-member is defined in terms of the spectral bands by reading down the columns of this matrix (column 1 gives the values on each spectral band for end-member 1 and so on).

3.   Number of scan lines in the input images (NL).
        Number of pixels per scan line in the input images (NPIX).
        Thus, each image in the set to be analysed has NL scanlines of NPIX pixels.

4.   This item consists of NBANDS filenames, one per line.
        Each filename is the name of an existing image file. These filenames must be in the same order as the rows of the end-member library matrix in item 2. Each must hold an image in raw format with the dimensions specified in item 3.

5.   This item consists of NEND filenames, one per line.
        Each filename is the name of an end-member proportions (image) file to be created by the program. The end-member files will be created in the order of the columns of the end-member library matrix in item 2.

6.   The name of a file to hold the residual image.

7.   Enter F as the first character on this line if the masking option is not to be used.
        Enter T if mask option is to be used.

8.   If T was entered at item 7 then:
        Enter T if the zero pixels in the mask image are to be ignored.
        Enter F if the non-zero pixels in the mask image are to be ignored.

9.   If T was entered at item 7 then give here the name of an existing mask file. This mask file contains zeros and ones only. If T is entered at item 8 then the zero pixels in this mask will be interpreted as signalling pixels that are not to be processed. These pixels will be given a proportion value of 0 in the output end-member images.
        If F is entered at item 8 then the non-zero pixels in the mask will be interpreted as pixels that are not to be processed. These pixels are given a proportion value of 0 in the end-member images.
        The mask can be created using a GIS, and may – for example – be based on geological boundaries or soil types. Alternatively, the MIPS display program can be used to generate a mask from the images themselves (for instance, using the NDVI to separate vegetated and non-vegetated areas) using the `Threshold` option (section A.4.9.3). The mask can be saved using the `Save Image` option (section A.4.9.1).

Example

A six-band image is to be processed, and four end-member spectra are provided. The image size is 1024 × 1024 pixels. A mask image is to be used and pixels in the images corresponding to zero pixels in the mask are to be ignored.

```
6 4
050   034   048   212
035   045   066   222
022   071   065   235
000   086   062   225
000   031   090   244
000   016   133   201
1024   1024
c:|images|tm-band.1
c:|images|tm-band.2
c:|images|tm-band.3
c:|images|tm-band.4
c:|images|tm-band.5
c:|images|tm-band.6
c:|images|end.1
c:|images|end.2
c:|images|end.3
c:|images|end.4
c:|images|residfile.img
T
T
c:|images|maskfile.img
```

B.3.4 CONSTMX1: constrained spectral unmixing

Program `constmx1` uses subroutine BVLS, which is a modified version of the routine NNLS that appeared in Lawson and Hansen (1995). The Fortran-90 code used here was downloaded from Netlib. It solves the least-squares problem $\mathbf{Ax} = \mathbf{b}$ subject to the constraint that $x_j \geq 0$ for all j. The elements of vector $\mathbf{x}$ are the mixture proportions. This program differs from `constmx` in that the elements of $\mathbf{x}$ are not constrained to sum to 1.0 or less. The method of use is exactly the same as that described above for `constmx` except, of course, that the program name is `constmx1`.

B.3.5 CUBCONV: illustrating cubic convolution resampling

`Cubconv` is a small program to illustrate the procedure of bicubic convolution (section 4.3.1.3). It takes a symmetric 4×4 grid and allows the user to specify a point in the central area of the grid (within the range rows 2–3, columns 2–3) and outputs the interpolated value at that point, together with the nearest-neighbour value, which is always 3. When you run this program the grid is displayed on the screen and you are asked to input the row and column coordinates of the test point. The interpolated value and the nearest-neighbour value are printed. Program `cubconv` is not a true Windows application and is best run from MS-DOS.

B.3.6 DIVERGEN: divergence analysis

The use of the transformed divergence J_T is described in section 8.9 as a method of feature selection. Program `divergen` takes the statistics file that is generated by the MIPS program `ml_train`. The average J_T is calculated over all pairs of classes for each band, and at stage 1 the highest average J_T is selected. Let the corresponding spectral band be denoted by i. At stage 2 the combinations of band i and all other bands in turn are used, and the highest average J_T is selected to give the best pair, say i and j. Now bands i and j are used together with each of the remaining bands in turn, and the procedure is repeated until all bands are included.

To use this program you must first collect training data coordinates using the MIPS display program's `Classify` option (section A.4.8). Secondly, this coordinates file is processed by `ml_train` (section B.3.20) to generate a statistics file. When you run `divergen` you will be asked for

- a log file name,
- the name of the statistics file, and
- the name of the image dictionary (INF) file to which the statistics file refers.

Output is directed to the log file, which can be printed or listed to the screen using the MS-DOS `type` command or via a word processor. Details of the J_T values for every calculation are listed, plus a summary of the best $1, 2, \ldots, n$ band combinations. Values of J_T are mapped on to a 0–100 scale. Note that the program reads all of the image files that are referenced by the nominated image dictionary (INF) file. If, for example, the INF file contains the names of seven files each holding a Landsat TM band then the divergence statistics will be computed for all seven bands. If you wish to select fewer files than are referenced in the INF file then edit the INF file and give the edited file a different name. The selected INF file must be the same as that selected in the `ml_train` program, which calculates statistics from the training data coordinates file that is output by the `Classify` option in the MIPS display program.

B.3.7 FOURIER: two-dimensional Fourier transform

Program `fourier` will carry out the discrete base-2 Fourier transform for a square image with a side length that is a power of 2 up to 512. The user is asked to supply the name of an INF file and then to select one image file from the list of images referenced by this INF file. The selected image is displayed after an automatic linear contrast stretch that takes the 5% and 95% points of the image histograms as the stretch limits (section 5.3.1). Press `Finish` on the menu bar once you have viewed the image (but note that the program keeps running while you are viewing the image). Next, the forward discrete Fourier transform of the image is computed and the log of the amplitude spectrum is displayed, again using the 5–95% contrast stretch. This image is automatically placed on the Windows Clipboard. Click on `Finish` as before to destroy the window. At this stage you can choose a frequency-domain filter from the following list:

- Ideal high-pass
- Ideal band-pass
- Ideal low-pass
- Exponential high-pass
- Exponential low-pass
- Butterworth high-pass
- Butterworth low-pass
- Trapezoidal high-pass
- Trapezoidal low-pass
- High-frequency boost
- Directional high-pass

These filters are described in general terms in section 7.5, and in more detail in Gonzales and Woods (1992). The high-frequency boost filter simply doubles the amplitudes with frequencies greater than F_{crit}.

All of the filters require the user to enter a value or values defining threshold frequency limits. For example, the ideal low-pass filter requires a single threshold frequency. Elements of the amplitude spectrum with a frequency greater than the threshold are set to zero. The ideal band-pass filter requires an upper and a lower threshold frequency to define the extent of the band. Frequencies below the lower threshold are set to zero, as are frequencies greater than the upper threshold. For a 512 × 512 image the permissible range of threshold frequencies is 1–255. Low frequencies have low wavenumbers. Some experience is necessary before a sensible choice can be made, and the reader is encouraged to experiment in order to familiarise him- or herself with the effects produced by varying the cut-off frequency. I have found that a threshold or cut-off frequency of around 25 effectively separates the low-frequency pattern and the high-frequency detail, though these values should be taken only as a rough guide.

A plot of the chosen filter's transfer function is drawn. Press the `OK` button when you have viewed it. Finally the filtered spectrum is transformed back to the spatial domain, and is scaled on to a 0–255 range and displayed following a 5–95% contrast stretch. The filtered image can be saved to a disk file. If you wish to save the image, answer `Yes` to the question 'Do you wish to save the filtered image?' and use the Windows 'File Open' dialogue box to specify the name of a file to hold the image. You may, for example, wish to filter the same image three times, using low-pass, band-pass and high-pass filters. Three filtered image files will thus be created, and can be displayed in red, green and blue as described in Chapter 6. The `make_inf` program (section B.3.16) must be run to generate image dictionary and histogram files before the MIPS image display program is used to view the result.

B.3.8   GEOMCORR: geometric correction using least-squares polynomials

Program `geomcorr` uses polynomials of order 3 or less, based on up to 100 ground control points (GCPs), in the registration of an image to a map. Only one selected band of an image set is registered. The nearest-neighbour interpolation procedure is used in the resampling process. The program is written to accept map coordinate input in the form of eastings and northings, which are measured in kilometres relative to some arbitrary datum (an example is the Ordnance Survey of Great Britain's National Grid).

Before using this program you must prepare a data file using a text editor (if you use a word processor make sure that the file is saved as plain text). The data file contains the easting and northing coordinate in kilometres and the image pixel number (counting the leftmost pixel as 1) and line number (counting the topmost line in the image as 1) for each GCP. The data for each GCP start a new record in the data file. An example data file is:

```
58.125   38.250   237.5    48.0
56.925   38.150   118.0    79.5
59.250   34.125   415.5   434.0
57.200   36.900   165.0   196.5
55.700   35.725    37.0   336.5
59.375   35.075   407.0   337.5
58.125   37.600   246.0   113.0
58.100   35.100   279.5   358.5
```

Note that the eastings and northings for these eight GCPs are not the true National Grid coordinates; it is sensible to subtract a constant from the actual values in order to avoid the accumulation of error as their powers and cross-products are calculated. Thus, if the true National Grid eastings for the GCPs in the list above are all preceded by the digits 67924 (so that the easting for the first point is 6792458.125) then it makes mathematical sense to remove the digits 67924. The result will be the same, in theory. In practice, the accumulation of squares and cubes of numbers as large as 6792 458.125 will result in rounding error.

Details of the transformation, as described in Chapter 4, are sent to a log file, the name of which is supplied in response to the first prompt. Give the full path associated with the file unless you want it to be created in your current directory. Next, give the maximum order of the polynomial surface to be applied. The program will accept values of 1, 2 or 3. If you choose a second- or a third-order surface then the statistics for the lower-order surfaces will be sent to the log file in addition to the statistics for the selected surface. The third prompt requests the name of the text file holding the GCP data as described above.

If you wish, you can stop the program once the statistics for each polynomial have been calculated. It is advisable to do this before proceeding to correct an image so that you can inspect the values of residual errors at each GCP. If you wish to stop the program, click on the NO button in response to the question 'Do you want to correct an image?' Otherwise, click YES to proceed. If you decide to proceed, you must select an INF file containing details of the image to be corrected, using the Windows file selection procedure. Next, a list of image files referenced by the INF file is listed on the screen, in numerical order $(1, 2, ..., n)$. Type the identification number of the file you want to correct. Make sure that the INF file refers to the image set for which the GCPs were collected.

Two values are required to answer the next question – the dimensions of the pixel in the output image. Give the east–west dimension first, then the north–south dimension. These values are specified in *metres*. The name of an output image file and an associated image dictionary (INF) file are required, and their names (including the full path) are requested next.

There are memory restrictions in the program, and an error message is printed if you exceed these limits. The program assigns 1 Mb of RAM to the input image so the product of the number of lines and pixels in the input image should not exceed 1 048 576. The output image

cannot have more than 2048 pixels per line. To speed up the correction procedure the image is resampled in 10×10 pixel blocks, so the output image width and height are always a multiple of 10. This may leave a black border around the corrected image.

Example

An example of the log file output for the program geomcorr is given below. The four equations *column(pixel)* $= f(E, N)$, *row(scan line)* $= f(E, N)$, $E = f(column, row)$ and $N = f(column, line)$ are calculated and the coefficients in the least-squares equation are printed. For example, the relationship *column(pixel)* $= f(E, N)$ are *column(pixel)* $= -4905.32 + 98.81E - 15.75N$. For each of the four relationships the actual and predicted values for each GCP are listed together with the individual residuals and the sum of squares of the residuals. Finally, the map coordinates of the image corners are printed, together with the coordinates of the bounding rectangle.

B.3.9   GLCM: grey-level co-occurrence matrix

The grey-level co-occurrence matrix (GLCM) is used in extracting texture features from the grey-level values forming an image (section 8.7.1). This program is not intended for operational use. It is provided to allow the reader to check his or her pencil-and-paper calculations of the grey-level transitions in a small synthetic image. The image size is restricted to a maximum of 80 rows and 80 columns, and the maximum number of grey levels is 10 (range 0–9). A data file must be created in text format before the program is used. You can use a standard editor to prepare this file. If you use a word processor be sure to save the file in text format. The file contains

- number of grey levels in the image (e.g. 10 for 0–9 range),
- number of rows and number of columns in the image, and
- image grey levels, row by row.

These data are separated by a space. The program calculates the GLCM for directions 0°, 45°, 90° and 135° and for a separation distance of 1 pixel. Output is sent to a text file, which can be printed (or inspected on-screen) once the program has run successfully.

FITTING EQUATION IMAGE PIXEL=F(EASTING, NORTHING)
FITTING POLYNOMIAL OF ORDER 1

Least-squares coefficients are:
```
 1  -0.490532E+04
 2   0.988100E+02
 3  -0.157471E+02
```

Explained sum of squares: 0.1219485E+06
Residual sum of squares : 0.3492578E+02
Total sum of squares   : 0.1219835E+06
Explained sum of squares as percent of total: 99.9714

EAST	NORTH	COL	PREDICTED	RESIDUAL
58.1250	38.2500	237.5000	235.6868	-1.8132
56.9250	38.1500	118.0000	118.6895	0.6895
59.2500	34.1250	415.5000	411.8048	-3.6952
57.2000	36.9000	165.0000	165.5462	0.5462
55.7000	35.7250	37.0000	35.8341	-1.1659
59.3750	35.0750	407.0000	409.1963	2.1963
58.1250	37.6000	246.0000	245.9225	-0.0775
58.1000	35.1000	279.5000	282.8198	3.3198

Standard error (standard deviation of residuals): 0.223370E+01

FITTING EQUATION IMAGE LINE=F(EASTING, NORTHING)
FITTING POLYNOMIAL OF ORDER 1

Least-squares coefficients are:
```
 1   0.478488E+04
 2  -0.170244E+02
 3  -0.979519E+02
```

Explained sum of squares: 0.1511574E+06
Residual sum of squares : 0.2928652E+01
Total sum of squares   : 0.1511603E+06
Explained sum of squares as percent of total: 99.9981

EAST	NORTH	ROW	PREDICTED	RESIDUAL
58.1250	38.2500	47.8800	48.6789	0.7989
56.9250	38.1500	79.3800	78.9032	-0.4768
59.2500	34.1250	434.0000	433.5781	-0.4219
57.2000	36.9000	196.7500	196.6614	-0.0886
55.7000	35.7250	336.7500	337.2918	0.5418
59.3750	35.0750	337.5000	338.3956	0.8956
58.1250	37.6000	113.0000	112.3478	-0.6522
58.1000	35.1000	358.2500	357.6532	-0.5968

Standard error (standard deviation of residuals): 0.646822E+00

FITTING EQUATION MAP EASTING=F (IMAGE PIXEL, LINE)
FITTING POLYNOMIAL OF ORDER 1

Least-squares coefficients are:
```
 1   0.558816E+02
 2   0.984416E-02
 3  -0.158171E-02
```

```
Explained sum of squares:   0.1039658E+02
Residual sum of squares  :  0.3419962E-02
Total sum of squares     :  0.1040000E+02
Explained sum of squares as percent of total:  99.9671
```

COL	ROW	EAST	PREDICTED	RESIDUAL
237.5000	47.8800	58.1250	58.1439	0.0189
118.0000	79.3800	56.9250	56.9176	-0.0073
415.5000	434.0000	59.2500	59.2854	0.0354
165.0000	196.7500	57.2000	57.1947	-0.0053
37.0000	336.7500	55.7000	55.7132	0.0132
407.0000	337.5000	59.3750	59.3543	-0.0207
246.0000	113.0000	58.1250	58.1245	-0.0005
279.5000	358.2500	58.1000	58.0664	-0.0336

Standard error (standard deviation of residuals): 0.221035E-01

FITTING EQUATION MAP NORTHING=F (IMAGE PIXEL, LINE)
FITTING POLYNOMIAL OF ORDER 1

Least-squares coefficients are:
```
  1    0.391369E+02
  2   -0.171115E-02
  3   -0.993391E-02
```

```
Explained sum of squares:   0.1724207E+02
Residual sum of squares  :  0.3619945E-03
Total sum of squares     :  0.1724243E+02
Explained sum of squares as percent of total:  99.9979
```

COL	ROW	NORTH	PREDICTED	RESIDUAL
237.5000	47.8800	38.2500	38.2548	0.0048
118.0000	79.3800	38.1500	38.1464	-0.0036
415.5000	434.0000	34.1250	34.1146	-0.0104
165.0000	196.7500	36.9000	36.9000	0.0000
37.0000	336.7500	35.7250	35.7283	0.0033
407.0000	337.5000	35.0750	35.0877	0.0127
246.0000	113.0000	37.6000	37.5934	-0.0066
279.5000	358.2500	35.1000	35.0998	-0.0002

Standard error (standard deviation of residuals): 0.719121E-02

THE MAP COORDINATES OF THE IMAGE CORNERS ARE:
```
BOTTOM LEFT  :  EASTING=  55.082  NORTHING=  34.049
TOP RIGHT:                60.920             38.251
BOTTOM RIGHT:             60.112             33.175
TOP LEFT:                 55.890             39.125
```

THE EXTREME COORDINATES ARE:
```
EASTINGS  :  MINIMUM=  55.082  MAXIMUM=  60.920
NORTHINGS:             33.175             39.125
```

Example data file

This data file refers to an image with four grey levels (0–3). The image matrix has 4 rows and 5 columns.

```
4
4  5
0  0  0  2  1
0  1  1  2  2
0  1  2  2  3
1  1  2  3  4
```

B.3.10   IMFACT: noise-reduced principal components analysis

This procedure is described in section 6.4.2. It is based on image factor analysis, hence the program name. This version allows the use of a mask image consisting of some zero pixels (perhaps indicating an area of interest) and some non-zero pixels, perhaps '1's or '255's. You can choose to process image pixels corresponding to the zero pixels in the mask, or to process image pixels corresponding to non-zero pixels. This feature could be used, for example, to analyse vegetated/non-vegetated or land/water areas separately, depending on the purpose of the project.

The only restriction on the use of the program is that the number of bands of imagery must be 12 or less. The inter-band correlation matrix is computed from a sample of pixels from the image set; every fourth line and fourth pixel are used. The following input is requested by the program:

1. The name of a log file, to which text output is directed.
2. The name of the image dictionary (INF) file which references the image set to be processed.
3. A flag or indicator: 1 means use all image files referenced by the INF file, and 2 means that only selected files should be used. If your INF file contains a mask image then select option 2 at this point. If you select option 2 then a list of image (and mask) files contained in the INF file is provided and you must indicate (by typing Y) which files to include. Do not select the mask file at this stage, even if you intend to use a mask.
4. A flag or indicator: Y means that you want to use a mask file, which must be referenced in the INF file that describes the image set that you are currently using. N means that you do not wish to mask the images. If your answer is Y then you select the mask file from the list of files referenced by the current INF file by typing its identification number, and then you

are asked whether the zero pixels or the non-zero pixels are to be used as the mask. The value 1 means that image pixels corresponding to zero pixels in the mask are to be ignored. The value 2 means that image pixels corresponding to non-zero pixels in the mask are to be ignored.

5. The number of principal components to be computed.
6. The names of the output image files to hold the principal component images.
7. The name of a new INF file to reference the principal component images.
8. The name of a file to hold the histograms of the principal component images.

B.3.11   ISODATA: unsupervised classification

The ISODATA algorithm is described in section 8.3. This version allows up to 12 image files with a maximum number of pixels per scan line of 8000. The program can generate up to 149 clusters. The cluster centres can either be generated randomly or input by the user. Control data are provided in a text file, which is created using a text editor. If you use a word processor to create this file, remember to save the file in text mode. The format of the control data file is:

Record   Contents

1. Number of starting cluster centres, N.

2. Flag – a value of 0 means 'generate the cluster centres randomly'. In this case, records 2* are omitted. A value of 1 means that the cluster centres are supplied by the user as records 2*.

2*. If the flag value entered on record 2 is equal to 1 then one record is supplied for each of the N cluster centres specified in record 1. Each cluster centre record consists of NBANDS values (where NBANDS is the number of spectral bands in the image set) separated by a space or a comma.

3. K, the desired number of clusters.

4. The minimum number of pixels allowed in a 'live' cluster.

5. The maximum standard deviation in any of the spectral bands before a cluster is split.

6. The minimum distance between cluster centres before clusters are merged.

7.   Maximum number of iterations to be carried out.

8.   Maximum number of classes that are allowed to merge at any given iteration.

9.   The minimum allowable change in average inter-cluster distance between one iteration and the next before the procedure is terminated.

10.   The absolute minimum value of average inter-cluster distance before the procedure is terminated.

11.   The largest squared distance between a pixel and the nearest cluster centre before the pixel is termed 'unclassified'.

The program asks for the name of a file containing these control data. It also asks for the name of a log file to which the results of the clustering procedure are written. You also supply the name of an output image file to hold the classified image (suffix cls) and the name of the INF file to be created to store the details of the classified image. Before running the MIPS display program to view the classified image you must run program make_his to create an image histogram file for the classified image. You may also run make_LUT to generate a LUT entry for each of the classes. This LUT can be read into the MIPS display program using Utilities|Read_LUT_file. Use of make_LUT is recommended rather than the use of the Enhance| Pseudocolour option in the MIPS display program. The number of classes can be read from the log file. If the number of classes is k then ensure that you provide values for LUT entries $0–k$, as the rows of the LUT array are numbered upwards from zero, not from one. Unclassified pixels are given the label $k + 1$.

The choice of suitable values for records 3–11 is a difficulty. However, the procedure is quite fast in operation if the maximum number of iterations (record 7) is set to a low value, say 3 or 4, so some experimentation is possible. The best advice is to run the program a number of times and inspect the log file to see what kinds of values are associated with parameters such as standard deviations of the spectral bands and inter-cluster distances.

B.3.12   JULDAY: Julian days calculator

Some remote sensing data sets contain the date of imaging expressed as a Julian day (the number of days elapsed since the start of the year) rather than the day, month and year. This program and the program in section B.3.13 (julday_inv) allow the conversion be-

tween these two methods of expressing the date.

Program julday calculates the number of days that have elapsed between the start of a reference year and a given date, expressed as DD MM YYYY. For example, if the given date is 22 11 1988 then you can calculate how many days (not including the given date) that have elapsed since the start of 1988 by using 1988 as the reference year. If you use another year, for example 1980, as the reference year then the program calculates the number of days between 1 1 1980 and the given date, 22 11 1988. The program does not work for dates before 1753 AD, as the Julian calendar was used in Britain before September 1752.

B.3.13   JULDAY_INV: inverse of Julian date

Program julday_inv takes the Julian day (days elapsed since the start of the year) plus the year number (e.g. 1988) and tells you the day and the month. The year number is required to determine whether or not the given year is a leap year.

B.3.14   KMEANS: k-means classification

The k-means classification procedure is described in Chapter 8. Program kmeans requires a control data file in the following format:

1.   Name of the image dictionary (INF) file describing the input image set.

2.   Name of a log file (which can be examined later using Windows Notepad, or any word processor).

3.   Number of cluster centres (must be 100 or less).

4.   Name of the image dictionary file to hold details of the classified image.

5.   Name of the file to hold the classified image.

6.   Name of the file to hold the image histogram.

7.   MAXIT: maximum number of iterations to be carried out (note that the number of iterations performed will, for technical reasons, always be MAXIT + 1). A value of 0 can be used to allocate image pixels to one of a set of user-provided cluster centres.

8.   MINREL: the clustering procedure terminates if fewer than MINREL pixels move from one cluster to another at any iteration.

9.   IPART is a flag. If its value is 1 then you must provide a set of cluster centre coordinates. The value IPART = 2 requests that cluster centres are generated randomly. No other value is allowed.

10. MAXD, the maximum average squared Euclidean distance between a pixel and a cluster centre. If the squared Euclidean distance (summed over all bands) between a cluster centre exceeds MAXD then the pixel is left unclassified and is allocated the identifier NC + 1 where NC is the number of clusters requested.

11A. If IPART (record 9) is equal to 1 (centres provided) then the centre coordinates for NC clusters are required, one per record. Each set of coordinates is NBANDS values in the range 0–255.

11B. If IPART is equal to 2 (random centres to be generated) then provide the upper bound for the range of randomly generated cluster centre coordinates. For example, if a value of 100 is specified, and if the number of bands in the input image set is four, then random centres will be set up as four-dimensional points with coordinates (c1, c2, c3, c4) generated in the range 0 to 100.

The value of 999999 for MAXD is used in the example below to ensure that all the image pixels are classified (since it is unlikely that any will have a Euclidean distance greater than 999999 from the nearest cluster centre). However, a careful choice of MAXD can ensure that only pixels that are 'reasonably' close to a cluster centre are labelled as members of that cluster. Any pixels further than MAXD from the nearest centre are declared 'unidentified' and given the label

NC + 1 (21 in the case of the example below). The name of the control data file is the only input requested by the program.

When clusters are generated randomly then the value provided in record 11B represents an estimate of the upper limit of the centre coordinates. It is more than likely that some of these randomly generated initial cluster centres will lie in empty areas of feature space. Earlier versions of this program simply recalculated the empty centre positions using the same algorithm as before, that is, generate random centroids in the specified range. This version uses a slightly more sophisticated algorithm that is borrowed from the ISODATA procedure. If there are nc clusters, and cluster i is empty, then look for a non-empty cluster in the range $\{i + 1, nc\}$. If there are none, look for a non-empty cluster in the range $\{i - 1, 1\}$. If there are none, then all clusters are empty. Otherwise, generate new coordinates for the empty cluster based on $x_{jk} \pm r/4$ where x is the coordinate of the non-empty cluster (j) on band k and r is the upper limit for cluster centres specified in record 11B. Thus, empty clusters are located near non-empty clusters, and may therefore collect some members at the next iteration.

Program kmeans has the following restrictions: maximum number of clusters is 100, maximum image scan line length is 8000 pixels, and maximum number of features is 12.

Example

An example of a data file is given below.

Record Number	Contents	Explanation
1.	c:\mips\images\egmar.inf	Name of input image dictionary file.
2.	c:\work\logfiles\kmeans.log	Name of log file.
3.	20	Number of cluster centres.
4.	c:\work\images\kmeans.inf	INF file for classified image.
5.	c:\work\images\kmeans.cls	Name of classified image file.
6.	c:\work\images\kmeans.hst	Name of histogram file.
7.	10	MAXIT.
8.	250	MINREL.
9.	2	IPART.
10.	999999	MAXD.
11B.	200	Upper limit for centre coordinates.

B.3.15   MAKE_HIS: generate a histogram file

The MIPS image dictionary (INF) file records the name of a file holding the histograms of the referenced image files in the order in which the image filenames are listed in the INF file. Some of the MIPS stand-alone programs automatically generate this histogram file and place an entry in the corresponding INF file. Others do not, and this is indicated in the program description. To check whether the INF file contains the histogram filename simply list the INF file using the MS-DOS `type` command, or Windows Notepad, and compare the contents of the INF file with the description given in section B.3.16 below.

The program requests two filenames as input. The first is the name of the relevant INF file, and the second is the name of the histogram file to be created.

B.3.16   MAKE_INF: generate a MIPS image dictionary file

`Make_inf` generates an image dictionary file and an associated image histogram file for a set of images. The INF file is used by all MIPS programs to access an image data set. The format of the INF file is as follows:

Record   Contents

1.   Number of bands of imagery (NBANDS). Number of scan lines in each image (NL). Number of pixels per scan line (NPIX).

2.   One filename per image band, each occupying a separate record.

3.   Sun elevation and azimuth values (use values of − 1 if these are unknown).

4.   Theoretical maximum pixel value in each band; one value per band.

5.   Day, month, and year of image acquisition (for example, 5 6 97).

6.   Name of file holding inter-band covariance matrix. Currently left blank.

7.   Name of file holding histograms.

8.   Name of file holding ground control point locations. Currently left blank.

For example, an image dictionary file describing three images making up a set could be:

```
3   512   512
c:\images\sicily.sp1
c:\images\sicily.sp2
c:\images\sicily.sp3
55 145
255   255   255
13   12   98
```

⚠ ⚠ ⚠

`c:\images\sicily.hst`

⚠ ⚠ ⚠

The symbol ⚠ ⚠ ⚠ means 'this record must be present, even if it is blank'. These records are included in order to allow for future development of MIPS.

Once the program is running you will be asked to press the `Return` key and enter a name for the INF file in the File Selection dialogue box. Remember to move to the disk and directory in which you store your image files before entering a filename. It is always good sense to keep all your images in one directory with several sub-directories, especially if you have a bad memory. Next, provide the answers to the following questions:

- How many bands of imagery?
- How many scan lines per image?
- How many pixels per scan line?

Press the `Return` key, when asked, and the File Selection dialogue box will appear once for each image in the image set. Use the mouse to double-click on an image filename, and this will be entered in the INF file in the order in which you select them. The following information will now be asked:

- Give the sun elevation angle.
- Give the sun azimuth angle.
- Give the maximum pixel value.
- Give the day of image acquisition.
- Give the month of image acquisition.
- Give the year of image acquisition.

You can type − 1 as the answer to any of these requests if you do not know the answer, except for the maximum pixel value. Use 255 if you do not know the theoretical upper limit of the range of pixel values (63 or 127 for some Landsat MSS images, 255 for Landsat TM and SPOT).

Finally, the program calculates the histograms for each band and stores these in an unformatted (binary) sequential file. You are asked to provide a filename for the histogram file, using the File Selection dialogue box. The program keeps you informed of progress as the histograms are calculated. These histograms are read by the MIPS display program and used to set the initial lookup tables (LUTs). When the program has finished you will be asked to press Return to close the output window and return to Windows.

You can check the consistency of an INF file using program test_inf, which is described in section B.3.37.

B.3.17 MAKE_LUT: manually generate a lookup table

Make_LUT allows you to create a lookup table (LUT) file directly. This may be useful, for instance, if you want to convert a classified image to colour. The LUT file can be read by the MIPS display program using Utilities|Read_LUT_file on the top menu bar (section A.4.9.6).

The first input item requested is the name of a file to hold the LUT. Then, for each of the 256 LUT entries (for pixel values 0–255) type an integer value in the range 0–255 for each colour. For example, if you wish to make the grey level '0' appear on the screen as brightest red, the first LUT entry would be 255 0 0 (maximum red, no green, no blue). You do not need to provide all 768 values. If your answer to each request 'provide RGB values' is −1 then the rest of the lookup table will be filled with zeros. For example, you may wish to display a classified image in colour, with eight classes. Input to make_LUT would be as shown below. Only the items in the left-hand column should be provided as input. The right-hand column contains helpful comments which are not part of the data. Once the program has completed, press Return to return to Windows. Note that the maximum likelihood classifier (B.3.21) uses the class label 'O' to indicate an unclassified pixel. This corresponds to entry number 1 in the LUT. Both ISODATA and *k* means use the class label NC + 1 to indicate an unclassified pixel.

RGB Values			Comment (*NOT* part of the data)
0	0	0	Class 0 is black.
255	0	0	Class 1 is maximum red.
0	255	0	Class 2 is maximum green.
0	0	255	Class 3 is maximum blue.
255	0	255	Class 4 is maximum magenta.
0	255	255	Class 5 is maximum cyan.
255	190	155	Class 6 is orange.
200	200	0	Class 7 is yellow.
0	127	0	Class 8 is mid-green.

−1	−1	−1	Signals end of input. The rest of the LUT is filled with zeros.

B.3.18 MASK: image masking

The mask program allows you to take a mask file containing pixel values of 0 or 255 and to apply this mask to one or more image files. The mask file is prepared using the MIPS image display program's Utilities|Threshold function (section A.4.9.3). After you have prepared the masked image, edit the image dictionary (INF) file referencing the image set to which the masked image refers, using a standard text editor. You need to increment the first number on the first line (the number of image bands referenced by the INF file), then add the name of the mask file to the end of the list of filenames, then add an additional '255' to the line containing maximum pixel values. Now run make_his (section B.3.15) to generate a new image histogram file. For example, an original image INF file is shown on the left and the modified file is shown on the right. The old histogram file, egmar.his, can be deleted. The new histogram file is egmar1.his.

Original INF file

```
     3       512       512
d:\mips\images\egmar2.img
d:\mips\images\egmar3.img
d:\mips\images\egmar4.img
    -1        -1
 255   255   255
   1     1     1
```

```
d:\mips\images\egmar.his
```

Modified INF file

```
     4       512       512
d:\mips\images\egmar2.img
d:\mips\images\egmar3.img
d:\mips\images\egmar4.img
d:\mips\images\eg4mask.img
    -1        -1
 255   255   255 255
   1     1     1
```

```
d:\mips\images\egmar.his
```

Remember that means: blank record. In the example, an INF file referencing three files (egmar2.img, egmar3.img and egmar4.img) is extended to reference the mask file (eg4mask.img). The number of bands of imagery (first number on line 1) is changed from 3 to 4, and the number of items on the 'theoretical maximum pixel value' line (containing three values of 255) is extended by one value. The structure of the INF file is described in more detail in the description of program make_inf (section B.3.16).

The mask program can handle up to 12 images, including the mask image, and a maximum of 2048 pixels per image scan line. The program first asks you to press Return then use the File Selection dialogue box to specify the name of a log file that will contain a record of your actions. You can print this later, or use a text editor to read it. Next you specify the name of the INF file containing the images and the mask file to be applied. Thirdly, from the list of files contained in the INF file, type the numerical ID of the mask file. For example, in the INF file listed above, the mask file is number 4. Now select the numerical ID of the image file to be masked, then provide the name of a file to hold the result of masking. The effect of the mask is to set to zero all pixel values in the image that correspond to pixel values of 0 in the mask image. The mask image may be imported from other software and might have a 'no mask' value of 1, or 255, or some other number in between.

You are now asked whether you want to repeat the operation on another image. Choose either the YES or the NO button. If you answer YES then the procedure described in the preceding paragraph is repeated. If you answer NO then you will be asked for the name of an INF file to reference the output image set. Finally, press Return to close the window and quit the program.

The consistency of an INF file can be checked using program test_inf, which is described in section B.3.37.

B.3.19   MIXMOD: spectral unmixing

This program performs spectral unmixing using unconstrained linear mixture modelling. The program uses the singular-value decomposition approach as described by Boardman (1989). Before running this program you must prepare a data file using a standard text editor. Note that the program can handle a maximum of 12 end-members and 12 spectral bands. The format of the data file is as follows:

1. Number of rows (image bands) and number of columns (end-members) in the end-member data matrix (separate these two numbers by a comma or a space).

2. The end-member spectral values, stored column-wise.

3. Number of image scan lines and number of pixels per image scan line in the images to be processed (separate these two numbers using a space or a comma).

4. One input image filename per input image band as specified on record 1, item 1.

5. One output filename per end-member, as specified on record 1, item 2.

6. Name of a file to hold the residual image.

Example

Given three end-members and four spectral bands, and with the end-member spectra of $\{12, 8, 5, 0\}$, $\{43, 35, 120, 130\}$ and $\{20, 42, 76, 88\}$ then a data file for mixmod might be as follows (the items following the exclamation mark (!) are comments and are not a part of the data file):

```
4  3                    ! 4 bands, 3 end-members
12  43   20             ! end-member spectra for band 1
 8  35   42             ! end-member spectra for band 2
 5 120   76             ! end-member spectra for band 3
 0 130   88             ! end-member spectra for band 4
c:\images\band1.img     ! filename for image band 1
c:\images\band2.img     ! filename for image band 2
c:\images\band3.img     ! filename for image band 3
c:\images\band4.img     ! filename for image band 4
c:\images\end1.img      ! filename for end-member 1
c:\images\end2.img      ! filename for end-member 2
c:\images\end3.img      ! filename for end-member 3
c:\images\resid.img     ! filename for (RMS) residuals
```

Once you start the program you will be asked to select the name of the data file using the File Selection dialogue box. When the data file has been opened successfully, you are asked to give the name of an output (log) file, again using the File Selection dialogue box. This file will contain details of the results, and can be printed or inspected using a text editor once the program run is completed.

In calculating the mixture proportions you can choose to use unconstrained proportions as calculated by the model or you can normalise the output for each pixel so that the proportions sum to 1.0. See section 8.5.1 for a discussion of the factors that might influence this choice. Answer either Y(es) or N(o) to the question concerning the method of generating output pixels. Note that if you answer N and compute the unconstrained proportions then, in order to accommodate overshoots, the proportion 0.0 is mapped to a pixel value of 100 and the proportion 1.0 is mapped to 200. Under-shoots occupy the range 0–99 and over-shoots are given values between 201 and 255.

The end-members are placed in the files specified in the data file. The RMS error for each pixel is computed from the unconstrained proportions whether or not the pixel normalisation option is selected.

B.3.20 ML__TRAIN: generate training statistics for maximum-likelihood classification

The sequence of operations required to generate a maximum-likelihood (ML) classified image in MIPS is as follows:

1. Run the MIPS display program.

2. Select an image set and display three bands in false colour.

3. Select the Classify option and choose to save the file of training class coordinates (see section A.4.8). Collect training data as described in section A.4.8, but do not go on to classify the image (the MIPS display program classifies only on the basis of the three bands displayed on the screen).

4. Run program ml_train to generate training class statistics from the coordinates file, using the same image set (INF file) as that used to collect the coordinates.

5. Optionally run program divergen to examine the contribution of each band (feature) to class separability.

6. Run program ml to classify the image set.

Program ml_train requires the following input:

1. Name of a log file to which text output is to be sent (this file can be examined once the program has terminated using the MS-DOS type command, Windows Notepad or any word processor).

2. Name of the image dictionary (INF) file (must be the same as that used in the training class coordinate collection procedure).

3. Name of file to hold training class statistics to be calculated by this program (this is a text file and can be examined using the MS-DOS type command, Windows Notepad or any word processor).

4. Name of training class coordinates file produced by the MIPS display program, as described above.

Program ml_train has the following restrictions: the maximum number of bands is 12 and the maximum number of pixels per scan line is 7000.

B.3.21 ML: maximum-likelihood classification

This program performs maximum-likelihood classification on a set of image files referenced by a single image dictionary (INF) file. Restrictions are: a maximum of 12 spectral bands, 60 classes and 7000 pixels per scan line. Before this program is run, the following steps must be performed:

1. Using the MIPS display program, select the Classify option and choose training area coordinates as described in section A.4.8.

2. Run ml_train using the coordinates file generated by the MIPS display program and the INF file of the image set to be classified.

The output from ml_train is a statistics (*.sts) file. This file can also be read by program diverg which calculates the divergence statistics (see section B.3.6).

Input to the program is, first, the name of a log file, to which details of the procedure are written. This log file can be printed or typed once the program run is finished. Then the names of the following files must be provided:

- the statistics file generated by ml_train,
- the INF file which references the images to be classified,
- the file to hold the classified image, and
- the INF file to hold details of the classified image.

The program allows you to vary the prior probabilities for each class as explained in section 8.4.2.3. You must also choose a 'critical chi-square value' for each class. For example, if you choose 10 (per cent) for class i this means that, in a multivariate-normal population with the same mean vector and covariance matrix as class i, 10% of all measurements will have a Mahalanobis distance that is equal to or less than the critical chi-square value at the 10% level; 90% will have a Mahalanobis distance that is larger than the critical value and will thus be rejected and labelled as zero (unclassified). Conversely, if you choose the 99% level then only 1% of the same multivariate-normal population will be rejected. If you select 0 then this test will not be applied and all pixels will be allocated to one of the specified classes.

The output from the program is an INF file referencing the classified image. You can use the MIPS display program to view the image. You can select the Enhance|Pseudocolour option in automatic mode (section A.4.4.4) to assign a colour to each class. The colours are chosen from the spectrum from red through green and blue and back to red. It is preferable, however, to create a lookup table for the classified image using the program `make_LUT` (section B.3.17). This lookup table can be read into the MIPS display program using the `Utilities|Read_LUT` option (section A.4.9.6). Remember, however, that class number 0 is used for unallocated pixels. It is sensible to set this to black (0, 0, 0) and allocate colours to LUT entries 1 to k, the number of classes.

B.3.22   NL__DESTR: non-linear de-striping after Horn and Woodham

This procedure is described in section 4.2.2.2, and this implementation will work only with Landsat MSS or TM images. It uses the overall image histogram as the 'target' histogram and fits the six (Landsat MSS) or 16 (Landsat TM) detector histograms to this target histogram in order to reduce the effects of striping in the image caused by unequal sensor calibrations. It is best to use a large portion of an image so that the target histogram is realistic. All the input and output use dialogue boxes, and are self-evident. The full image histograms and the histograms of the individual sensors are written to the log file, plus the lookup table for each image pixel value (0–63, 0–127 or 0–255). The lookup table shows the values assigned to the pixels imaged by each of the six or 16 individual detectors – for example, the row labelled '51' may contain six values (49, 50, 53, 52, 51, 54) for an MSS image. This indicates that the value 49 replaces all values of 51 in the input image for lines scanned by detector 1, while the value 50 replaces all values of 51 in the input image for lines scanned by detector 2, and so on. The de-striped image is written to a file specified by the user. Note that `make_inf` (section B.3.16) must be run to generate a new INF file before the de-striped image can be displayed. The program can handle images of any size up to the limits of the virtual memory of the machine on which it is running.

B.3.23   NLMAP: non-linear mapping

Non-linear mapping is discussed in section 8.4.1. Before you run the program you must create a text file containing the data to be analysed. This text file has a simple structure. The first record has two numbers separated by a space or a comma. The first is the number of data items (pixels) and the second is the number of measurements on each pixel, for example, the number of spectral bands. If the number of pixels is denoted by n and the number of measurements by m then the following n records each contain the m values or measurements. The data set may, for example, consist of the centroids of clusters generated by the `isodata` or `kmeans` programs. These centroids will generally be m-dimensional vectors, and the non-linear mapping program may be a way of projecting these vectors on to a subspace of dimension 2 or 3.

When the `nlmap` program is run it first asks whether the data file described above has been prepared. If not, you can exit gracefully. The name of the data file and of a log file (which can be read later using the MS-DOS editor, Windows Notepad or any word processor) are requested. Next, the number of dimensions in the subspace (2 or 3) is selected by clicking the appropriate button on the window. The next input window asks for the value of the 'magic factor', which is basically a step length. A value of around 0.35 is generally recommended. The program will reduce the value of the 'magic factor' as the number of iterations increases, so that it does not oscillate around a minimum. The number of iterations, which is requested next, depends on the relationship between the dimensionality (m) of the input data and that of the subspace, as well as on the number of data items (pixels). Often convergence is reached in a few iterations whereas on other occasions the program requires several hundred iterations to converge. The process is quite fast so it is possible to experiment. Requesting too many iterations will do no harm. The procedure stops when the difference between the 'mapping error' on successive iterations is less than the value of the convergence criterion, which is specified next.

Once the process has converged the data values are plotted, two dimensions at a time. Click on the `Plot` menu item to see the plot. If you have chosen a dimensionality of 3 for the subspace then you must choose the two axes of the plot from a drop-down list (you can plot axes 1 and 2, 1 and 3 or 2 and 3). Once the plot appears a message is presented to ask whether you wish to identify any of the points on the graph using the mouse. If you answer `YES` then the numerical identifier of the corresponding data item is listed when you click the mouse left button close to a point on the graph. The selected point then turns green. This facility is useful if you want to discover the relationships between the data items in the subspace. Click the right mouse button to exit from the point identification procedure. The plot can be sent to the Windows Clipboard by clicking on the `Write_Clipboard` menu item. It can then be pasted into another Windows application, such as a word processor. Click the menu item `Exit` and confirm your choice to terminate the program.

B.3.24   NADIRCOR: cross-scan nadir correction

This program implements a correction procedure to compensate for cross-scan variations in illumination, as described in section 4.5. Input requested by the program is:

1. The name of the file to be corrected. This filename is chosen using the Windows File Selection dialogue box.
2. The number of scan lines and the number of pixels per scan line in the image. This is entered in a dialogue box.
3. The name of a log file.
4. The nadir pixel number. This may not be the central pixel in the image if a sub-image has been selected.
5. The name of a file to hold the corrected image.

Note that this program does not generate an INF file and operates separately on the individual bands of an image set. Thus, to correct an image set containing NBANDS images, it is necessary to run the program NBANDS times, once per image, and then to run program make_inf before using the MIPS display program to view the results.

A plot showing the second-order polynomial function and the means of the columns of the corrected and uncorrected images is written to the Clipboard (Figure B.1) in yellow, purple and dark blue respectively. You can paste it into a suitable Windows application.

B.3.25   PCA: principal components analysis

The input to program pca is almost identical to that for program imfact, which is described above (section B.3.10). The only difference is that the user is asked to choose whether to base the analysis on the correlations or covariances between the bands. The same restrictions apply – a maximum of 12 bands can be used (this is a property of the INF file rather than program pca) and, since allocatable arrays are used, the maximum image size depends only on available memory (physical and virtual).

B.3.26   PIPED: parallelepiped classification

The parallelepiped classifier is described in Chapter 8. Program piped reads all its input data from a control data file that must be prepared using a text editor before the program is run. If a word processor is used to create the control data file then remember to save the file in text mode. The contents of this control data file are as follows:

1. Name of a log file for text output (include full path).
2. Number of classes (maximum 30).
3. Name of the INF file describing the images to be classified. It is assumed that all images are used in the classification.
4. Name of the output classified image to be created (include full path).

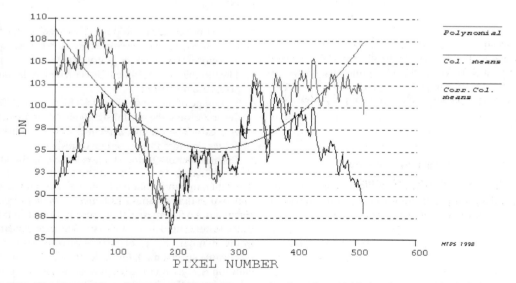

Figure B.1   Plot of output from program nadircor. The second-order polynomial fitting the corrected column means is drawn in yellow. The purple curve shows the corrected column means, and the dark blue curve is the column means after correction (original in colour).

5. Name of the INF file to describe the classified image (include full path).
6. Lower and upper limits on each band in turn for each class in turn. These values define the parallelepipeds in feature space.

An example data file is shown below.

```
d:\mydata\piped.log
5
d:\mips\images\kenya.inf
d:\mips\images\piped.cls
d:\mips\images\piped.inf
    0    20
   30    70
   80   120
   20    70
   30    90
   40   120
   30    80
   90   120
    0   100
   80    90
  100   110
  110   120
   35    55
   75   105
   20    90
```

The program does not create a histogram file for the output image. Before you try to display the classified image using the MIPS display program, run `make_his` (section B.3.15) to generate a histogram file. The input to `make_his` is the name of the INF file that describes the classified image.

B.3.27   PLOT__PIX: plot pixel spectra

Program `plot_pix` takes a text file containing 15 or fewer pixel spectra and produces a plot. The pixel spectra are presented row-wise, for example:

```
   10    20    30
   13    24    35
   90    70    40
   90    40    90
  130   180    22
```

This data file provides details of five pixel spectra measured on three bands. The first pixel has values (10, 20, 30) on bands 1–3. The text file is prepared using a standard editor. If a word processor is used, make sure that you save the file in text mode. The program first requests the name of the data file, then the number of bands in the spectra. A plot window appears – click on the `Plot` button. The plot is automatically copied to the Windows Clipboard. Click the `Exit` button to leave the program. The spectral plot for the data file shown above is illustrated in Figure B.2. Program `plot_pix` is useful in examining, for instance, the end-member spectra in mixture modelling or the centroids of clusters produced by the `isodata` or `kmeans` classification programs.

B.3.28   QMODE: Q-mode factor analysis

In Chapter 8 the topic of selection of suitable end-members for mixture modelling is discussed. This is an experimental program that follows a procedure described by Imbrie (1963) for the analysis of sediment data in terms of a small number of components. It is not feasible to apply this procedure to all of the pixels in the image, so a sampling procedure is adopted. Program `sel_pix` (section B.3.32) allows the random sampling of an image and the output from `sel_pix` can be read directly by `qmode`. The `qmode` program requires the following data to be provided by the user:

1. The name of a file to hold the results (a log file). If the file that you specify already exists then you are asked to confirm that it can be overwritten.
2. The number of spectral bands m on which the sample pixels are measured.
3. The name of the input data file. This is an ASCII text file containing the digital signal levels for n pixels on m bands. The file has n records each containing m values separated by a comma or a space.
4. The number of Q-mode factors to be calculated.

The first stage in the calculation is the computation of the $\cos\theta$ matrix of similarities among the n pixels. The eigenvalues of the matrix that exceed 0.00001 in magnitude are computed and listed. The user selects an appropriate number, say p, of factors on the basis of this eigenvalue spectrum. A Q-mode principal components solution is written to the log file. This matrix shows the relationships between the n pixels and the p principal components. Next, the principal components matrix is transformed using the varimax procedure. The pixels

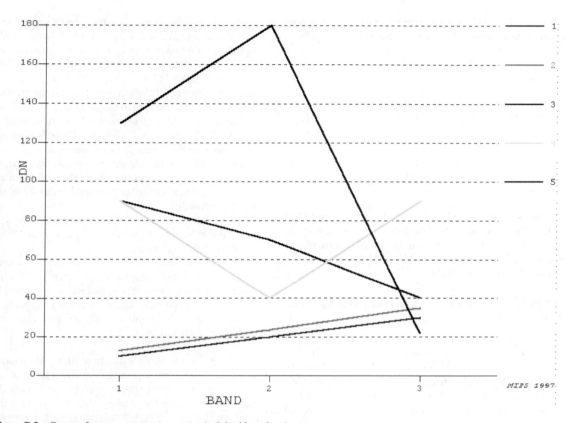

Figure B.2   Output from program `plot_pix` (original in colour).

with the highest loadings on each of the *p* varimax-rotated components are selected as the end-members, and an oblique rotation of the varimax axes is applied. Finally, this oblique solution is standardised and the column order is rearranged.

Imbrie (1963) generated a test data file with 10 measurements on 10 sediment samples. His 10 measurements were generated from three pure types or end-members. The data file is:

```
 5.00   25.00   15.00    5.00    5.00   20.00   10.00   5.00   5.00   5.00
10.00   30.00   17.00   17.00    8.00    8.00    5.00   4.00   1.00   0.00
 3.00    6.00   10.00   13.00   25.00   15.00   13.00   8.00   5.00   2.00
 7.50   27.50   16.00   11.00    6.50   14.00    7.50   4.50   3.00   2.50
 4.60   21.20   14.00    6.60    9.00   19.00   10.60   5.60   5.00   4.40
 3.80   13.60   12.00    9.80   17.00   17.00   11.80   6.80   5.00   3.20
 8.30   26.60   15.90   14.20    9.10   11.10    6.80   4.60   2.20   1.20
 6.10   22.70   14.60   10.20    9.90   15.40    9.10   5.30   3.80   2.90
 7.60   24.20   15.20   13.80   10.80   11.80    7.60   5.00   2.60   1.40
 3.90   10.30   11.20   12.60   21.30   14.80   11.90   7.30   4.60   2.10
```

The standardised oblique-rotated Q-mode solution computed by qmode for three factors is:

```
 1  1.000  0.000  0.000
 2  0.000  1.000  0.000
 3  0.000  0.000  1.000
 4  0.500  0.500  0.000
 5  0.800  0.000  0.200
 6  0.400  0.000  0.600
 7  0.200  0.700  0.100
 8  0.500  0.300  0.200
 9  0.200  0.600  0.200
10  0.100  0.100  0.800
```

Limitations are: a maximum of 1024 scan lines and 1024 pixels per scan line in the image set, a maximum of 20 Q-mode factors and a maximum of 20 bands in the image set. This program is an MS-DOS application and is best run from the DOS prompt.

B.3.29   READ__ESATM__CD: read TM image from Eurimage CD

This program reads a sub-image from one or more of the seven Landsat TM bands stored on a CD. The program has been tested only with CDs in the European Space Agency (ESA) format, as supplied by Eurimage. Each CD contains a single (multi-band) TM image. It has a root directory and a subdirectory called \scene1. The images normally have 5760 scan lines each containing 6920 pixels including left and right fill pixels. To use the program, place the Landsat TM CD in the drive, run read_esatm_cd, and answer the following questions:

1. What is the drive letter of the CD? (Your answer could be D or E (without a colon) depending on the number of hard drives on your PC.)
2. What is the drive letter for the disk on which the extracted image files are to be placed? (Usually C or D, again depending on the number of disk drives. Do not include the final colon.)
3. What is the pathname for the extracted files on the disk? (Your answer could, for example, be \mydata\myimages\ noting the final '\'.)
4. What is the log filename? (Include the full specification, for example, c:\mydata\logfiles\tm_logfile.log.)
5. Give start scan line, number of scan lines and line skip factor. (The scan lines are numbered 1–*n* from the top of the image, as shown in Figure A.1. The line skip factor is one more than the number of lines to omit between each retrieved line, so to read every line

then line skip factor is 1. To read every second line the line skip factor is 2. For example, to extract 512 scan lines starting at scan line number 196 in the full image, and reading every fourth scan line in the full image, enter 196 512 4.)
6. Give the number of the leftmost pixel, total number of pixels and the pixel skip factor. (Pixels making up a scan line are numbered from the left-hand side of the image, as shown in Figure A.1. Other values are analogous to the values for scan line selection.)
7. Enter FP to have the file pointer records printed, or press Return to suppress the printing of these records. (See the ESA format manual for details of file pointer records.)
8. How many bands of imagery to read? (The CD contains data for seven bands; type a number between 1 and 7 inclusive.)
9. Now give the numeric IDs for these bands. (If your answer to the last question was *N* then you should provide *N* band identifiers at this point; for example, if *N* = 3 then you have asked for three TM bands to be read, so at this point you should give the band numbers, such as 1, 4 and 7.)

Each image is associated with three files – the leader, image and trailer files. For each of the images that you have selected you must specify whether you want the contents of the leader file to be printed. You request this by typing the character L in response to the prompt Enter L for printout of leader file. If you do not want the contents of the trailer file to be printed then simply press the Return key.

When the imagery file is encountered on the CD you are asked to provide the name of a file to hold the extracted image. Give the filename only, not the drive letter and path (for example, enter band1.tm rather than c:\mydata\images\band1.tm). The program assumes that these files will be stored on the disk and path that you specified in response to items 2 and 3 in the above list. The imagery file header record provides details of the number of scan lines per image and the number of pixels per scan line. The number of left and right border pixels is also printed. Enter T in response to the prompt Enter T to get printout of trailer file. If you do not require a printout of the trailer file, just press the Return key. The questions relating to the leader, imagery and trailer files are repeated for each of the bands that you identified for extraction.

Once the image data have been successfully read from the CD you are asked to provide the name of an image dictionary (INF) file that is to contain the details of the extracted images. It is assumed that the INF file will be stored on the hard drive and path that you specified in response to items 2 and 3 on the list above. You are also asked for the name of a file to hold the image histograms. Again, supply the filename only, not the drive letter or path.

The contents of the volume directory record and the leader, imagery and trailer files are summarised below.

Volume directory record	Provides details of sensor, image path and row on the World Reference System, and date of imaging.
Leader file	Contains details of the day, month and year of image acquisition, image centre latitude and longitude, system processing operations, radiometric and geometric corrections, Sun angles, offset and gain coefficients, the lookup tables for each detector, and map projection details.
Imagery file	Provides the image data plus additional information relating to each scan line of the image.
Trailer file	Contains the histograms for the raw data for each of the 16 TM detectors for the given band.

B.3.30   READ_SPOT_CD: read a SPOT image from CD

Place the SPOT CD containing the image you want to access and determine the CD drive letter (in the following description it is assumed that D: is the CD drive). Go to the MS-DOS prompt and type dir D: (assuming that the drive letter of your CD is D:). Alternatively, use Windows Explorer to navigate to the CD drive and determine the path names of the various sub-directories of the CD's root directory. These will be normally D:|scene01, D:|scene02 and so on. The details accompanying the SPOT CD will indicate which images are contained in these directories. Run the read_SPOT_cd program, which will first ask you for the CD drive letter (if the return key is pressed then D: is assumed). The second item requested is the scene number, which is the final digit in the sub-directory

name (1 for sub-directory D:|scene01, for example).

The descriptive data contained on the CD are written to a text file so that you can use a standard word processor, editor or the MS-DOS type command to read it. You provide the name of this text file as the third input item. The next three input items are the number of scan lines you wish to read, the starting line number, and the line sampling rate. To help you, the number of scan lines and pixels per line in the image are printed on the screen immediately before this input is requested. If you give an unacceptable answer then a message is printed and you are asked to re-input. No check is made for negative values of these three input items, as it is assumed that the user is making an attempt to provide sensible values. The line sampling rate is interpreted as: '1' means every line, '2' means every second line, and so on. It is therefore possible to request 512 scan lines to be read, starting at line 329 of the full image and using a sampling rate of 3. This would cause scan lines 329, 332, 335, ... and so on to be written to the output image files.

The next three items of input are the start pixel number (counting the leftmost pixel as number 1), the number of pixels to be written to the output file, and the pixel sampling rate, which is interpreted in the same way as the line sampling rate. Lastly, the names of the files to be created on your hard disk are requested. If the SPOT image you are reading is panchromatic then you provide one filename. If it is a multispectral image then three filenames are needed. The Windows File Selection box is used to select these filenames. It appears after you have clicked OK on the information box that tells you that filenames are required. Note that you will have to run make_inf before you can use the MIPS display program to view the SPOT images that are read by this program.

Example

An example of the text file that is generated by this program is listed below. Note that some of the fields are not filled in; this is because the relevant data were not written to the CD by the ground receiving station. The file below was generated when the image set described by the images referenced in the dictionary file camarg1.inf (Appendix C) were generated.

```
Volume Directory File
Logical Volume ID: SP1 X 1BBIL 1
File pointer record #1
---------------------
Referenced file number              1
                name        SP1 X1B LEADBIL
                class       LEADER FILE
                class code  LEAD
                data type   MIXED BINARY AND ASCII
                data type code  MBAA
No. of records in referenced file:   27
1st record length:    3960
max record length:    3960
length type:        FIXED LENGTH
length type code:   FIXD
File pointer record #2
---------------------
Referenced file number              2
                name        SP1 X1B IMGYBIL
                class       IMAGERY FILE
                class code  IMGY
                data type   BINARY ONLY
                data type code  BINO
no. of records in referenced file:   8983
1st record length:    5400
max record length:    5400
length type:        FIXED LENGTH
length type code:   FIXD
File pointer record #3
---------------------
Referenced file number              3
                name        SP1 X1B TRAIBIL
                class       TRAILER FILE
                class code  TRAI
                data type   MIXED BINARY AND ASCII
                data type code  MBBA
no. of records in referenced file:    3
1st record length:    1080
max record length:    1080
length type:        FIXED LENGTH
length type code:   FIXD
Leader File
Scene ID: S1H1870117103104
Scene centre latitude:    N0434026
Scene centre longitude:   E0043151
Spectral Mode (XS or PAN): XS
Preprocessing level identification: 1B
Radiometric calibration designator: 1
Deconvolution designator: 1
Resampling designator: CC
Pixel size along line:    20
Pixel size along column:  20
Map projection ID:
```

```
Image size in Map Projection along Y axis:    059854
Image size in Map Projection along X axis:    070397
Number of GCP:
Number of registration control points:
GRS Designator for Reference Scene:
Reference Scene centre time:
Mission ID for reference scene:
Sun calibration operation date: 19861115
This is a multispectral image
Absolute calibration gains: 00.86262   00.79872   00.89310
Absolute calibration offsets: 00.00000   00.00000   00.00000
Opening imagery file with record length 5400
Imagery File Descriptor
Length of this record:
No. of image records:     8983
Image record length:     5400
Number of images:          3
Number of lines/image excluding border lines:    2994
Number of left border pixels per line:    0
Number of image pixels per line:    3522
Number of right border pixels per line:    1778
Number of top border lines:                0
Number of bottom border pixels:            0
Interleaving indicator:   BIL
maximum image lines: 2994
maximum pixels per line: 3522
top border     0
left border    0
```

B.3.31   SCATTER: scatter plot

Program scatter produces a scatter diagram showing the relationship between pixels in two selected spectral bands of an image set. When the program is started, an information box appears asking for the name of a log file. Click OK or press the Return key, and the standard Windows File Selection dialogue box appears. Either type in a new filename or choose an existing log file from the list. If you choose the latter option then the existing log file will be overwritten. Information relating to your choice of image files and their histograms is written to the log file, which can be viewed later by using the DOS type command or a Windows text editor such as Notepad (so make a note of the name you give). Another information box appears, telling you to choose an image INF file. Click OK or press Return and the File Selection dialogue box reappears. This time select an image dictionary file (INF file). The images referenced by the INF file will be listed, with a numerical identifier, such as 1, 2, 3. Select the image band to be plotted as the x-axis by typing its numerical identifier. Then select the image band to be plotted as the y-axis in the same way.

For example, you might select to plot image files numbered 1 and 4.

A graphical window now appears on the screen. Its size depends upon the dimension of the screen. Click Plot on the menu bar (or press Alt and P). The axes of the plot are drawn in red. The plot area is filled with white. The rest of the window is filled with your selected Windows background colour. The cursor should turn to the 'busy' state and two scan lines of pixels are plotted at each flicker. The reason the cursor flickers is that scatter is yielding control occasionally in order to allow Windows to catch up with any other processing. You will notice that the plot produces different coloured points. These colours are related to the number of pixels that fall at the given point, as follows:

1	black	7	green
2	light blue	8	yellow
3	dark blue	9	light red
4	light cyan	10	light magenta
5	cyan	11	magenta
6	light green	12	red

Once the plot is complete (the cursor reverts to a cross) you can click on the menu item `Write-Clipboard` to transfer the plot to the Windows Clipboard (Figure B.3). It is then accessible to other Windows applications such as word processors and image display and manipulation programs. The menu item `Exit` terminates the program and returns you to Windows. Plotting image files much larger than 512 × 512 is rather time-consuming, unless you are using a powerful machine with a fast graphics card.

B.3.32   SEL_PIX: randomly sample the pixels in an image set

This program is provided so that input for program `qmode` can be generated from an image set. Images of up to 1024 scan lines with 1024 pixels per scan line can be handled. When the program is started it asks for the name of the file to which the sampled pixel values are to be written in ASCII text form. Next, a log filename must be specified, then an INF filename. The name of each file is output to the screen, and the file is selected by answering Y to the question 'Select this file?' Note that the

program can sample using a mask, and if your INF file references an image mask file do not select the mask file at this stage. After the image files have been chosen, answer the question 'Do you want to use a mask?' with either Y or N. If you use a mask – which must have been previously prepared using, for example, the `Threshold` function in the MIPS display program (Appendix A) – then the identification number of the mask file must be entered (the program provides a list of all the files referenced by your INF file). The mask is a binary image and contains pixel values that are either 0 or 1. You can choose to sample in areas of the image corresponding to values of zero in the mask, or in areas corresponding to values of one in the mask. If you choose not to use a mask the samples are selected from the full image. Finally, enter the number of pixels to be sampled. A random sampling procedure is used. Note that the `qmode` program has an input limit of 1000 pixels.

B.3.33   SPECINDX: spectral indices

Jackson (1983) describes a method of calculating the coefficients of linear combinations of spectral bands

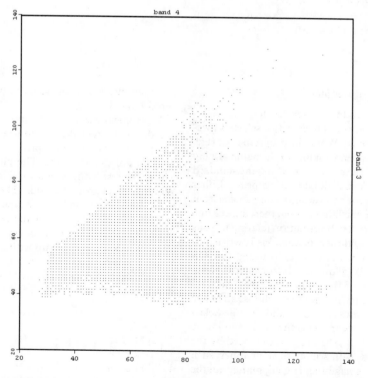

Figure B.3   Sample output from program `scatter`. This output was obtained by writing the plot to the Windows Clipboard, and pasting it into a word processor document (original in colour).

using Gram–Schmidt orthogonal polynomials (mentioned in section 6.3.2). If n is the number of spectral bands and m the number of samples of land cover types then $m - 1$ linear combinations can be extracted, though not all of them will necessarily be meaningful. The data matrix required as input to the program contains the values of m and n as the first record, followed by m records of reflectance values, each representing a different cover type. The first and second records are the reflectance values for dry soil and wet soil, respectively. The third and fourth records are the reflectances for green and senescent vegetation. The data for the wet and dry soils are described by Jackson (1983, p. 417) as 'critical'. However, the vegetation data are less important. Four cover types provide the data for three indices to be calculated. The multispectral tasselled cap function shows only two meaningful band combinations, representing greenness and brightness. The additional spectral bands of the Landsat TM provide the information necessary to define a third function, wetness. Hence it is likely that only two or three functions will be of interest. Jackson (1983) notes that the greenness value for a bare soil is not necessarily zero, and that the greenness value computed for bare soil should be subtracted from the greenness values of all pixels. The program requires a data file, as noted above. A typical data file is:

```
  4   4
15.10   20.32   28.73   32.45
 7.59   11.79   15.52   17.65
 3.45    2.80   28.51   43.82
11.58   17.59   25.71   31.36
```

These data are taken from table 3 of Jackson (1983). The first row contains the values for the number of spectral bands (4) then the number of samples (4). The next four rows hold the reflectances in the four bands for dry soil, wet soil, green vegetation and senescent vegetation. Results are written to a log file, which is specified by the user. The log file for the example is:

```
COEFFICIENTS OF THE ORTHOGONAL FUNCTIONS
COLUMNWISE
1  0.32848   -0.44797   -0.61251
2  0.37310   -0.68965    0.61249
3  0.57780    0.06700   -0.39302
4  0.64734    0.56499    0.30859
```

The coefficients for the brightness function are given in column 1, those for greenness in column 2 and those for the 'yellowness' function in column 3.

B.3.34   SPEC__ANG: spectral angle mapping

This procedure is described in section 8.5.1 as an alternative to spectral unmixing. The program requires a control data file as input. This is a text file and can be prepared using any text editor. If a word processor is used then care must be taken to save the file in text mode. The control data file contains the following information:

1. Number of lines in each image, number of pixels per scan line, number of spectral bands (images) and number of end-members. These four numbers are separated by commas or spaces.
2. One input filename per image band.
3. One output filename per end-member.
4. Band 1 values of all end-members, followed by the band 2 values of all end-members, and so on for all bands.

Example

To apply spectral angle mapping to a four-band input image set (held in files d:\myimages\band.1 to band.4), each of which is composed of 512 lines with 512 pixels per line, and to compute three end-member images (to be stored in d:\myimages\end.1 to end.3), with end-member spectra (13, 10, 7, 0), (46, 55, 127, 143) and (89, 112, 111, 121), the following data file would be prepared:

```
512  512  4  3
c:\myimages\band.1
c:\myimages\band.2
c:\myimages\band.3
c:\myimages\band.4
c:\myimages\specang.1
c:\myimages\specang.2
c:\myimages\specang.3
13   46    89
10   55   112
 7  127   111
 0  143   121
```

The program does not generate an INF file for the end-member images, so program make_inf should be run, giving the names of the end-member images as input, before the MIPS display program is used.

B.3.35   SUB__EXTR: extract sub-image from source image

Some of the images on the CD are too large to be displayed in a convenient fashion using the MIPS display program (though you can specify a start line and start pixel for image reading). However, you may prefer to use `sub_extr` to extract a sub-image from a source image. When you run this program you must supply the following information:

1. Name of a log file (which will hold a summary of the operation performed).
2. Name of the source file.
3. Number of lines and number of pixels per line in the source file.
4. Name of the new sub-image file.
5. Number of lines and number of pixels per line in the new sub-image file.
6. Starting line and starting pixel in the source image at which extraction is to begin.

Items 1, 2 and 4 use the Windows File Open dialogue box. The other items require numbers to be typed at the keyboard. If your specification is illogical the program will stop. Note that before you can view the sub-image using the MIPS display program you must run `make_inf` to generate an INF file and a histogram file.

B.3.36   TASSCAP: tasselled cap transform

The tasselled cap transform computes four output images from six Landsat TM reflective bands (1–5 and 7) (see section 6.3.2). This program requires a control data file containing the following data items:

1. Line/pixel sampling rates for estimating max and min 'raw' function values. Put 0, 0 if standard mapping used (see item 7 below).
2. Name of log file.
3. Name of dictionary file (`*.inf`) referencing the six TM reflective bands.
4. TM band numbers with 0 indicating 'not this band'. Thus, an entry of 1 2 3 4 5 0 6 would indicate that referenced files 1–5 are TM bands 1–5 and referenced file 6 is TM band 7. This feature allows you to store the TM bands in the INF file in any order, and to select a specific file. For example, a mask file may be the first referenced file and the thermal band (6) may be present so you could then use the ordering 1 2 3 4 5 0 6.
5. Four filenames to hold the four tasselled cap functions. You *must* include the full path, for example `c:\myfiles\myimages\tasscap.1`.
6. Name of a text file holding tasselled cap coefficients. The standard coefficients are provided on the CD in a file called `tasscoef.dat`, which is located in the root directory of the CD. Again, you must provide

the full path. Enter the coefficients as a matrix of six rows (one per TM reflective band) and four columns (one per tasselled cap function).

7. Enter (literally) either `STANDARD` or `ESTIMATE`. `STANDARD` means: use the built-in values of the maximum and minimum of the raw tasselled cap function to scale the output image values to the range 0–255.

   `ESTIMATE` means: take the pixels in the full image corresponding to the line and pixel sampling rates given in step 1 and compute raw function values from these pixels. Use these values to estimate the maximum and minimum tasselled cap values for the entire image.

8. Name of dictionary (`*.inf`) file for output data set. Again, supply the full path name.
9. Name of file to hold the histograms of the calculated images. Supply the full path name.

Example

Note that the full path (e.g. `c:\work\images\`) is provided for each of the files.

```
0  0
c:\work\logfiles\tcap.log
c:\mips\images\littlept.inf
1  2  3  4  5  6
c:\mips\images\tass.1
c:\mips\images\tass.2
c:\mips\images\tass.3
c:\mips\images\tass.4
c:\mips\programs\tasscoef.dat
standard
c:\work\images\tass.inf
c:\work\images\tass.hst
```

B.3.37   TEST_INF: check the consistency of an INF file

Before running a MIPS program that requires an image dictionary (INF) file, it is a good idea to check that the file is correctly structured. Run `test_inf` and provide the name of the INF file to be checked. First, the number of referenced image files is checked. The maximum is 12. Then each image file is opened and read, using the information in record 1 of the image file (number of bands, number of lines per image and number of pixels per line). The histogram file is also checked and read. The other information held in the INF file (pixel upper value, date, and Sun angles) is displayed for visual checking. If an error is discovered, you can use a text editor or a word processor to amend the INF file, but if you use a word processor remember to save the amended file as a text file.

B.3.38   UNPACK__GB: unpack a generic binary file

Many commercial image processing software packages allow the export of 'generic binary' files in band-sequential (BSQ) format. The description 'generic binary' means in raw format. 'Band-sequential' means that one generic binary file contains all of the bands 1, 2,..., n in order. Generally, you can choose BSQ as an output option. If you wish to use `unpack_gb` then make sure

that the exported file is generic binary without headers. Start the program using one of the procedures described at the beginning of this appendix. Give the name of the generic binary file containing the n bands of imagery, and specify the number of scan lines and pixels per scan line in each image band (e.g. 512, 512) as well as the number of bands. Finally, supply a filename for each of the bands as they are unpacked. Run `make_inf` to generate a dictionary file for the unpacked (raw) images before using the MIPS display program (Appendix A) to view the result. Check the INF file for consistency before you use it by running `test_inf` (section B.37).

Appendix C

Installing the MIPS Software

C.1 INTRODUCTION

The CD-ROM that accompanies this book contains (i) the MIPS display program (Appendix A), (ii) the MIPS stand-alone programs (Appendix B), (iii) a selection of digital images, described in Appendix D, and (iv) an HTML file containing links to remote sensing and GIS WWW sites (Appendix E). This Appendix tells you how to transfer these files to your computer. If – and only if! – your CD drive letter is D: you can run all the programs and access the images directly from the CD – you do not need to copy the programs. This is because the image dictionary files (INF files) on the CD specify that the drive on which the images are located is D:. If the CD drive letter is E: then the images referenced by the INF file will not be found, and you will need to copy the image dictionary (INF) files to your hard disk using the procedures described below.

C.2 COPYING SOFTWARE AND IMAGES FROM THE CD

If your CD drive letter is D: you can run the programs and access the sample images simply by placing the CD in the drive, starting Explorer, setting the drive to D: and clicking on the name of the program that you want to run. You can click the Type button in Explorer and the files will be listed in order of their function. Executable files are known to Windows as 'Applications'.

If the CD drive letter is not D: then you need to copy the software and the image dictionary (INF) files from the CD to your hard drive using a program called MIPS_Install, which was kindly provided by Michael McCullagh, to whom I am most grateful. The image files can be left on the CD. The steps are as follows:

1. Place the CD in the CD drive and wait for it to become ready.
2. Start Windows Explorer, and change the drive de-

signator to the drive letter of your CD drive (when you click the down-arrow at the top left of the Explorer window you will see a list of all your disk drives. Double-click on the CD drive letter and a list of files and folders contained on the drive will appear in the main Explorer window).

3. Click the Type button at the head of the third column from the left in the main Explorer window. The 'Application' files will be listed together as a group.

4. Look for an application called MIPS_Install. exe and double-click on its icon. The program displays a window, as shown in Figure C.1. At the top of the window is an input box headed Current Installation Base Directory Choice. This will initially contain the drive letter of your hard drive, e.g. C:\. You can change the drive letter by clicking on the middle column headed Disk. Remember that this is the disk on to which the files will be written.

5. The Base Directory is initially C:\. You can change the drive letter as explained above. You can also change the Base Directory by clicking on a directory name in the left-hand window, headed Add directory. The programs will be installed in a directory called Mips\Programs\, which is a subdirectory of your Base Directory (i.e. C:\Mips\Programs\ if your Base Directory is C:\). The image dictionary files will be transferred to a subdirectory of your Base Directory called \Mips\Images\. Normally the Base Directory will be C:\ so there is no need to change the initial settings. You can change either Mips\Programs\ or \Mips\Images\ by pointing and clicking with the mouse and then manually editing.

6. At the lower right of the MIPS_Install window, you will see the message Image files will be left on CD. If you want to copy all of the images

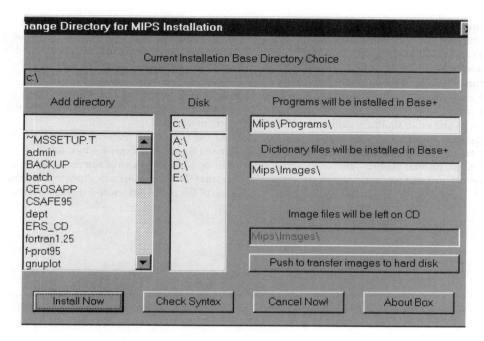

Figure C.1 `MIPS_Install` window.

to your hard drive, then press the button `Push to transfer images to hard disk`. The button is a toggle switch, so pressing it again will cause it to revert to its initial state. Normally the image files will be left on the CD.

7. Finally, press the `Install Now` button (bottom left of the window). The files should now be transferred to your hard disk.
8. You can start any of the programs described in Appendices A and B by using Explorer to locate them in their new location on your hard disk, then double clicking on the appropriate icon.
9. If you leave the settings as they are, then your working directory in the MIPS display program (Appendix A) is `C:\Mips\Images`.
10. If you leave the images on the CD, as recommended, then you must have the CD-ROM in its drive when you run any of the programs described in Appendices A and B.

C.3   USING THE USEFUL__LINKS.HTML FILE

The file `Useful_links.html` is located on the CD in the directory `\www`. It contains several hundred WWW links to remote sensing and GIS sites. To access this file you need a Web Browser such as Internet Explorer or Netscape Communicator, plus a connection to the Internet. Note that the following paragraphs assume that your CD drive letter is `D:`. If your CD drive letter is not `D:` then simply change the drive letter to `E:` or whatever your CD drive letter happens to be.

If your browser is Internet Explorer then simply choose `File|Open`, then select your CD drive letter, and finally navigate to the file `D:\www\Useful_links.html`. Double-click on the filename, which will then be displayed in a window. Click `OK`.

If you are using Netscape Communicator, click `File|Open Page`. Click `Choose File` and navigate to the file `D:\www\Useful_links.html` on the CD drive. Ensure that the 'Navigate' radio button is checked. Click `Open`.

The WWW page has a simple hierarchical format. At the top of the page is a list of 22 headings, as follows:

1. Data and metadata services, including document search
2. Organisations
3. Programmes
4. Satellite Sensors/Systems
5. GIS
6. DEMs
7. Data formats

8. Techniques and methods
9. Earth and environmental science
10. Computing and algorithms, etc.
11. Odds and ends
12. Instruments (radiometers, etc.)
13. People
14. WWW tools
15. Companies and other similar organisations (including software suppliers)
16. Other remote sensing and GIS WWW pages
17. Publications, tutorial material, etc.
18. Conferences, proceedings
19. Bibliographic databases
20. Applications
21. Cartography, map projections, etc.
22. Sources of digital maps

Simply click on the heading of interest, and you will be taken to the corresponding list of links. Double-click on a link to activate it.

At the top of the useful links page is a link to the Remote Sensing Society. Follow this link to find the most up-to-date version of the useful links file. Remember that WWW addresses change unpredictably, so some of the useful links will disappear in time.

Appendix D

Description of Sample Image Data Sets

D.1 ACKNOWLEDGEMENTS

I am grateful to the following organisations and individuals for permission to use copyrighted material:

- NASA's Observatorium (Jennifer McCullough, Curator), is thanked for permission to include the images marked [nasa]. Some of these images were made available to the NASA Observatorium by Intermountain Digital Imaging, LC and CORE Software Technology, whose assistance is acknowledged. NASA Observatorium can be found at http://observe.ivv.nasa.gov
- Images marked [esa] are copyright © ESA 1994, distributed by Eurimage.
- Mr R.D. Freeman, DERA RADARSAT Marketing Manager, kindly supplied the RADARSAT image extracts that are indicated in the list below by [rsat]. All of these images are © Canadian Space Agency/Agence spatiale canadienne, and were received by the Canada Centre for Remote Sensing (CCRS) and processed and distributed by RADARSAT International. These images, plus three full RADARSAT scenes and over 30 other image extracts, are contained on the CD-ROM 'RADARSAT Images', produced by RADARSAT International in 1996.
- The SPOT images contained on this CD (and marked with [spot]) are copyright © CNES 1986/1998 – Spot Image Distribution, and permission to use the data was kindly provided by Spot Image, 5 rue des Satellites, BP 4359, F-31030 Toulouse, France. I am grateful to Mme Isabelle Guidolin for her patience. Spot Image can be found at http://www.spotimage.fr/
- Images marked [usgs] were obtained from the US Geological Survey, EROS Data Center, Sioux Falls, South Dakota, USA.

D.2 USING THE IMAGE DATA SETS

The images listed below are contained in the root directory of the CD-ROM. You can read these image files directly from the CD by selecting drive D or E (depending on the setup of your computer) in the Windows File Selection dialogue box, and then changing the working directory to `D:\` or `E:\`. Note that if you copy these files to your computer's hard disk, you must include all of the files listed in the associated INF file. In addition, you must change all of the drive and path references contained in the INF file. The INF file is in text format so you can access it directly via Windows Notepad. Currently the drive, path and directory are listed as `D:\`. If you copy an image set to `C:\mips\images` (assuming that your hard drive is `C:`) you must change the INF file so that ALL the files that are referenced in that INF file have the drive, path and directory `C:\mips\images`. If you copy the image files to any other drive and/or directory, say `D:\images`, then the INF files must be edited so that the drive, path and directory of all of the referenced files are amended, otherwise the programs will not work.

The program `MIPS_Install`, which is described in Appendix C, will copy either the image dictionary files (*.inf) or the image dictionary files plus the image files on to your hard disk, and will automatically update the dictionary files. The default is to copy the dictionary files, and to leave the image files on the CD. This is recommended, as the image files occupy around 50Mb of disk space. See program `test_inf` (B.3.37), which checks the consistency of a specified image dictionary file.

D.3 IMAGE DATA SETS

The image data sets contained on the CD-ROM and listed in the following subsections are as follows (see section C.1 for the meanings of the various source symbols):

D.3.1   Mississippi River [nasa]
D.3.2   Little Colorado River [nasa]
D.3.3   London [nasa]
D.3.4   Paris [nasa]
D.3.5   San Joaquin Valley, California [nasa]
D.3.6   Morro Bay, California [nasa]
D.3.7   Candlewood Lake, Connecticut [nasa]
D.3.8   Rio de Janeiro, Brazil [nasa]
D.3.9   Nottingham [spot]
D.3.10 Chott el Guettar, Tunisia [usgs]
D.3.11 Gregory Rift Valley, Kenya [usgs]
D.3.12 Littleport, Cambridgeshire [esa]
D.3.13 Los Monegros, NE Spain [esa]
D.3.14 Fry Canyon, Arizona/Utah [usgs]
D.3.15 Tanzanian Coast [usgs]
D.3.16 The Camargue, France [spot]
D.3.17 Red Sea Hills, Sudan [usgs]
D.3.18 RADARSAT images[rsat] – East Anglia, The Netherlands, Indonesia and western Canada
D.3.19 AVHRR browse data, Australia

D.3.1   Mississippi River [nasa]

The centre latitude and longitude of this Landsat TM image set are 34°46′N and 90°27′W, a point to the south-west of Memphis, Tennessee. The date of acquisition is 13 January 1983. The image has 512 rows, 512 columns and seven bands. The image set is referenced by the INF file `missis.inf`.

D.3.2   Little Colorado River [nasa]

The centre latitude and longitude of this Landsat TM image set are 36°12′N and 111°47′W, to the east of the San Francisco peaks, Arizona. Here the Little Colorado River flows NNE across the Painted Desert. The date of acquisition is 24 August 1985. The image has 512 rows, 512 columns and seven bands. The image set is referenced by the INF file `litcolorado.inf`.

D.3.3   London [nasa]

This Landsat TM image set covers the centre of London (51°30′N, 0°20′W). The date of acquisition is 18 August 1984. The image has 512 rows, 512 columns and seven bands. The image set is referenced by the INF file `london.inf`.

D.3.4   Paris [nasa]

This Landsat TM image set covers the centre of Paris (48°50′N, 2°20′E), and was acquired on 9 May 1987. The

size of the images is 512 lines of 512 pixels. The image set is referenced by the INF file `paris.inf`.

D.3.5   San Joaquin Valley, California [nasa]

The centre of this Landsat TM image set is at the point 35°11′N, 119°06′W, near Fresno, California. The date of acquisition is 15 September 1986. The image has 512 rows, 512 columns and seven bands. The image set is referenced by the INF file `san-joaq.inf`.

D.3.6   Morro Bay, California [nasa]

The centre of this Landsat TM image set is at the point 35°21′N, 120°49′W, to the north of San Luis Obispo, California. The date of acquisition is 19 November 1984. The image has 512 rows, 512 columns and seven bands. The image set is referenced by the INF file `morrobay.inf`.

D.3.7   Candlewood Lake, Connecticut [nasa]

The centre of this Landsat TM image set is at the point 41°30′N, 73°30′W, in the state of Connecticut. Two image sets are provided. The first was acquired in the summer, on 10 June 1984, and the second in autumn, on 9 October 1986. Both images have 512 rows, 512 columns and seven bands. The summer image set is referenced by the INF file `cans.inf` and the autumn (fall) data are referenced in `canf.inf`.

D.3.8   Rio de Janeiro, Brazil [nasa]

This Landsat TM image set shows the city of Rio de Janeiro, Brazil (23°S, 43°W), and its surroundings. The date of acquisition is 5 August 1985. The image has 512 rows, 512 columns and seven bands. The image set is referenced by the INF file `rio.inf`.

D.3.9   Nottingham [spot]

This multispectral SPOT image covers the area around the city of Nottingham, UK (52°50′N, 1°10′W). The image has 1024 rows and 1024 columns. The date of acquisition is unknown. The name of the INF file that references this image set is `notspot1.inf`. Because this image is greater than 512 × 512 pixels in size it cannot be displayed directly by the MIPS display program (Appendix A). One of the following options must be used: (i) specify a start line and start pixel giving the coordinates of the top left corner of the sub-image to be displayed, relative to the top left of the full image (which has coordinates 1, 1) and display a 512 × 512 extract, or (ii) specify a start line and start pixel of 1, and subsample the image using a line-skip and pixel-skip factor of 2 to display every second line and second pixel.

D.3.10 Chott el Guettar, Tunisia [usgs]

The Chott el Guettar is a dry saline lake located near 33°50′N, 8°30′E. The area shown on this 512 × 512 pixel sub-image covers part of the salt lake bed, with the southern slopes of the Djebel Ortaba to the north of the lake. The main road from Gabes to Gafsa runs along the northern side of the saline lake in a general NW–SE direction. Three Landsat TM bands (2, 3 and 4) are provided, and these are referenced in the INF file `egmar.inf`. The date of image acquisition is not known.

D.3.11 Gregory Rift Valley, Kenya [usgs]

This image set contains three Landsat TM bands (3, 5 and 7) and is 512 × 512 pixels in size. The Gregory Rift Valley is part of the East African Rift system. The area shown on this image lies to the north of Lake Baringo and south of Silali volcano, at approximately 1°N, 36°E. This area is sparsely vegetated and several lava flows of different ages and compositions are clearly apparent. The eroded area to the west of the small volcanic cone visible in the lower right of the image is a pyroclastic deposit. Alluvial areas extend on either side of the river that flows eastwards across the northern part of the image area. The image acquisition date is 30 July 1984. The three TM bands are referenced by the INF file `kenya.inf`.

D.3.12 Littleport, Cambridgeshire [esa]

The small town of Littleport is located about 5 km north of the city of Ely, Cambridgeshire, UK, at 52°25′N, 0°20′E. The main features of the area shown on the image are the River Ouse, running NE on the right-hand side of the image, and the parallel Old and New Bedford Rivers, also running in a NE direction. This area is low-lying (around sea-level) and flat. The blue areas on the band 432 colour composite are either ploughed fields or towns and villages. The main crops in this fertile region of the Fens are wheat, barley and sugar-beet. The image area contains a few small clouds and associated shadows. Six of the seven Landsat TM bands are included (band 6 is excluded). Image size is

512 × 512 pixels. The INF file for this image set is `littlept.inf`.

D.3.13 Los Monegros, N.E. Spain [esa]

The Los Monegros region lies in the Spanish province of Aragon. It is a semi-arid and sparsely populated upland region, with an average altitude of 350 m. Recent developments include the building of irrigation canals to bring water for agriculture from the Flumen river system to the north of the area shown on this Landsat TM image, acquired on 7 July 1997. The image covers the area between the village of Bujaraloz (in the top right corner of the image) and the northern slopes of the Ebro Valley. The approximate latitude and longitude of the image centre are 41°30′N and 0°15′W. The Laguna la Playa, which is generally dry in summer, lies to the south of Bujaraloz. The southern part of the area is mainly dryland farming (wheat and barley). The north-west part of the image covers a dissected upland area covered by low coniferous trees. Bands 1–7 of this Landsat TM image are referenced by the file `LosMoneg97.inf`. Bands 1–5 and band 7 of a Landsat TM sub-image, acquired in 1991 and covering approximately the same area, are referenced in `LosMoneg91.inf`. Differences between the 1991 and 1997 images are due to changes in agriculture, including crop rotation and extension of the ploughed area, and differences in the weather conditions in the days preceding image acquisition. The period before the 1997 image was acquired was particularly wet.

D.3.14 Fry Canyon, Arizona/Utah [usgs]

This four-band Landsat MSS image has been subsampled (taking every sixth pixel on every fourth line). It covers the area from the north of the Grand Canyon in Arizona (around the latitude of the city of Page) to the Canyonlands National Park, south of Moab, Utah (approximate centre position 36°N, 111°W). Lake Powell is a prominent water feature in the south of the region. The main tributary of the Colorado River, the San Juan, joins near the northern end of Lake Powell. The upland areas to the west of the Colorado include the snow-covered Henry Mountains, well-known to readers of Grove Karl Gilbert famous monograph on their geology and geomorphology, the Waterpocket Fold, the Kaiparowitz Plateau, and Smoky Mountain. Navajo Mountain lies to the south of the Colorado–San Juan confluence. The mountain ridges are vegetated, mainly by low trees, and the lowland areas are semi-arid. The file `fry.inf` references four Landsat images, the MSS bands labelled 4, 5, 6 and 7.

D.3.15   Tanzanian Coast [usgs]

This three-band Landsat MSS image set covers an area of the Tanzanian coast south of Dar-es-Salaam and north of Mafia Island, at a latitude and longitude of 7°30'S, 39°25'W. The northern part of the coast is fringed with coral reefs, and a small coral island is apparent in the lower centre of the image. The reddish colours over the land indicate forest. Brownish and whitish areas have been cleared. The coastal waters are clear, and the colour variations here refer to water depth as the light in MSS bands 4 and 5 is reflected from the sea bed. The edge of the continental shelf is clearly visible. This image set is 512 × 512 pixels in size and is referenced by tanzcoas.inf. Note that Landsat MSS bands 4, 5 and 7 are provided, and that the dynamic range of these bands is 7 bits for bands 4 and 5, and 6 bits for band 7.

D.3.16   The Camargue, France [spot]

Two SPOT multispectral images, each of 512 × 512 pixels, show the area to the north of the town of Tarascon in the Camargue area of the Rhône delta in the south of France (43°50'N, 4°45'E). The first image set (referenced by camarg.inf) was collected on 12 January 1987. The second (camarg1.inf) was collected on 17 January of the same year. The two image sets show the changes that occurred between these two dates, largely the result of a comprehensive snowfall. These images are extracts from a demonstration CD ('SPOT Scene') produced by SPOT Image, and were read using the read_SPOT_cd program described in Appendix B.

D.3.17   Red Sea Hills, Sudan [usgs]

The Red Sea Hills are located in eastern Sudan (between 18° and 19°N latitude and around 36°–38°E longitude). The area is described by Koch and Mather (1997; see also Mather et al., 1998) as a 200 km wide range of mountains, rising steeply from the coastal plain to elevations of 1000 m and more. The sandy Nubian Desert lies to the west. The upland area is heavily dissected by a network of drainage channels called *khors*. The underlying rock is volcanic and granitic, with some limestone areas, and is heavily faulted with the main directions of faulting being N–S, with a subsidiary E–W trend. Two image sets are provided. One is a degraded version of a Landsat TM image from 1984, which has been resampled at a six-pixel interval in both rows and columns

and then padded with zeros to give an image size of 1024 × 1024 pixels. The second is a full resolution extract from the same image, and is also 1024 × 1024 pixels in size. The degraded image is referenced by the dictionary file sudanlo.inf, while the full resolution image is referenced by sudanhi.inf.

D.3.18   RADARSAT images [rsat]

Four RADARSAT image sets are included on this CD. Each of the images is © Canadian Space Agency/Agence spatiale canadienne, and was received by the Canada Centre for Remote Sensing (CCRS) and processed and distributed by RADARSAT International. The image dictionary files hengelo.inf, indonesia.inf and okanagan.inf each reference a single RADARSAT SAR image, which is 512 × 512 pixels in size. The *hengelo* image covers an agricultural area near the town of Hengelo, in The Netherlands. The *indonesia* image shows a volcanic peak on the island of Java, while the *okanagan* image is of a forested area around Okanagan in British Columbia, Canada. The eastanglia.inf image dictionary file references four RADARSAT images, each 512 × 512 pixels in size, covering an area of East Anglia (UK) to the south-west of Lakenheath. These four images were collected at different times during the crop-growing season. They have been geometrically corrected, and can therefore be overlaid. Colour differences then indicate differential growth, which affects the backscattering properties of the surface. I am grateful to Mr G. Gill for providing these images.

D.3.19   AVHRR, Australia

This is a browse image of a five-band NOAA-14 scene bcovering parts of New Guinea, Northern Territory and Queensland. The area of the image is approximately 2000 km by 4500 km, giving a pixel size of around 5 km across the scan lines (columns) and 4 km in the along-track (row) direction. The date of image capture was 20 July 1996 at 04:31 GMT. I am grateful to Susan Camp-bell of the CSIRO Earth Observation Centre (EOC), Canberra, Australia, for her help in providing the image data. The data were collected by the Australian Institute of Marine Science (AIMS) in Townsville, Queensland. The scale of this image is roughly that of the Global Area Coverage (GAC) data, and shows the problems involved in obtaining cloud-free NDVI composites from daily AVHRR data. The image dictionary file for this image set is avhrr.inf.

Appendix E

World Wide Web Bookmarks

The CD contains a file called `useful_links.html`, in directory www. This file holds over 900 links to World Wide Web (WWW) sites across the world. Undoubtedly some of the addresses contained in this file will become out of date, and readers should check with the WWW home page of the Remote Sensing Society for the most recent version of this file. The bookmarks are sub-divided into categories for ease of access. You can also use the `Find` facility on your browser in order to locate items of interest.

You will need an Internet browser to read this file. The most commonly-used browsers are `Netscape` and `Internet Explorer`. If your preferred browser is Netscape, then choose `File|Open File in Browser` and the select the file `D:\www\useful_links.`

`html`. If you use Internet Explorer choose `File|Open` and enter the filename `D:\www\useful_links.html` in the box. If your CD drive letter is not `D:` then do not use `D:` in the path – use your CD drive letter instead.

The bookmarks contain references to data suppliers, organisations, software providers, tutorial materials, algorithms, and applications of remote sensing and GIS. There are many thousands of WWW sites that provide remote-sensing-related information. However, the bookmarks provided here should give you a flying start. Check the WWW page `http:/www.geog.nottingham.ac.uk/~mather/booknews.html` for news of updates to the `useful_links.html` file.

References

Abramowitz, M. and Stegun, I.A. (eds), 1972, *Handbook of Mathematical Functions*. New York: Dover Books.

Adams, J.B., Smith, M.O. and Gillespie, A.R., 1989, Simple models for complex natural surfaces: a strategy for the hyperspectral era of remote sensing. *Proceedings of the IEEE International Geoscience and Remote Sensing Symposium (IGARSS'89)*, Vancouver, British Columbia, Canada, 10–14 July 1989. New York: IEEE Press, vol. 1, 16–21.

Adams, J.B., Smith, M.O. and Gillespie, A.R., 1993, Imaging spectroscopy: interpretation based on spectral mixture analysis. In Pieters, C.M. and Englert, P.A.J. (eds) (1993), *Remote Geochemical Analysis: Elemental and Mineralogical Composition*. Cambridge: Cambridge University Press, 145–166.

Aleksander, I. and Morton, J., 1990, *An Introduction to Neural Computing*. London: Chapman and Hall.

Alföldi, T.T., 1982, Remote sensing for water quality monitoring. In Johanssen, C.J. and Sanders, J.L. (eds) (1982), *Remote Sensing for Resource Management*. Ankeny, Iowa: Soil Conservation Society of America, 317–328.

Al-Hinai, K.G., Khan, M.A. and Canaas, A.A., 1991, Enhancement of sand dune texture from Landsat imagery using difference of Gaussian filter. *International Journal of Remote Sensing*, **12**, 1063–1069.

Allan, J.A., 1984, The role and future of remote sensing. *Proceedings of the Tenth Anniversary International Conference of the Remote Sensing Society*. Nottingham: The Remote Sensing Society, 23–30.

Alley, R.E., 1995, *Algorithm Theoretical Basis Document*, version 2.0, 1 March 1995. Jet Propulsion Laboratory, Pasadena, Calif.

Alparone, L., Baronti, S., Carla, R. and Pugilisi, C., 1996, An adaptive order-statistics filter for SAR images. *International Journal of Remote Sensing*, **17**, 1357–1365.

Anderberg, M.R., 1973, *Cluster Analysis for Applications*. New York: Academic Press.

Andreadis, I., Glavas, E. and Tsalides, Ph., 1995, Image enhancement using colour information. *International Journal of Remote Sensing*, **16**, 2285–2289.

Anuta, P.E., 1970, Spatial registration of multispectral and multitemporal digital imagery using Fast Fourier Transform techniques. *IEEE Transactions on Geoscience Electronics*, **8**, 353–368.

Ardö, J., Pilesjö, P. and Skidmore, A., 1997, Neural networks, multitemporal Landsat Thematic Mapper data and topographic data to classify forest damage in the Czech Republic. *Canadian Journal of Remote Sensing*, **23**, 217–219.

Arnaud, M., 1994, The SPOT programme. In Mather, P.M. (ed.) (1994), 29–39.

Askne, J. (ed.), 1995, *Sensors and Environmental Applications of Remote Sensing. Proceedings of the 14th EARSeL Symposium*, Göteborg, Sweden, 6–8 June 1994. Rotterdam: A.A. Balkema

Asrar, G. (ed.), 1989, *Theory and Applications of Optical Remote Sensing*. New York: Wiley-Interscience.

Atkinson, P.M. and Curran, P.J., 1997, Choosing an appropriate spatial resolution. *Photogrammetric Engineering and Remote Sensing*, **63**, 1345–1351.

Austin, J., Harding, S., Kanellopoulos, I., Lees, K., McNaughton, H., Roli, F., Vernazza, G. and Wilkinson, G., 1997, *Connectionist Computation in Earth Observation*. Joint Research Centre, European Commission, Report EUR 17314 EN, Brussels, Belgium.

Babey, S.K. and Anger, C.D., 1989, A Compact Airborne Spectrographic Imager (CASI). *Proceedings of the IEEE International Geoscience and Remote Sensing Symposium (IGARSS'89)*, Vancouver, British Columbia, Canada, 10–14 July 1989. New York: IEEE Press, 1028–1031.

Bannari, A., Morin, D., Bénié, G.B. and Bonn, F.J., 1995a, A theoretical review of different mathematical models of geometric corrections applied to remote sensing images. *Remote Sensing Reviews*, **13**, 27–47.

Bannari, A., Morin, D., Bonn, F. and Huete, A., 1995b, A review of vegetation indices. *Remote Sensing Reviews*, **13**, 95–120.

Baret, F. and Guyot, G., 1991, Potential and limits of vegetation indices for LAI and PAR assessment. *Remote Sensing of Environment*, **35**, 161–173.

Baret, F., Jacquemond, S. and Hanocq, J.F., 1993, The soil line concept in remote sensing. *Remote Sensing Reviews*, **7**, 65–82.

Barker, J.L. and Gunther, F.L., 1983, Landsat-4 sensor performance. *Proceedings of the Eighth W.T. Pecora Memorial Remote Sensing Symposium, 'Land Remote Sensing Advances for the Eighties'*. Sioux Falls, SD: US Geological Survey, 46–74.

Barnea, D.I. and Silverman, H.F., 1972, A class of algorithm for fast digital image registration. *IEEE Transactions on Computers*, **21**, 179–186.

Barnsley, M.J., 1983, The implications of view angle effects on the use of multispectral data for vegetation studies. *Proceedings of the International Conference on Remote Sensing for Rangeland Monitoring and Management*, Silsoe, Bedfordshire, England. Nottingham: The Remote Sensing Society, 173–177.

Barnsley, M.J., 1994, Environmental monitoring using multiple-view-angle (MVA) remotely-sensed data. In Foody, G. and Curran, P. (eds) (1994), *Environmental Remote Sensing from Regional to Global Scales*. Chichester: Wiley, 181–201.

Barnsley, M.J. and Kay, S.A.W., 1990, The relationship between sensor geometry, vegetation-canopy geometry and image variance. *International Journal of Remote Sensing*, **11**, 1075–1083.

Bastin, L., 1997, Comparison of fuzzy *c*-mean classification, linear mixture modelling and MLC probabilities as tools for unmixing coarse pixels. *International Journal of Remote Sensing*, **18**, 3629–3648.

Basu, J.P. and Odell, P.L., 1974, Effects of intraclass correlation among training samples on the misclassification probabilities of Bayes' procedure. *Pattern Recognition*, **6**, 13–16.

Bateson, A. and Curtiss, B., 1996, A method for manual endmember selection and spectral unmixing. *Remote Sensing of Environment*, **55**, 229–243.

Beale, R. and Jackson, T., 1990, *Neural Computing: An Introduction*. Bristol: Adam Hilger.

Beauchemin, M., Thomson, K.B.P. and Edwards, G., 1996, Edge detection and speckle adaptive filtering based on a second-order textural measure. *International Journal of Remote Sensing*, **17**, 1751–1759.

Begni, G., 1986, Absolute calibration of SPOT data. *SPOT Newsletter*, **10**, 2–3.

Begni, G., Dinguirard, M.C., Jackson, R.D. and Slater, P.N., 1988, Absolute calibration of the SPOT-1 HRV cameras. *SPIE*, **660**, 66–76.

Belward, A., 1991, Spectral characteristics of vegetation, soil and water in the visible, near-infrared and middle-infrared wavelengths. In Belward, A. and Valenzuela, C.R. (eds) (1991), 31–53.

Belward, A. and Valenzuela, C.R. (eds), 1991, *Remote Sensing and Geographical Information Systems for Resource Management in Developing Countries*. Dordrecht: Kluwer Academic.

Ben-Dor, E. and Kruse, F.A., 1995, Surface mineral mapping of the Makhtesh Ramon Negev, Israel, using GER 63 channel scanner data. *International Journal of Remote Sensing*, **16**, 3529–3553.

Benediktsson, J.A. and Sveinsson, J.R., 1997, Feature extraction for neural network classifiers. In Kanellopoulos, I. *et al.* (eds) (1997), 97–104.

Benediktsson, J.A., Swain, P.H. and Ersoy, O.K., 1990, Neural network approaches versus statistical methods in the classification of multisource remote sensing data. *IEEE Transactions of Geoscience and Remote Sensing*, **28**, 540–552.

Benny, A.H., 1981, Automatic relocation of ground control points in Landsat imagery. *Proceedings of the International Conference on Matching Remote Sensing Technologies and their Applications*. Nottingham: The Remote Sensing Society, 307–315.

Bergland, G.D., 1969, A guided tour of the Fast Fourier Transform. *IEEE Spectrum*, **6**, 41–45.

Bergland, G.D. and Dolan, M.T., 1979, Fast Fourier transform algorithms. *Programs for Digital Signal Processing*. New York: IEEE Acoustics, Speech and Signal Processing Society, IEEE Press/Wiley, Section 1.2–1.

Bernard, A.C., Kanellopoulos, I. and Wilkinson, G.G., 1996, Neural net classification of mixtures. In Binaghi, E. *et al.* (eds) (1996), 53–58.

Bernstein, R., Lotspiech, J.B., Myers, J., Kolsky, H.G. and Lees, R.D., 1984, Analysis and processing of Landsat-4 sensor data using advanced image processing techniques and technologies. *IEEE Transactions on Geoscience and Remote Sensing*, **22**, 192–221.

Bezdek, J.C., 1993, Editorial: Fuzzy models – what are they, and why? *IEEE Transactions on Fuzzy Systems*, **1**, 1–6.

Bezdek, J.C., 1994, The thirsty traveler visits Gamont: A rejoinder to 'Comments on fuzzy sets – what are they and why?' *IEEE Transactions on Fuzzy Systems*, **2**, 43–45.

Bezdek, J.C., Ehrlich, R. and Full, W., 1984, FCM: The fuzzy *c*-means clustering algorithm. *Computers & Geosciences*, **10**, 191–203.

Billingsley, F.C. 1983, Data processing and reprocessing. In Colwell, R.N. (ed.) (1983), 719–792.

Binaghi, E., Brivio, P.A. and Rampini, A. (eds), 1996, *Proceedings of the 14th International Workshop on Soft Computing in Remote Sensing Data Analysis*, Milan, 4–5 December 1995. Singapore: World Scientific.

Bird, A.C., 1991a, Principles of remote sensing: electromagnetic radiation, reflectance and emissivity. In Belward, A. and Valenzuela, C.R. (eds) (1991), 1–15.

Bird, A.C., 1991b, Principles of remote sensing: interaction of electromagnetic radiation with the atmosphere and the Earth. In Belward, A. and Valenzuela, C.R. (eds) (1991), 17–30.

Bishof, H., Schneider, W. and Pinz, A.J., 1992, Multispectral classification of Landsat images using neural networks. *IEEE Transactions on Geoscience and Remote Sensing*, **30**, 482–490.

Bishop, C.M., 1995, *Neural Networks for Pattern Recognition*. Oxford: Clarendon Press.

Bishop, Y.M., Fienberg, S.E. and Holland, P.W., 1975, *Discrete Multivariate Analysis: Theory and Practice*. Cambridge, Mass.: MIT Press.

Blamire, P., 1996, The influence of relative sample size in training artificial neural networks. *International Journal of Remote Sensing*, **17**, 223–230.

Blaser, T.J. and Caloz, R., 1991, Digital ortho-image registration from a SPOT panchromatic image using a digital elevation model. *IEEE Transactions on Geoscience and Remote Sensing*, **29**, 2431–2434.

Blom, R.G., 1988, Effects of variations in look angle and wavelength in radar images of volcanic and aeolian terrains, or now you see it, now you don't. *International Journal of Remote Sensing*, **9**, 945–965.

Blom, R.G. and Daily, M., 1982, Radar image processing for rock-type discrimination. *IEEE Transactions on Geoscience Electronics*, **20**, 343–351.

Blonda, P.N. and Pasquariello, G., 1991, An experiment for the interpretation of multitemporal remotely sensed images based on a fuzzy logic approach. *International Journal of Remote Sensing*, **12**, 463–476.

Boardman, J., 1989, Inversion of imaging spectrometer data using singular value decomposition. *Proceedings of the IEEE International Geoscience and Remote Sensing Symposium (IGARSS'89)*, Vancouver, British Columbia, Canada, 10–14 July 1989. New York: IEEE Press, vol. 4, 2069–2072.

Bodechtel, J. and Zilger, J., 1996, MOMS – History, concepts, goals. *Proceedings of the MOMS-02 Symposium*, Cologne, Germany, 5–7 July 1995. Paris: European Association of Remote Sensing Laboratories (EARSeL), 12–25.

Bolstad, P.V. and Lillesand, T.M., 1992, Semi-automated training approaches for spectral class definition. *International Journal of Remote Sensing*, **13**, 3157–3166.

Bolstad, P.V. and Stowe, T., 1994, An evaluation of DEM accuracy: elevation, slope and aspect. *Photogrammetric Engineering and Remote Sensing*, **60**, 1327–1332.

Bolstad, P.V., Gessler, P. and Thomas, M.L., 1990, Positional uncertainty in manually-digitised map data. *International Journal of Geographical Information Systems*, **4**, 39–42.

Bonham-Carter, G.F., 1994, *Geographic Information Systems for Geoscientists*. Oxford: Pergamon/Elsevier Science.

Bonhomme, R., 1993, The solar radiation: characteristics and distribution in the canopy. In Varlet-Grancher, C., Bonhomme, R. and Sinoquet, H. (eds) (1993), *Crop Structures and Light Microclimate, Characteristics and Applications*. Paris: INRA Editions, 17–28.

Bouman, B.A.M., 1992, Accuracy of estimating the Leaf Area Index from vegetation indices derived from crop reflectance characteristics: a simulation study. *International Journal of Remote Sensing*, **13**, 3069–3084.

Bow, S.-T., 1992, *Pattern Recognition and Image Preprocessing*. New York: Marcel Dekker.

Box, E.O., Holben, B.N. and Kalb, V., 1989, Accuracy of the AVHRR Vegetation Index as a predictor of biomass, primary productivity and net CO_2 flux. *Vegetatio*, **80**, 71–89.

Brady, M., 1982, Computational approaches to image understanding. *Association of Computer Manufacturers' (ACM) Computing Surveys*, **14**, 3–71.

Brown, S.R. and Scholz, C.H., 1985, Broad bandwidth study of the topography of natural rock surfaces. *Journal of Geophysical Research*, **90**, 12575–12582.

Brownrigg, D.R.K., 1984, The weighted median filter. *Communications of the Association of Computer Manufacturers (ACM)*, **27**, 807–818.

Brush, R.J.H., 1985, A method for real-time navigation of AVHRR imagery. *IEEE Transactions on Geoscience and Remote Sensing*, **23**, 876–887.

Bruzzone, L., Conese, C., Maselli, F. and Roli, F., 1997, Multisource classification of complex rural areas using statistical and neural-network approaches. *Photogrammetric Engineering and Remote Sensing*, **63**, 523–533.

Bryant, R.G., 1996, Validated linear mixture modelling of Landsat TM data for mapping evaporite minerals on a playa surface: methods and applications. *International Journal of Remote Sensing*, **17**, 315–330.

Buckingham, W.F. and Sommer, S.E., 1983, Mineralogical characterization of rock surfaces formed by hydrothermal alteration and weathering: application to remote sensing. *Economic Geology*, **78**, 664–674.

Buckley, J.J. and Hayashi, Y., 1994, Fuzzy neural networks: a survey. *Fuzzy Sets and Systems*, **66**, 1–13.

Budkewitsch, P., Newton, G. and Hynes, A.J., 1994, Characterisation and extraction of linear features from digital images. *Canadian Journal of Remote Sensing*, **20**, 268–279.

Burrough, P.A. and Frank, A., 1996, *Geographic Objects with Indeterminate Boundaries*. London: Taylor and Francis.

Calder, N., 1991, *Spaceship Earth*. London: Viking Books/Channel Four Television.

Campbell, J.B., 1981, Spatial correlation effects upon accuracy of supervised classification of land cover. *Photogrammetric Engineering and Remote Sensing*, **47**, 355–357.

Campbell, N.A., 1980, Robust procedure in multivariate analysis. I: Robust covariance estimation. *Applied Statistics*, **29**, 231–237.

Campbell, N.A., 1996, The decorrelation stretch transform. *International Journal of Remote Sensing*, **17**, 1939–1949.

Cannon, R.J., Dave, J.A., Bezdek, J.C. and Trivedi, M.M., 1986, Segmentation of a Thematic Mapper image using the fuzzy c-means clustering algorithm. *IEEE Transactions on Geoscience and Remote Sensing*, **24**, 400–408.

Canters, F., 1997, Evaluating the uncertainty of area estimates derived from fuzzy land cover classification. *Photogrammetric Engineering and Remote Sensing*, **63**, 403–414.

Cappellini, V., Chiuderi, A. and Fini, S., 1995, Neural networks in remote sensing multisensor data processing. In Askne, J. (ed.) (1995), 457–462.

Carara, A., Bitelli, G. and Carla, T., 1997, Comparison of techniques for generating digital terrain models from contour lines. *International Journal of Geographical Information Science*, **11**, 451–473.

Carlson, G.E. and Ebell, W.J., 1995, Co-occurrence matrices for small region texture measurements and comparison. *International Journal of Remote Sensing*, **16**, 1417–1423.

Carpenter, G.A., Grossberg, S., Markuzon, N., Reynolds, J.H. and Rosen, D.B., 1992, Fuzzy ARTMAP: a neural network architecture for incremental supervised learning for analog multidimensional maps. *IEEE Transactions on Neural Networks*, **3**, 698–712.

Carpenter, G.A., Gjaja, M.N., Gopal, S. and Woodcock, C.E., 1997, ART neural networks for remote sensing: vegetation classification from Landsat TM and terrain data. *IEEE Transactions on Geoscience and Remote Sensing*, **35**, 308–325.

Chalmers, A.I. and Harris, R., 1981, Band ratios in multispectral analysis of Landsat digital data. *Proceedings of the 8th Annual Conference of the Remote Sensing Society*. Nottingham: The Remote Sensing Society, 139–146.

Chang, W.C., 1983, On using principal components before separating a mixture of two multivariate normal distributions. *Applied Statistics*, **32**, 267–275.

Chapman, R.E., 1995, *Physics for Geologists*. London: UCL Press.

Chavez, P.S., Jr, 1986, Processing techniques for digital sonar images from GLORIA. *Photogrammetric Engineering and Remote Sensing*, **52**, 1133–1145.

Chavez, P.S., Jr, 1988, An improved dark-object subtraction technique for atmospheric scattering correction of multi-spectral data. *Remote Sensing of Environment*, **24**, 459–479.

Chavez, P.S., Jr, 1996, Image-based atmospheric corrections – revisited and improved. *Photogrammetric Engineering and Remote Sensing*, **62**, 1025–1036.

Chavez, P.S., Jr and Bauer, B., 1982, An automatic kernel-size selection technique for edge enhancement. *Remote Sensing of Environment*, **12**, 23–38.

Che, N. and Price, J.C., 1992, Survey of radiometric calibration results and methods for visible and near-infrared channels of NOAA-7, -9, and -11 AVHRRs. *Remote Sensing of Environment*, **41**, 19–27.

Chilar, J., St-Laurent, L. and Dyer, J.A., 1991, Relation between the normalised vegetation index and ecological variables. *Remote Sensing of Environment*, **35**, 279–298.

Chittenini, C.B., 1983, Edge and line detection in multidimensional noisy imagery. *IEEE Transactions on Geoscience and Remote Sensing*, **21**, 163–174.

Chiuderi, A. and Cappellini, V., 1996, A Kohonen's Self Organising Map for land cover classification. In Parlow. E. (ed.) (1996a), 107–112.

Clark, C. and Cañas, A., 1995, Spectral identification by artificial neural network and genetic algorithm. *International Journal of Remote Sensing*, **16**, 2255–2275.

Clavet, D.M., Lassere, M. and Pouliot, J., 1993, GPS control for 1:50,000 scale topographic mapping from satellite images. *Photogrammetric Engineering and Remote Sensing*, **59**, 107–111.

Cliff, A.D. and Ord, J.K., 1973, *Spatial Autocorrelation*. London: Pion Press.

Collins, J.B. and Woodcock, C.E., 1996, Assessment of several linear change detection techniques for mapping forest mortality using multitemporal Landsat TM data. *Remote Sensing of Environment*, **56**, 66–77.

Colwell, R.N. (ed.), 1983, *Manual of Remote Sensing*, 2 vols. Falls Church, Va.: American Society of Photogrammetry.

Conese, C., Gilabert, M.A., Maselli, F. and Bottai, L., 1993, Topographic normalisation of TM scenes through the use of an atmospheric correction method and digital terrain models. *Photogrammetric Engineering and Remote Sensing*, **59**, 1745–1753.

Congalton, R.G., 1991, A review of assessing the accuracy of classifications of remotely-sensed data. *Remote Sensing of Environment*, **37**, 35–46.

Congalton, R.G., Oderwald, R. and Mead, R., 1983, Landsat classification accuracy using discrete multivariate analysis statistical techniques. *Photogrammetric Engineering and Remote Sensing*, **49**, 1671–1678.

Cook, A.E. and Pinder, J.E., III, 1996, Relative accuracy of rectifications of coordinates determined from maps and the Global Positioning System. *Photogrammetric Engineering and Remote Sensing*, **62**, 73–77.

Cracknell, A.P., 1997, *The Advanced Very High Resolution Radiometer*. London: Taylor and Francis.

Cracknell, A.P., 1998, Review Article: Synergy in remote sensing – what's in a pixel? *International Journal of Remote Sensing*, **19**, 2025–2074.

Craig, R.G., 1979, Autocorrelation in Landsat data. *Proceedings of the 13th International Symposium on Remote Sensing of Environment*. Ann Arbor, Mich.: Environmental Research Institute of Michigan, 1517–1524.

Crawford, P.S., Brooks, A.R. and Brush, R.J.H., 1996, Fast navigation of AVHRR images using complex orbital models. *International Journal of Remote Sensing*, **17**, 197–212.

Crippen, R.W., 1989, A simple spatial filtering routine for the cosmetic removal of scan line noise from Landsat TM P-tape imagery. *Photogrammetric Engineering and Remote Sensing*, **55**, 327–331.

Crist, E.P., 1983, The TM tasseled cap – a preliminary formulation. *Proceedings of the Symposium on Machine Processing of Remotely-Sensed Data 1983*, Purdue University, West Lafayette, Ind., 357–364.

Crist, E.P. and Cicone, R.C., 1984a, A physically-based transformation of Thematic Mapper data – the TM Tasseled Cap. *IEEE Transactions on Geoscience and Remote Sensing*, **22**, 256–263.

Crist, E.P. and Cicone, R.C., 1984b, Comparison of the dimensionality and features of simulated Landsat-4 MSS and TM data. *Remote Sensing of Environment*, **14**, 235–246.

Crist, E.P. and Kauth, R.J., 1986, The Tasseled Cap de-mystified. *Photogrammetric Engineering and Remote Sensing*, **52**, 81–86.

Cross, A.M., Settle, J.J., Drake, N.A. and Paivinen, R.T.M., 1991, Subpixel measurement of tropical forest cover using AVHRR data. *International Journal of Remote Sensing*, **12**, 1119–1129.

Curran, P.J., 1980, Multispectral remote sensing of vegetation amount. *Progress in Physical Geography*, **4**, 315–341.

Curran, P.J., 1983, Multispectral remote sensing for estimation of green leaf area index. *Philosophical Transactions of the Royal Society of London*, **309A**, 257–270.

Curran, P.J., 1994, Imaging spectroscopy. *Progress in Physical Geography*, **18**, 247–266.

Curran, P.J. and Atkinson, P.M., 1998, Geostatistics and remote sensing. *Progress in Physical Geography*, **22**, 61–78.

Czaplewski, R.L., 1992, Misclassification bias in areal estimates. *Photogrammetric Engineering and Remote Sensing*, **58**, 189–192.

Daily, M., 1983, Hue–saturation–intensity split-spectrum processing of Seasat radar images. *Photogrammetric Engineering and Remote Sensing*, **49**, 349–355.

Dale, P.E.R., Chandica, A.L. and Evans, M., 1996, Using image subtraction and classification to evaluate change in sub-tropical inter-tidal wetlands. *International Journal of Remote*

Sensing, **17**, 703–719.

Danielsson, P.E., 1981, Getting the median faster. *Computer Graphics and Image Processing*, **15**, 71–78.

Davis, J.C., 1973, *Statistics and Data Analysis in Geology*. New York: Wiley.

Dayhoff, J.E., 1990, *Neural Network Architectures: An Introduction*. New York: Van Nostrand Reinhold.

Decell, H.P. and Guseman, L.F., 1979, Linear feature selection with applications. *Pattern Recognition*, **11**, 55–63.

de Cola, L., 1994, Simulating and mapping spatial complexity using multi-scale techniques. *International Journal of Geographical Information Systems*, **8**, 411–427.

de Jong, S.M. and Burrough, P.A., 1995, A fractal approach to the classification of Mediterranean vegetation types using remotely sensed data. *Photogrammetric Engineering and Remote Sensing*, **61**, 1041–1063.

Deschamps, P.Y., Herman, M. and Tanré, D., 1983, Definitions of atmospheric radiance and transmittances in remote sensing. *Remote Sensing of Environment*, **13**, 89–92.

Desnos, Y.-L. and Matteini, V., 1993, Review on structural detection and speckle filtering on ERS-1 images. *EARSeL Advances in Remote Sensing*, **2**, 52–65. Paris: European Association of Remote Sensing Laboratories.

de Souza Filho, C.R., Drury, S.A., Denniss, A.M., Carlton, R.W.T. and Rothery, D.A., 1996, Restoration of corrupted optical Fuyo-1 (JERS-1) data using frequency domain techniques. *Photogrammetric Engineering and Remote Sensing*, **62**, 1037–1047.

Dikshit, O., 1996, Textural classification for ecological research using ATM images. *International Journal of Remote Sensing*, **17**, 887–915.

Diner, D.J., Bruegge, C.J., Martonchik, J.V., Bothwell, G.W., Danielson, E.D., Floyd, E.L., Ford, V.G., Hovland, L.E., Jones, K.L. and White, M.L., 1991, A Multi-Angle Imaging SpectroRadiometer for terrestrial remote sensing from the Earth Observing System. *International Journal of Imaging Systems and Technology*, **3**, 92–107.

Dobbertin, M. and Biging, G.S., 1996, A simulation study of the effect of scene autocorrelation, training size and sampling method on classification accuracy. *Canadian Journal of Remote Sensing*, **22**, 360–367.

Dowman, I., 1992, The geometry of SAR images for geocoding and stereo applications. *International Journal of Remote Sensing*, **13**, 1609–1617.

Dowman, I., Laycock, J. and Whalley, J., 1993, Geocoding in the UK. In Schreier, G. (ed.) (1993a), 373–387.

Drewniok, C., 1994, Multi-spectral edge detection. Some experiments on data from Landsat-TM. *International Journal of Remote Sensing*, **15**, 3743–3765.

Drury, S.A., 1993, *Image Interpretation in Geology*, 2nd edn. London: Allen and Unwin.

Du, L. and Lee, J.S., 1996, Fuzzy classification of earth terrain covers using complex polarimetric SAR data. *International Journal of Remote Sensing*, **17**, 809–826.

Duda, O. and Hart, P.E., 1973, *Pattern Classification and Scene Analysis*. New York: Wiley.

Duggin, M.J., 1985, Factors influencing the discrimination and quantification of terrestrial features using remotely-sensed radiance. *International Journal of Remote Sensing*, **6**, 3–28.

Duguay, C.R. and Peddle, D.R., 1996, Comparison of evidential reasoning and neural network approaches in a multi-source classification of alpine tundra vegetation. *Canadian Journal of Remote Sensing*, **22**, 433–440.

Dymond, J.R., 1992, How accurately do image classifiers estimate area? *International Journal of Remote Sensing*, **13**, 1735–1742.

Eberlein, R.B. and Wezka, J.S., 1975, Mixtures of derivative operators as edge detectors. *Computer Graphics and Image Processing*, **4**, 180–185.

Egbert, D.D. and Ulaby, F.T., 1972, Effect of angles on reflectivity. *Photogrammetric Engineering*, **29**, 556–564.

Eghbali, H.J., 1979, A K-S test for detecting changes from Landsat imagery data. *IEEE Transactions on Systems, Man and Cybernetics*, **9**, 17–23.

Ehleringer, J.R. and Field, C.B. (eds) (1991), *Scaling Physiological Processes – Leaf to Globe*. San Diego: Academic Press.

Elachi, C., 1987, *Introduction to the Physics and Techniques of Remote Sensing*. New York: Wiley.

Eliason, E.M. and McEwen, A.S., 1990, Adaptive box filters for the removal of random noise from digital images. *Photogrammetric Engineering and Remote Sensing*, **56**, 453–458.

Elvidge, C.D. and Lyon, R.J.P., 1985, Influence of rock–soil spectral variation on the assessment of green biomass. *Remote Sensing of Environment*, **17**, 265–279.

Elvidge, C.D., Yuan, D., Weerackoon, R.D. and Lunetta, R.S., 1995, Relative radiometric normalisation of Landsat Multispectral Scanner (MSS) data using an automatic scattergram-controlled regression. *Photogrammetric Engineering and Remote Sensing*, **61**, 1255–1260.

Evans, D.L., Plant, J.J. and Stofan, E.R., 1997, Overview of the Spaceborne Imaging Radar-C/X-Band Synthetic Aperture Radar (SIR-C/X-SAR) missions. *Remote Sensing of Environment*, **59**, 135–140.

Farag, A.A., 1992, Edge-based image segmentation. *Remote Sensing Reviews*, **6**, 95–122.

Ferrari, M.C., 1992, Improved decorrelation stretching of TM data for geological applications: first results in Northern Somalia. *International Journal of Remote Sensing*, **13**, 841–851.

Filella, I. and Penuelas, J., 1994, The red edge position and shape as indicators of plant chlorophyll content, biomass and hydric status. *International Journal of Remote Sensing*, **15**, 1459–1470.

Fioravanti, S., 1994, Multifractals: theory and application to image texture recognition. In Wilkinson, G.G., Kanellopoulos, I. and Mégier, J. (eds) (1994), *Fractals in Geoscience and Remote Sensing: Proceedings of a Joint JRC/EARSeL Expert Meeting*, Ispra, Italy, 14–15 April 1994. Report EUR 16092. Luxembourg: Office for Official Publications of the European Communities, 152–175.

Fisher, P., 1997, The pixel: a snare and a delusion. *International Journal of Remote Sensing*, **18**, 679–685.

Fitzgerald, R.W. and Lees, B.G., 1994, Assessing the classification accuracy of multisource remote sensing data. *Remote*

Sensing of Environment, **47**, 362–368.

Flygare, A.-M., 1997, A comparison of contextual classification methods using Landsat TM. *International Journal of Remote Sensing*, **18**, 3835–3842.

Foley, J.D., van Dam, A., Feiner, S.K. and Hughes, J.F., 1990, *Computer Graphics – Principles and Practice, Second Edition in C*. Reading, Mass.: Addison-Wesley.

Foley, J.D., van Dam, A., Feiner, S.K., Hughes, J.F. and Phillips, R.L., 1994, *Introduction to Computer Graphics*. Reading, Mass.: Addison-Wesley.

Foody, G.M., 1995, Using prior knowledge in artificial neural net classification with a minimal training set. *International Journal of Remote Sensing*, **16**, 301–312.

Foody, G.M., 1996, Approaches for the production and evaluation of fuzzy land cover classification from remotely-sensed data. *International Journal of Remote Sensing*, **17**, 1317–1340.

Foody, G.M. and Arora, M.K., 1996, Fuzzy thematic mapping: incorporating mixed pixels in the training, allocation and testing stages of supervised image classification, In Binaghi. E. *et al.* (eds) (1996), 43–52.

Foody, G.M., McCullagh, M.B. and Yates, W.B., 1995, The effect of training set size and composition on artificial neural net classification. *International Journal of Remote Sensing*, **16**, 1707–1723.

Foody, G.M., Lucas, R.M., Curran, P.J. and Honzak, M., 1996, Estimation of the areal extent of land-cover classes that only occur at the sub-pixel level. *Canadian Journal of Remote Sensing*, **22**, 428–432.

Foody, G.M, Lucas, R.M., Curran, P.J. and Honzak, M., 1997, Non-linear mixture modelling without end members using an artificial neural network. *International Journal of Remote Sensing*, **18**, 937–953.

Forshaw, M.R.B., Haskell, A., Miller, P.F., Stanley, D.J. and Townshend, J.R.G., 1983, Spatial resolution of remotely-sensed imagery: a review paper. *International Journal of Remote Sensing*, **4**, 497–520.

Frank, T.D., 1985, Differentiating semiarid environments using Landsat reflectance data. *Professional Geographer*, **37**, 36–46.

Franklin, S.E. and Peddle, D.R., 1987, Texture analysis of digital image data using spatial co-occurrence. *Computers and Geosciences*, **13**, 293–311.

Frankot, R.T. and Chellappa, R., 1987, Lognormal random-field models and their applications to radar image synthesis. *IEEE Transactions on Geoscience and Remote Sensing*, **25**, 195–207.

Freeman, A., Villasenor, J., Klein, J.D., Hoogeboom, P. and Groot, J., 1994, On the use of multi-frequency and polarimetric radar backscatter features for classification of agricultural crops. *International Journal of Remote Sensing*, **15**, 1799–1812.

Frei, U., Graf, K.Chr. and Meier, E., 1993, Cartographic reference systems. In Schreier, G. (ed.), (1993a), 213–234.

Frulla, L.A., Milovich, J.A. and Gagliardini, D.A., 1995, Illumination and observational geometry for NOAA-AVHRR imagery. *International Journal of Remote Sensing*, **16**, 2233–2253.

Fusco, L. and Trevese, D., 1985, On the reconstruction of lost data in images of more than one band. *International Journal of Remote Sensing*, **6**, 1535–1544.

Gamon, J.A., Field, C.B., Roberts, D.A., Ustin, S.L. and Valentini, R., 1993, Functional patterns in an annual grassland during an AVIRIS overflight. *Remote Sensing of Environment*, **44**, 239–253.

García-Haro, F.J., Gilabert, M.A. and Melia, J., 1996, Linear spectral mixture modelling to estimate vegetation amount from optical spectral data. *International Journal of Remote Sensing*, **17**, 3373–3400.

Gardner, B.P., Blad, B.L., Thompson, D.R. and Henderson, K., 1985, Evaluation and interpretation of Thematic Mapper ratios for estimating corn growth parameters. *Remote Sensing of Environment*, **18**, 225–234.

Gauthier, Y., Bernier, M. and Fortin, J.-P., 1998, Aspect and incidence angle sensitivity in ERS-1 SAR data. *International Journal of Remote Sensing*, **19**, 2001–2006.

Gens, R. and van Genderen, J.L., 1996, SAR interferometry – issues, techniques, applications. *International Journal of Remote Sensing*, **17**, 1803–1835.

German, G.W.H. and Gahegan, M.N., 1996, Neural network architectures for the classification of temporal image sequences. *Computers and Geosciences*, **22**, 969–979.

Gillespie, A.R., Kahle, A.B. and Walker, R.E., 1986, Color enhancement of highly correlated images. I: Decorrelation and HSI contrast stretches. *Remote Sensing of Environment*, **20**, 209–235.

Goetz, A.F.H., 1984, High spectral resolution remote sensing of the land. *Proceedings of the Society of Photo-Optical Instrumentation Engineers (SPIE)*, **268**, 56–68.

Goetz, A.F.H., 1989, Spectral remote sensing in geology. In Asrar, G. (ed.) (1989), 491–526.

Goetz, A.F.H. and Rowan, L.C., 1981, Geologic remote sensing. *Science*, **211**, 781–791.

Goetz, A.F.H., Rock B.N. and Rowan, L.C., 1983, Remote sensing for exploration: an overview. *Economic Geology*, **78**, 573–589.

Goetz, A.F.H., Vane, G., Solomon, J.E. and Rock, B.N., 1985, Imaging spectroscopy for earth remote sensing. *Science*, **228**, 1147–1153.

Gong, P. and Howarth, P.J., 1990, An assessment of some factors influencing multispectral land-cover classification. *Photogrammetric Engineering and Remote Sensing*, **56**, 597–603.

Gong, P., Ledrew, E.F. and Miller, J.R., 1992, Registration noise reduction in difference images for change detection. *International Journal of Remote Sensing*, **13**, 773–739.

Gong, P., Pu, R. and Chen, J., 1996, Mapping ecological land systems and classification uncertainties from digital elevation and forest-cover data using neural networks. *Photogrammetric Engineering and Remote Sensing*, **62**, 1249–1260.

Gonzales, R.C. and Woods, R.E., 1992, *Digital Image Processing*. Reading, Mass.: Addison-Wesley.

Goodenough, D.G., Narendra, P.M. and O'Neill, K., 1978, Feature subset selection in remote sensing. *Canadian Journal of Remote Sensing*, **4**, 143–148.

Gopal, S. and Woodcock, C., 1994, Theory and methods for accuracy assessment of thematic maps using fuzzy sets. *Photogrammetric Engineering and Remote Sensing*, **60**, 181–188.

Gopal, S. and Woodcock, C., 1996, Remote sensing of forest change using artificial neural networks. *IEEE Transactions on Geoscience and Remote Sensing*, **34**, 398–403.

Green, A.A., Berman, M., Switzer, P. and Craig, M.D., 1988, A transformation for ordering multispectral data in terms of image quality with implications for noise removal. *IEEE Transactions on Geoscience and Remote Sensing*, **26**, 65–74.

Green, W.B., 1983, *Digital Image Processing: A Systems Approach*. New York: Van Nostrand Reinhold.

Greenfield, S., 1997, *The Human Brain: A Guided Tour*. London: Weidenfeld and Nicolson.

Grossman, Y.L., Ustin, S.L., Jacquemond, S., Sanderson, E.W., Schmuck, G. and Verdebout, J., 1996, Critique of stepwise multiple linear regression for the extraction of leaf biochemistry information from leaf reflectance data. *Remote Sensing of Environment*, **56**, 182–193.

Guo, L.J. and Moore, J.McM., 1996, Direct decorrelation stretch technique for RGB colour composition. *International Journal of Remote Sensing*, **17**, 1005–1018.

Gupta, R.P., 1991, *Remote Sensing in Geology*. Berlin: Springer-Verlag.

Gurney, C.M., 1980, Threshold selection for line detection algorithms. *IEEE Transactions on Geoscience and Remote Sensing*, **18**, 204–211.

Gurney, C.M. and Townshend, J.R.G., 1983, The use of contextual information in the classification of remotely sensed data. *Photogrammetric Engineering and Remote Sensing*, **49**, 55–64.

Gutman, G. and Ignatov, A., 1995, Global land monitoring from AVHRR: potential and limitations. *International Journal of Remote Sensing*, **16**, 2301–2309.

Guyot, G. and Gu, X.-F., 1994, Effect of radiometric corrections on NDVI determined from SPOT-HRV and Landsat-TM data. *Remote Sensing of Environment*, **49**, 169–180.

Hall, F.G., Strebel, D.E., Nickeson, J.R. and Goetz, S.J., 1991, Radiometric rectification: toward a common radiometric response among multidate, multisensor images. *Remote Sensing of Environment*, **35**, 11–27.

Hansen, M., Dubayah, R. and Defries, R., 1996, Classification trees: an alternative to traditional land cover classifiers. *International Journal of Remote Sensing*, **17**, 1075–1081.

Haralick, R.M., 1980, Edge and region analysis of digital image data. *Computer Graphics and Image Processing*, **12**, 50–73.

Haralick, R.M. and Fu, K.-S., 1983, Pattern recognition and classification. In Colwell, R.N. (ed.) (1983), 793–805.

Haralick, R.M. and Shanmugam, K.S., 1974, Combined spectral and spatial processing of ERTS imagery data. *Remote Sensing of Environment*, **3**, 3–13.

Haralick, R.M., Shanmugam, K.S. and Dinstein, I., 1973, Textural features for image classification. *IEEE Transactions on Systems, Man and Cybernetics*, **3**, 610–622.

Harman, H.H., 1967, *Modern Factor Analysis*, 2nd edn.

Chicago: Chicago University Press.

Harries, J., Llewellyn-Jones, D.T., Mutlow, C., Murray, M.J., Barton, I.J. and Prata, A.J., 1994, The ATSR programme: instruments, data and science. In Mather, P.M. (ed.) (1994), 20–28.

Harris, J.R., Murray, R. and Hirose, T., 1990, IHS transform for the integration of radar imagery and other remotely sensed data. *Photogrammetric Engineering and Remote Sensing*, **56**, 1631–1641.

Harris, R., 1981, An experiment in probabilistic relaxation for terrain cover classification of Kuwait from Landsat imagery. *Proceedings of the 15th International Conference on Remote Sensing of Environment*. Ann Arbor, Mich.: Environmental Research Institute of Michigan, 1245–1252.

Harris, R., 1985, Contextual classification post-processing of Landsat data using a probabilistic relaxation model. *International Journal of Remote Sensing*, **6**, 847–866.

Hartigan, J.A., 1975, *Clustering Algorithms*. New York: Wiley.

Hay, J.C. and Mackay, D.C., 1985, Estimating solar irradiances on inclined surfaces: a review and assessment of methodologies. *International Journal of Solar Energy*, **3**, 203–240.

Healy, M.J.R., 1968, Multivariate normal plotting. *Applied Statistics*, **17**, 157–161.

Hearn, D. and Baker, M.P., 1994, *Computer Graphics*, 2nd edn. London: Prentice-Hall International.

Henebry, G.M., 1997, Advantages of principal components analysis for land cover segmentation from SAR image series. *Proceedings of the 3rd ERS Symposium*, Florence, Italy, 18–21 March. Available via WWW from:
`http://earthnet.esrin.esa.it/pub/florence/`
`program-details/data/henebry3/index.html`

Hepner, G.F., Logan, T., Ritter, N. and Bryant, N., 1990, Artificial neural network classification using a minimal training set: comparison to conventional supervised classification. *Photogrammetric Engineering and Remote Sensing*, **56**, 469–473.

Hill, I.D., 1973, The Normal integral. Algorithm AS66. *Applied Statistics*, **22**, 424.

Hill, J., 1991, A quantitative approach to remote sensing: sensor calibration and comparison. In Belward, A.S. and Valenzuela, C.R. (eds) (1991), 97–110.

Hill, J., 1993–4, Monitoring land degradation and soil erosion in Mediterranean environments. *ITC Journal*, **1993–4**, 323–331.

Hill, J. and Aifadopoulou, D., 1990, Comparative analysis of Landsat-5 TM and SPOT HRV-1 data for use in multiple sensor approaches. *Remote Sensing of Environment*, **34**, 55–70.

Hill, J. and Horstert, P.O., 1996, Monitoring the growth of a Mediterranean metropolis based on the analysis of spectral mixtures – a case study on Athens. In Parlow, E. (ed.) (1996a), 21–31.

Hill, J., Mehl, W. and Radeloff, V., 1995, Improved terrain mapping by combining correction of atmospheric and topographic effects in Landsat TM imagery. In Askne, J. (ed.) (1995), 143–151.

Hobbs, R.J. and Mooney, H.A. (eds), 1990, *Remote Sensing of*

Biospheric Functioning. New York: Springer-Verlag.

Hoffer, R.M., 1978, Biological and physical considerations in applying computer-aided analysis techniques to remote sensor data. In Swain, P.H. and Davis, S.M. (eds) (1978), *Remote Sensing – the Quantitative Approach*. New York: McGraw-Hill, chapter 5.

Holben, B.N. and Fraser, R.S., 1984, Red and near-infrared sensor response to off-nadir viewing. *International Journal of Remote Sensing*, **5**, 145–160.

Holben, B.N. and Kimes, D., 1986, Directional reflectance response in AVHRR red and near-infrared bands for three cover types and varying atmospheric conditions. *Remote Sensing of Environment*, **19**, 213–226.

Holm, R.G., Moran, M.S., Jackson, R.D., Slater, P.N., Yuan, B. and Biggar, S.F., 1989, Surface reflectance factor retrieval from Thematic Mapper data. *Remote Sensing of Environment*, **27**, 47–57.

Horn, B.K.P. and Woodham, R.J., 1979, Destriping Landsat MSS images by histogram modification. *Computer Graphics and Image Processing*, **10**, 69–83.

Huang, T.S., Yang, G.J. and Tang, G.Y., 1979, A fast two-dimensional median filtering algorithm. *IEEE Transactions on Acoustics, Speech and Signal Processing*, **27**, 13–18.

Huete, A., 1989, Soil influences in remotely sensed vegetation-canopy spectra. In Asrar, G. (ed.) (1989), 107–141.

Hung, C.C., 1993, Competitive learning networks for unsupervised training. *International Journal of Remote Sensing*, **14**, 2411–2415.

Hunt, G.R., 1977, Spectral signatures of particulate minerals in the visible and near-infrared. *Geophysics*, **42**, 501–513.

Hunt, G.R., 1979, Near-infrared (1.3–2.4 μm) spectra of alteration minerals: potential for use in remote sensing. *Geophysics*, **44**, 1974–1986.

Hunt, G.R. and Ashley, R.P., 1979, Spectra of altered rocks in the visible and near infrared. *Economic Geology*, **74**, 1613–1629.

Hunt, G.R. and Salisbury, J.W., 1970, Visible and near-infrared spectra of minerals and rocks: I. Silicate minerals. *Modern Geology*, **1**, 283–300.

Hunt, G.R. and Salisbury, J.W., 1971, Visible and near-infrared spectra of minerals and rocks: II. Carbonates. *Modern Geology*, **2**, 23–30.

Hunt, G.R., Salisbury, J.W. and Lenhoff, D.J., 1971, Visible and near-infrared spectra of minerals and rocks: III. Oxides and hydroxides. *Modern Geology*, **2**, 195–205.

Hunter, G.J. and Goodchild, M.F., 1997, Modeling the uncertainty of slope and aspect estimates derived from spatial databases. *Geographical Analysis*, **29**, 35–49.

Hussein, H.A., Rabie, S.I. and Abdel Nabie, S., 1996, Structural interpretation of aeromagnetic data, Gabel Zubeir area, North Eastern Desert, Egypt. *International Journal of Remote Sensing*, **17**, 1997–2012.

Hutchinson, C.F., 1982, Techniques for combining Landsat and ancillary data for digital classification improvement. *Photogrammetric Engineering and Remote Sensing*, **48**, 123–130.

Hyman, A.H. and Barnsley, M.J., 1997, On the potential for land cover mapping from multiple-view-angle (MVA) remotely sensed images. *Proceedings of the 23rd Annual Conference and Exhibition of the Remote Sensing Society*, University of Reading, 2–4 September 1997. Nottingham: The Remote Sensing Society, 135–140.

Hyppänen, H., 1996, Spatial autocorrelation and optimum spatial resolution of optical remote sensing data in boreal forest environment. *International Journal of Remote Sensing*, **17**, 3441–3452.

Ichoku, C. and Karnieli, A., 1996, A review of mixture modeling techniques for sub-pixel land cover estimation. *Remote Sensing Reviews*, **13**, 161–186.

Ichoku, C., Karnieli, A., Meisels, A. and Chorowicz, J., 1996, Detection of drainage channel networks on satellite imagery. *International Journal of Remote Sensing*, **17**, 1659–1678.

Imbrie, J., 1963, *Factor and Vector Analysis Programs for Analyzing Geologic Data*. Technical Report, 6, Office of Naval Research Geography Branch, Task No. 389-135, Contract No. 1228 (26), Northwestern University, Evanston, Ill., October 1963.

Irons, J.R. and Petersen, G.W., 1981, Texture transforms of remotely sensed data. *Remote Sensing of Environment*, **11**, 359–370.

Irons, J.R., Weismiller, R.A. and Petersen, G.W., 1989, Soil reflectance. In Asrar, G. (ed.) (1989), 66–106.

Itten, K.I. and Meyer, P., 1993, Geometric and radiometric correction of TM data of mountainous forested areas. *IEEE Transactions on Geoscience and Remote Sensing*, **31**, 764–770.

Jackson, R.D., 1983, Spectral indices in *n*-space. *Remote Sensing of Environment*, **13**, 409–421.

Jackson, R.D., Slater, P.N. and Pinter, P.J., Jr, 1983, Discrimination of growth and water stress in wheat by various vegetation indices through clear and turbid atmospheres. *Remote Sensing of Environment*, **13**, 187–208.

Jarvis, C.H. and Stuart, N., 1996, The sensitivity of a neural net for classifying remotely sensed imagery. *Computers and Geosciences*, **22**, 959–967.

Jensen, J.R., 1986, *Introductory Digital Image Processing – A Remote Sensing Perspective*. Englewood Cliffs, NJ: Prentice-Hall.

Jensen, J.R. and Toll, D.L., 1982, Detecting residential land-use development at the urban fringe. *Photogrammetric Engineering and Remote Sensing*, **48**, 629–643.

Johnsen, H., Lauknes, L. and Guneriussen, T., 1995, Geocoding of fast-delivery SAR image mode products using DEM data. *International Journal of Remote Sensing*, **16**, 1957–1968.

Johnson, R.D. and Kasischke, E.S., 1998, Change vector analysis: a technique for the multispectral monitoring of land cover and condition. *International Journal of Remote Sensing*, **19**, 411–426.

Jones, A.R., Settle, J.J. and Wyatt, B.K., 1988, Use of digital terrain data in the interpretation of SPOT-1 HRV data. *International Journal of Remote Sensing*, **9**, 669–682.

Joseph, G., 1996, Imaging sensors for remote sensing. *Remote Sensing Reviews*, **13**, 257–342.

Jupp, D.L.B. and Mayo, K.K., 1982, The use of residual images in Landsat image analysis. *Photogrammetric Engineering and Remote Sensing*, **48**, 595–604.

Jupp, D.L.B., Strahler, A.H. and Woodcock, C.E., 1988, Autocorrelation and regularisation in digital images. I: Basic theory. *IEEE Transactions on Geoscience and Remote Sensing*, **26**, 463–473.

Jupp, D.L.B., Strahler, A.H. and Woodcock, C.E., 1989, Autocorrelation and regularisation in digital images. II: Simple image models. *IEEE Transactions on Geoscience and Remote Sensing*, **27**, 247–258.

Justice, C.O., Townshend, J.R.G., Holben, B.N. and Tucker, C.J., 1985, Analysis of the phenology of global vegetation using meteorological satellites. *International Journal of Remote Sensing*, **6**, 1271–1318.

Jutz, S.L. and Chorowicz, J., 1993, Geological mapping and detection of oblique extensional structures in the Kenyan Rift Valley with a SPOT/Landsat data merge. *International Journal of Remote Sensing*, **14**, 1677–1688.

Kahle, A.B. and Rowan, L.C., 1980, Evaluation of multispectral infrared aircraft images for lithological mapping in the East Tintic Mountains, Utah. *Geology*, **8**, 234–239.

Kailath, T., 1967, The divergence and Bhattacharyya distance measures in feature selection. *IEEE Transactions on Communication Theory*, **15**, 52–60.

Kalkhan, M.A., Reich, R.M. and Czaplewski, R.L., 1997, Variance estimates and confidence intervals for the kappa measure of classification accuracy. *Canadian Journal of Remote Sensing*, **23**, 210–216.

Kanal, L., 1974, Patterns in pattern recognition: 1968–1974. *IEEE Transactions on Information Theory*, **20**, 697–722.

Kaneko, T., 1976, Evaluation of Landsat image registration accuracy. *Photogrammetric Engineering and Remote Sensing*, **42**, 1285–1299.

Kanellopoulos, I., Varas, A., Wilkinson, G.G. and Mégier, J., 1992, Land-cover discrimination in SPOT HRV imagery using an artificial neural network – a 20-class experiment. *International Journal of Remote Sensing*, **13**, 917–924.

Kanellopoulos, I., Wilkinson, G.G., Roli, F. and Austin, J. (eds) (1997), *Neurocomputation in Remote Sensing Data Analysis*. Heidelberg: Springer-Verlag.

Kardoulas, N.G., Bird, A.C. and Lawan, A.I., 1996, Geometric correction of SPOT and Landsat imagery: a comparison of map and GPS-derived control points. *Photogrammetric Engineering and Remote Sensing*, **62**, 1173–1177.

Katawa, Y., Ueno, S. and Kusaka, T., 1988, Radiometric correction for atmospheric and topographic effects on Landsat MSS images. *International Journal of Remote Sensing*, **9**, 729–748.

Kaufmann, H., Berger, M., Meissner, M. and Süssenguth, G., 1996, MOMS-02/STS-55 design and validation of spectral and panchromatic modules. *Proceedings of the MOMS Symposium*, Cologne, Germany, 5–7 July 1995. Paris: European Association of Remote Sensing Laboratories (EARSeL), 26–38.

Kauth, R.J. and Thomas, G., 1976, The tasselled cap – a graphic description of the spectral–temporal development of agricultural crops as seen by Landsat. *Proceedings of the Symposium on Machine Processing of Remotely-Sensed Data* 1976, Purdue University, West Lafayette, Ind., vol. 4B, 41–51.

Keller, J.M., Chen, S. and Crownover, R.M., 1989, Texture description and segmentation through fractal geometry. *Computer Graphics, Vision and Image Processing*, **45**, 10–166.

Kerdiles, H. and Grondona, M.O., 1995, NOAA-AVHRR NDVI decomposition and sub-pixel classification using linear mixing in the Argentine Pampa. *International Journal of Remote Sensing*, **16**, 1303–1325.

Kess, B.L., Steinwand, D.R. and Reichenbach, S.E., 1996, Compression of Global Land 1-km AVHRR dataset. *International Journal of Remote Sensing*, **17**, 2955–2969.

Key, J.R., Maslanik, J.A. and Barry, R.G., 1989, Cloud classification from satellite data using a fuzzy sets algorithm: a polar example. *International Journal of Remote Sensing*, **10**, 1823–1842.

Khan, B., Hayes, L.W.B. and Cracknell, A.P., 1995, The effects of higher-order resampling on AVHRR data. *International Journal of Remote Sensing*, **16**, 147–163.

Kim, H. and Swain, P.H., 1995, Evidential reasoning approach to multisource data classification in remote sensing. *Proceedings of the International Geoscience and Remote Sensing Symposium (IGARSS'89)*, Vancouver, British Columbia, Canada, 10–14 July 1989. New York: IEEE Press, vol. 2, 829–832.

Kimble, G., 1951, The inadequacy of the regional concept. In Dudley Stamp, L. and Wooldridge, S.W (eds) (1951), *London Essays in Geography: Rodwell Jones Memorial Volume*. London: Longmans Green, 151–174.

Kingsley, S. and Quegan, S., 1992, *Understanding Radar Systems*. London: McGraw-Hill.

Kittler, J., 1983, Image processing for remote sensing. *Philosophical Transactions of the Royal Society of London*, **309A**, 323–335.

Kittler, J. and Föglein, J., 1984, Contextual classification of multispectral pixel data. *Image and Vision Computing*, **2**, 13–29.

Kneizys, F.X., Shettle, E.P., Gallery, W.O., Chetwynd, J.H., Abreu, L.W., Selby, J.E.A., Clough, S.A. and Fenn, R.W., 1983, *Atmospheric Transmittance/Radiance: Computer Code LOWTRAN-6*. US Air Force Geophysics Laboratory, AFGL-TR-83-0187.

Koch, M. and Mather, P.M., 1997, Lineament mapping for groundwater resource assessment: a comparison of digital Synthetic Aperture Radar imagery and stereoscopic Large Format Camera photographs in the Red Sea Hills, Sudan. *International Journal of Remote Sensing*, **18**, 1465–1482.

Kohl, H.G. and Hill, J., 1988, Geometric registration of multitemporal TM data over mountainous areas by use of a low resolution digital elevation model. *Proceedings of the 8th EARSeL Symposium on Alpine and Mediterranean Areas: A Challenge for Remote Sensing*, Capri, Italy, 17–20 May 1998. Paris: European Association of Remote Sensing Laboratories (EARSel), 323–355.

Kontoes, C.C. and Rokos, D., 1996, The integration of spatial context information in an experimental knowledge-based system and the supervised relaxation algorithm – two successful approaches to improving SPOT-XS classification. *International Journal of Remote Sensing*, **17**, 3093–3106.

Kowalik, W.S., Lyon, R.J.P. and Switzer, P., 1983, The effects of additive radiance terms on ratios of Landsat data. *Photogrammetric Engineering and Remote Sensing*, **49**, 659–669.

Kramer, H.J., 1994, *Observation of the Earth and its Environment – Survey of Missions and Sensors*. Berlin: Springer-Verlag.

Krishnamurthy, J., Manalavan, P. and Saivasan, V., 1992, Application of digital enhancement techniques for groundwater exploration in hard rock terrains. *International Journal of Remote Sensing*, **13**, 2925–2942.

Kropatsch, W.G. and Strobl, D., 1990, The generation of SAR layover and shadow maps from digital elevation models. *IEEE Transactions on Geoscience and Remote Sensing*, **28**, 90–107.

Kruse, F.A., Lefkoff, A.B., Boardman, J.W., Heidebrecht, K.B., Shapiro, A.T., Barloon, P.J. and Goetz, A.F.H., 1993, The Spectral Image Processing System (SIPS) – Interactive visualization and analysis of imaging spectrometer data. *Remote Sensing of Environment*, **44**, 145–163.

Kumar, R., 1979, Comparison of feature selection techniques for Earth resources data. In Gardner, W.E. (ed.) (1979), *Proceedings of an International Conference on Applications of Machine-Aided Image Analysis*, Oxford, 1979, Bristol: Institute of Physics, Conference Series, vol. 44, 238–243.

Kwok, R., Curlander, J.C. and Pang, S.S., 1987, Rectification of terrain-induced distortions in radar images. *Photogrammetric Engineering and Remote Sensing*, **53**, 507–513.

Labovitz, M.L. and Matsuoko, E.J., 1984, The influence of autocorrelation on signature extraction – an example from a geobotanical investigation of Cotter Basin, Montana. *International Journal of Remote Sensing*, **5**, 315–332.

Labovitz, M.L., Toll, D.L. and Kennard, R.E., 1982, Preliminary evidence for the influence of physiography and scale upon the autocorrelation function of remotely sensed data. *International Journal of Remote Sensing*, **3**, 13–30.

Lambin, E.F., 1996, Change detection at multiple temporal scales: seasonal and annual variations in landscape variables. *Photogrammetric Engineering and Remote Sensing*, **62**, 931–938.

Lambin, E.F. and Strahler, A.H., 1994, Indicators for land cover change for change-vector analysis in multitemporal space at coarse spatial scales. *International Journal of Remote Sensing*, **15**, 2099–2119.

Land, E.H., 1977, The retinex theory of color vision. *Scientific American*, December 1977, 108–128.

Landgrebe, D.A., 1980, The development of a spectral–spatial classifier for Earth observation data. *Pattern Recognition*, **12**, 165–175.

Landgrebe, D.A. and staff, 1975, *A Study of the Utilisation of ERTS-1 Data from the Wabash River Basin*. Laboratory for the Applications of Remote Sensing, Purdue University, West Lafayette, Ind., Information Note 052375.

Lark, R.M., 1995, A reappraisal of unsupervised classification,

I: Correspondence between spectral and conceptual classes. *International Journal of Remote Sensing*, **16**, 1425–1443.

Lark, R.M., 1996, Geostatistical description of texture on an aerial photograph for discriminating classes of land cover. *International Journal of Remote Sensing*, **17**, 2115–2133.

Lark, R.M., 1998, Forming spatially-coherent regions by classification of multi-variate data: an example from the analysis of maps of crop yield. *International Journal of Remote Sensing*, **19**, 83–98.

Lawrence, R.L. and Ripple, W.J., 1996, Determining patch perimeters in raster image processing and geographic information systems. *International Journal of Remote Sensing*, **17**, 1255–1259.

Lawson, C.L. and Hansen, R.J. (eds), 1995, *Solving Least Squares Problems*. Englewood Cliffs, NJ: Prentice Hall.

Leachtenauer, J.C., 1977, Optical power spectrum analysis: scale and resolution effects. *Photogrammetric Engineering and Remote Sensing*, **43**, 1117–1125.

Leberl, F.W., 1990, *Radargrammetric Image Processing*. Dedham, Mass.: Artech House.

Le Cun, Y., Denker, J.S. and Solla, S.A., 1990, Optimal brain damage. In Touretsky, D.S. (ed.) (1990), *Advances in Neural Information*. San Mateo, Calif.: Morgan Kaufmann, 598–605.

Lee, C. and Landgrebe, P.A., 1993, Decision boundary feature extraction for non-parametric classifiers. *IEEE Transactions on Systems, Man, and Cybernetics*, **23**, 433–444.

Lee, J.S., 1983a, Digital image smoothing and the sigma filter. *Computer Graphics, Vision and Image Processing*, **17**, 24–32.

Lee, J.S., 1983b, A simple speckle smoothing algorithm for synthetic aperture radar images. *IEEE Transactions on Systems, Man and Cybernetics*, **13**, 85–89.

Lee, J.S., Jurkevich, I., Dewaele, P., Wambacq, P. and Oosterlinck, M., 1994, Speckle filtering of synthetic aperture radar images: a review. *Remote Sensing Reviews*, **8**, 313–340.

Lee, T. and Richards, J.A., 1985, A low-cost classifier for multitemporal applications. *International Journal of Remote Sensing*, **6**, 1405–1418.

Lee, T., Richards, J.A. and Swain, P.H., 1987, Probabilistic and evidential approaches for multisource data analysis. *IEEE Transactions on Geoscience and Remote Sensing*, **25**, 283–293.

Lei, Q., Henkel, J., Frei, M., Mehl, H., Lörchner, G. and Bodechtel, J., 1996, Radiometric noise correction of panchromatic high resolution data of MOMS-02. *Proceedings of the MOMS Symposium*, Cologne, Germany, 5–7 July 1995. Paris: European Association of Remote Sensing Laboratories (EARSeL), 303–313.

Leprieur, C., Kerr, Y.H. and Pichon, J.M., 1996, Critical assessment of vegetation indices from AVHRR in a semi-arid environment. *International Journal of Remote Sensing*, **17**, 2549–2563.

Levin, S.A., 1991, Concepts of scale at the local level. In Ehleringer, J.R. and Field, C.B. (eds) (1991), 7–19.

Li, M., Daels, L. and Antrop, M., 1996, Lambertian and Minnaert relation simulation for topographic normalization. *Proceedings of the 11th Thematic Conference and Workshops on Applied Geologic Remote Sensing*, Las Vegas, Nevada,

27–29 February 1996. Ann Arbor, Mich.: Environmental Research Institute of Michigan, vol. 2, 133–141.

Li, W.-H., Weeks, R. and Gillespie, A.R., 1998, Multiple scattering in the remote sensing of natural surfaces. *International Journal of Remote Sensing*, **19**, 1725–1740.

Li, Z., 1993, Theoretical models of the accuracy of Digital Terrain Models: an evaluation and some observations. *Photogrammetric Record*, **14**, 651–660.

Lillesand, T.M. and Keifer, R.W., 1994, *Remote Sensing and Image Interpretation*, 3rd edn. New York: Wiley.

Lobo, A., Chic, O. and Casterad, A., 1996, Classification of Mediterranean crops with multi-sensor data: per-pixel versus per-object statistics and image segmentation. *International Journal of Remote Sensing*, **17**, 2385–2400.

Logan, T.L. and Strahler, A.H., 1983, Optimal Landsat transforms for forest applications. *Proceedings of the Symposium on Machine Processing of Remotely Sensed Data* 1983, Purdue University, West Lafayette, Ind., 146–153.

Lopes, A., Nezry, E., Touzi, R. and Laur, H., 1993, Structure detection and statistical adaptive speckle filtering in SAR images. *International Journal of Remote Sensing*, **14**, 1735–1758.

Lowitz, G.E., 1978, Stability and dimensionality of Karhunen-Loeve multispectral image expansions. *Pattern Recognition*, **10**, 359–363.

Lowman, P.D., Jr, 1994, Radar geology of the Canadian Shield: a 10-year review. *Canadian Journal of Remote Sensing*, **20**, 198–209.

Lynn, P.A., 1982, *An Introduction to the Analysis and Processing of Signals*, 2nd edn. London: Macmillan.

McCullagh, M.J. and Davis, J.C., 1972, Optical analysis of two-dimensional patterns. *Annals of the Association of American Geographers*, **62**, 561–577.

McDonnell, M.J., 1981, Box filtering techniques. *Computer Graphics and Image Processing*, **17**, 65–70.

MacFarlane, N. and Thomas, M.H.B., 1984, Speckle reduction algorithms and their application to SAR imagery. *Proceedings of the Tenth Anniversary International Conference of The Remote Sensing Society: Satellite Remote Sensing – Review and Preview*. Nottingham: The Remote Sensing Society, 391–398.

Maguire, D.J., Goodchild, M.F. and Rhind, D.W. (eds), 1991, *Geographic Information Systems – Principles and Applications*. Harlow, Essex: Longman Scientific and Technical.

Mannan, B., Roy, J. and Ray, A.K., 1998, Fuzzy ARTMAP supervised classification of multi-spectral remotely-sensed images. *International Journal of Remote Sensing*, **19**, 767–774.

Marceau, J., Howarth, P.J., Dubois, J.M. and Gratton, D.J., 1990, Evaluation of the grey-level co-occurrence matrix method for land cover classification using SPOT imagery. *IEEE Transactions on Geoscience and Remote Sensing*, **28**, 513–519.

Markham, B.L. and Barker, J.L., 1987, Thematic Mapper bandpass solar exoatmospheric irradiances. *International Journal of Remote Sensing*, **8**, 517–523.

Markham, B.L., Halthore, R.N. and Goetz, S.J., 1992, Surface reflectance retrieval from satellite and aircraft sensors: result of sensor and algorithm comparisons during FIFE. *Journal of Geophysical Research*, **97**, 18785–18795.

Marr, D. and Hildreth, E.C. 1980, Theory of edge detection. *Proceedings of the Royal Society of London*, **B209**, 187–217.

Martin, F.J. and Turner, R.W., 1993, SAR speckle reduction by weighted filtering. *International Journal of Remote Sensing*, **14**, 1759–1774.

Maselli, F., Conese, C., de Filippis, T. and Norcini, S., 1995a, Estimation of forest parameters through fuzzy classification. *IEEE Transactions on Geoscience and Remote Sensing*, **33**, 77–84.

Maselli, F., Conese, C., de Filippis, T. and Romani, M., 1995b, Integration of ancillary data into a maximum likelihood classification with nonparametric priors. *ISPRS Journal of Photogrammetry and Remote Sensing*, **50**, 2–11.

Maselli, F., Rodolfi, A. and Conese, C., 1996, Fuzzy classification of spatially degraded TM data for the estimation of sub-pixel components. *International Journal of Remote Sensing*, **17**, 537–551.

Mason, D.C., Corr, D.G., Cross, A., Hoggs, D.C., Lawrence, D.H., Petrou, M. and Tailor, A.M., 1988, The use of digital map data in the segmentation and classification of remotely-sensed images. *International Journal of Remote Sensing*, **9**, 195–205.

Massonet, D., 1993, Geoscientific applications at CNES. In Schreier, G. (ed.) (1993a), 397–415.

Mather, P.M., 1976, *Computational Methods of Multivariate Analysis in Physical Geography*. Chichester: Wiley.

Mather, P.M., 1991, *Computer Applications in Geography*. Chichester: Wiley.

Mather, P.M. (ed.), 1994, *TERRA 2: Understanding the Terrestrial Environment – Remote Sensing Data Systems and Networks*. Chichester: Wiley.

Mather, P.M., 1995, Map–image registration using least-squares polynomials. *International Journal of Geographical Information Systems*, **9**, 543–554.

Mather, P.M., Tso, B. and Koch, M., 1998, An evaluation of Landsat TM spectral data and SAR-derived textural information for lithological discrimination in the Red Sea Hills, Sudan. *International Journal of Remote Sensing*, **19**, 587–604.

Maxwell, E.L., 1976, Multivariate system analysis of multispectral images. *Photogrammetric Engineering and Remote Sensing*, **42**, 1173–1186.

Meadows, P., 1995, The calibration of ERS-1 synthetic aperture radar images using UK-PAF imagery. In Askne, J. (ed.) (1995), 423–430.

Menenti, M., Azzali, S., Verhoef, W. and van Swol, R., 1993, Mapping agroecological zones and time-lag in vegetation growth by means of Fourier analysis of time series of NDVI images. *Advances in Space Research*, **13**, 233–237.

Milovich, J.A., Frulla, L.A. and Gagliardini, D.A., 1995, Environmental contribution to the atmospheric correction for Landsat-MSS images. *International Journal of Remote Sensing*, **16**, 2515–2537.

Milton, E.J., 1986, Principles of field spectroscopy and its role in remote sensing. In Bradbury, P.A. and Rollin, E.A. (eds)

(1986), *Ground Truth for Remote Sensing, Proceedings of a Remote Sensing Workshop*, Department of Geography, The University of Nottingham, May 1986. Nottingham: The Remote Sensing Society, 14–39.

Misra, P.N. and Wheeler, S.G., 1978, Crop classification with Landsat Multispectral Scanner data. *Pattern Recognition*, **10**, 1–13.

Moik, J., 1980, *Digital Processing of Remotely-Sensed Images*. Washington DC: National Aeronautics and Space Administration (NASA), Special Publication 431.

Montserud, R.A. and Leamans, R., 1992, Comparing global vegetation maps with the kappa statistic. *Ecological Modelling*, **62**, 275–293.

Moon, W.M., Li, B., Singhroy, V., So, C.S. and Yamaguchi, Y., 1994, Notes on JERS-1 SAR data characteristics for geological applications. *Canadian Journal of Remote Sensing*, **20**, 329–332.

Moore, G.K. and Waltz, F.A., 1983, Objective procedures for lineament enhancement and extraction. *Photogrammetric Engineering and Remote Sensing*, **49**, 641–647.

Morad, M., Chalmers, A.I. and O'Regan, P.R., 1996, The role of root-mean-square error in the geo-transformation of images in GIS. *International Journal of Geographical Information Systems*, **10**, 347–353.

Moran, M.S., Jackson, R.D., Hart, G.F., Slater, P.N., Bartell, R.J., Biggar, S.F., Gellman, D.I. and Santer, R.P., 1990, Obtaining surface reflectance factors from atmospheric and view angle corrected SPOT-1 HRV data. *Remote Sensing of Environment*, **32**, 203–214.

Moran, M.S., Jackson, R.D., Slater, P.N. and Teillet, P.M., 1992, Evaluation of simplified procedures for retrieval of land surface reflectance factors from satellite sensor output. *Remote Sensing of Environment*, **41**, 169–184.

Moran, M.S., Jackson, R.D., Clarke, T.R., Qi, J., Cabot, F., Thome, K.J. and Markham, B.L., 1995, Reflectance factor retrieval from Landsat TM and SPOT HRV data for bright and dark targets. *Remote Sensing of Environment*, **52**, 218–230.

Moreno, J.E. and Melia, J., 1993, A method for accurate geometric correction of NOAA AVHRR HRPT data. *IEEE Transactions on Geoscience and Remote Sensing*, **31**, 204–226.

Motrena, P. and Rebordão, J.M., 1998, Invariant models for ground control points in high-resolution images. *International Journal of Remote Sensing*, **19**, 1359–1375.

Muasher, M.J. and Landgrebe, D.A., 1984, A binary tree feature selection technique for limited training sample size. *Remote Sensing of Environment*, **16**, 183–194.

Mulder, N.J., 1980, A view on digital image processing. *ITC Journal*, **1980–3**, 452–476.

Muller, E., 1993, Evaluation and correction of angular anisotropic effects in multidate SPOT and Thematic Mapper data. *Remote Sensing of Environment*, **45**, 295–309.

Murphy, R.J. and Wadge, G., 1994, The effects of vegetation on the ability to map soils using imaging spectrometer data. *International Journal of Remote Sensing*, **15**, 63–86.

Myneni, R.B., Hall, F.G., Sellers, P.J. and Marshak, A.L., 1995, The interpretation of spectral vegetation indices. *IEEE Transactions on Geoscience and Remote Sensing*, **33**, 481–486.

Nagao, M. and Matsuyama, T., 1979, Edge preserving smoothing. *Computer Graphics and Image Processing*, **9**, 394–407.

Nalbant, S.S. and Alptekin, Ö., 1995, The use of Landsat Thematic Mapper imagery for analysing lithology and structure of Korucu–Duğla area in Western Turkey. *International Journal of Remote Sensing*, **16**, 2357–2374.

Narendra, P.M. and Fukunaga, K., 1977, A branch and bound algorithm for feature subset selection. *IEEE Transactions on Computers*, **26**, 917–921.

NASA, 1988, *Earth Observing System Instrument Panel Report*, Volume IIf, *Synthetic Aperture Radar*. Washington DC: National Aeronautics and Space Administration (NASA).

Nasrabadi, N.M., Ibikunle, J.O. and King, R.A., 1984, A new line detection technique in noisy images. *Proceedings of the Tenth Anniversary International Conference of the Remote Sensing Society*. Nottingham: Remote Sensing Society, 237–246.

Novak, K., 1992, Rectification of digital imagery. *Photogrammetric Engineering and Remote Sensing*, **58**, 339–344.

Oliver, C. and Quegan, S., 1998, *Understanding Synthetic Aperture Radar*. Boston: Artech House.

O'Leary, D.W., Friedmann, J.D. and Pohn, H.A., 1976, Lineament, linear, lineation: some proposed new standards for old terms. *Bulletin of the Geological Society of America*, **87**, 1463–1469.

Olsson, H., 1993, Regression functions for multitemporal relative calibrations of Thematic Mapper data over boreal forests. *Remote Sensing of Environment*, **46**, 89–102.

Olsson, H., 1995, Radiometric calibration of Thematic Mapper data for forest change detection. *International Journal of Remote Sensing*, **16**, 81–96.

Olsson, L. and Ekhlund, L., 1994, Fourier series for analysis of temporal sequences of satellite sensor imagery. *International Journal of Remote Sensing*, **15**, 3735–3741.

Ormsby, J., 1992, Evaluation of natural and man-made features using Landsat TM data. *International Journal of Remote Sensing*, **13**, 303–318.

Orti, F., Garcia, A. and Martin, M.A., 1979, Geometric correction of Landsat MSS images using a ground control point library. In Allan, J.A. and Harris, R. (eds) (1979), *Proceedings of the 5th Annual Conference of the Remote Sensing Society*, Durham University, 1978. Nottingham: The Remote Sensing Society, 17–26.

Overheim, R.D. and Wagner, D.L., 1982, *Light and Color*. New York: Wiley.

Palà, V. and Pons, X., 1995, Incorporation of relief in polynomial-based geometric corrections. *Photogrammetric Engineering and Remote Sensing*, **61**, 935–944.

Pan, J.-J. and Chang, C.-I., 1992, Destriping of Landsat MSS images by filtering techniques. *Photogrammetric Engineering and Remote Sensing*, **58**, 1417–1423.

Paola, J. and Schowengerdt, R.A., 1995a, A review and analysis of backpropagation neural networks for classification of remotely-sensed multi-spectral imagery. *International Journal of Remote Sensing*, **16**, 3033–3058.

Paola, J. and Schowengerdt, R.A., 1995b, A detailed comparison of backpropagation neural networks and maximum-likelihood classifiers for urban land use classification. *IEEE Transactions on Geoscience and Remote Sensing*, **33**, 981–996.

Paola, J. and Schowengerdt, R.A., 1997, The effect of neural network structure on multispectral land-use/land-cover classification. *Photogrammetric Engineering and Remote Sensing*, **63**, 535–544.

Parlow, E. (ed.), 1996a, *Progress in Environmental Remote Sensing Research and Applications, Proceedings of the 15th EARSeL Symposium*, Basle, Switzerland, 4–6 September 1996. Rotterdam: A.A. Balkema.

Parlow, E., 1996b, Correction of terrain controlled illumination effects in satellite data. In Parlow, E. (ed.) (1996a), 139–145.

Pavlidis, T., 1982, *Algorithms for Graphics and Image Processing*. Berlin: Springer-Verlag.

Peddle, D.R., 1993, An empirical comparison of evidential reasoning, linear discriminant analysis, and maximum likelihood algorithms for land cover classification. *Canadian Journal of Remote Sensing*, **19**, 31–44.

Peddle, D.R., 1995a, Knowledge foundation for supervised evidential classification. *Photogrammetric Engineering and Remote Sensing*, **61**, 409–417.

Peddle, D.R., 1995b, Mercury⊕: an evidential reasoning classifier. *Computers and Geosciences*, **21**, 1163–1176.

Peddle, D.R., Foody, G.M., Zhang, A., Franklin, S.E. and LeDrew, E.F., 1994, Multi-source image classification. II: An empirical comparison of evidential reasoning and neural network approaches. *Canadian Journal of Remote Sensing*, **20**, 396–407.

Peleg, S., 1980, A new probabilistic relaxation scheme. *IEEE Transactions on Pattern Analysis and Machine Intelligence*, **2**, 362–369

Peli, T. and Malah, D., 1982, A study of edge-detection algorithms. *Computer Graphics and Image Processing*, **20**, 1–21.

Perry, C.R. and Lautenschlager, L.F., 1984, Functional equivalence of spectral vegetation indices. *Remote Sensing of Environment*, **14**, 169–182.

Pinter, P.J., Jackson, R.D., Idso, S.B. and Reginato, R.J., 1983, Diurnal patterns of wheat spectral reflectance. *IEEE Transactions on Geoscience and Remote Sensing*, **21**, 156–163.

Pinty, B., Leprieur, C. and Verstraete, M.M., 1993, Towards a quantitative interpretation of vegetation indices. Part I: Biophysical canopy properties and classical indices. *Remote Sensing Reviews*, **7**, 127–150.

Pitas, I., 1993, *Digital Image Processing Algorithms*. Englewood Cliffs, NJ: Prentice-Hall.

Pohl, C. and van Genderen, J.L., 1998, Multisensor image fusion in remote sensing: concepts, methods and application. *International Journal of Remote Sensing*, **19**, 823–854.

Popp, T., 1995, Correcting atmospheric masking to retrieve the spectral albedo of land surfaces from satellite measurements. *International Journal of Remote Sensing*, **16**, 3843–3508.

Prasad, L. and Iyengar, S.S., 1997, *Wavelet Analysis with Applications to Image Processing*. Boca Raton, FL: CRC Press.

Pratt, W.K., 1978, *Digital Image Processing*. New York: Wiley.

Price, J.C., 1987, Calibration of satellite radiometers and the comparison of vegetation indices. *Remote Sensing of Environment*, **21**, 15–27.

Price, J.C., 1988, An update on visible and near infrared calibration of satellite instruments. *Remote Sensing of Environment*, **24**, 419–422.

Price, J.C., 1994, How unique are spectral signatures? *Remote Sensing of Environment*, **49**, 181–186.

Proy, C., Tanré, D. and Deschamps, P.Y., 1989, Evaluation of topographic effects in remotely sensed data. *Remote Sensing of Environment*, **30**, 21–32.

Quegan, S. and Rhodes, I., 1994, Relating polarimetric SAR data to surface properties – the MAC-Europe experiment. In Mather, P.M. (ed.) (1994), 159–174.

Quegan, S. and Wright, A., 1984, Automatic segmentation techniques for satellite-borne synthetic aperture radar (SAR) images. *Proceedings of the Tenth Anniversary Conference of the Remote Sensing Society*. Nottingham: Remote Sensing Society, 161–167.

Ramirez, R.W., 1985, *The Fast Fourier Transform: Fundamentals and Concepts*. Englewood Cliffs, NJ: Prentice-Hall.

Ranchin, T. and Wald, L., 1993, The wavelet transform for the analysis of remotely sensed images. *International Journal of Remote Sensing*, **14**, 615–619.

Ranson, K.J., Biehl, L.L. and Bauer, M.E., 1985, Variation in spectral response of soybeans with respect to illumination, view and canopy geometry. *International Journal of Remote Sensing*, **6**, 1827–1842.

Rao, C.R.N. and Chen, J., 1996, Post-launch calibration of the visible and near-infrared channels on the Advanced Very High Resolution Radiometer on the NOAA-14 spacecraft. *International Journal of Remote Sensing*, **17**, 2743–2747.

Rao, K.S. and Rao, Y.S., 1995, Frequency dependence of polarisation phase difference. *International Journal of Remote Sensing*, **16**, 3605–3617.

Ray, T.W. and Murray, B.C., 1996, Nonlinear spectral mixing in desert vegetation. *Remote Sensing of Environment*, **55**, 59–74.

Rayner, J.N., 1971, *An Introduction to Spectral Analysis*. London: Pion Press.

Rees, W.G., 1990, *Physical Principles of Remote Sensing*. Cambridge: Cambridge University Press.

Rees, W.G. and Satchell, M.J.F., 1997, The effect of median filtering on synthetic aperture radar images. *International Journal of Remote Sensing*, **18**, 2887–2893.

Riazanoff, S., Cervelle, B. and Chorowicz, J., 1990, Parametrizable skeletonisation of binary and multi-level images. *Pattern Recognition Letters*, **11**, 25–33.

Ribed, P.S. and Lopez, A.M., 1995, Monitoring burnt areas by principal components analysis of multitemporal TM data. *International Journal of Remote Sensing*, **16**, 1577–1587.

Richards, J.A., 1993, *Remote Sensing Digital Image Analysis: An Introduction*, 2nd edn. Berlin: Springer-Verlag.

Richardson, A.J. and Wiegand, C.L., 1977, Distinguishing vegetation from soil background information. *Photogrammetric Engineering and Remote Sensing*, **43**, 1541–1552.

Richter, R., 1996, A spatially-adaptive fast atmospheric correction algorithm. *International Journal of Remote Sensing*, **17**, 1201–1214.

Riou, R. and Seyler, F., 1997, Texture analysis of tropical rain forest infrared satellite imagery. *Photogrammetric Engineering and Remote Sensing*, **63**, 515–521.

Roberts, D., Adams, J.B. and Smith, M.O., 1993, Discriminating green vegetation, non-photosynthetic vegetation and soils in AVIRIS data. *Remote Sensing of Environment*, **44**, 1–25.

Roger, R.E., 1996, Principal components transform with simple, automatic noise adjustment. *International Journal of Remote Sensing*, **17**, 2719–2727.

Rondeaux, G., 1995, Vegetation monitoring by remote sensing: a review of biophysical indices. *Photo-Interpretation*, No. 1995/3, 197–216.

Rondeaux, G., Steven, M.D. and Baret, F., 1996, Optimisation of Soil-Adjusted Vegetation Indices. *Remote Sensing of Environment*, **55**, 95–107.

Rosenfeld, A., 1976, Iterative methods in image analysis. *Pattern Recognition*, **10**, 181–187.

Rosenfeld, A. and Kak, A.C., 1982, *Digital Picture Processing*, 2nd edn, vol. 1. New York: Academic Press.

Rothery, D.A. and Hunt, G.A., 1990, A simple way to perform decorrelation stretching and related techniques on menu-driven image processing systems. *International Journal of Remote Sensing*, **11**, 133–137.

Rowan, L.C., Wetlaufer, P., Goetz, A.F.H., Billingsley, F. and Stewart, J., 1974, *Discrimination of Rock Types and Detection of Hydrothermally-Altered Areas in South-Central Nevada by the Use of Computer-Enhanced ERTS Images*. US Geological Survey, Professional Paper 883. Washington DC: US Government Printing Office.

Running, S.W., Justice, C.O., Salomonson, V.V., Strahler, A.H., Huete, A.R., Muller, J.-P., Vanderbilt, V., Wan, Z.M., Teillet, P.M. and Carneggie, D., 1994, Terrestrial remote sensing science and algorithms planned for EOS/MODIS. *International Journal of Remote Sensing*, **15**, 3587–3620.

Russ, J.C., 1995, *The Image Processing Handbook*, 2nd edn. Boca Raton, FL: CRC Press, 88–100.

Samet, H., 1990, *The Design and Analysis of Spatial Data Structures*. Reading, Mass.: Addison-Wesley.

Sammon, J.W., Jr, 1969, A nonlinear mapping algorithm for data structure analysis. *IEEE Transactions on Computers*, **18**, 401–409.

Sarkar, N. and Chaudhuri, B.B., 1994, An efficient differential box-counting approach to compute the fractal dimension of an image. *IEEE Transactions on Systems, Man and Cybernetics*, **24**, 155–120.

Sawter, R., Deuze, J.L., Devaux, G., Vermote, E., Guyot, G., Tu, X.F., Verbrugge, M. and Leroy, M., 1991, SPOT calibration on the test site at La Crau, France. *Cinquième Colloque International Mesures Physiques et Signatures en Télédétection*, Courcheval, 14–18 January 1991. ESA-SP-319. Paris: European Space Agency, 77–80.

Schaale, M. and Furrer, R., 1995, Land surface classification by neural networks. *International Journal of Remote Sensing*, **16**, 3003–3031.

Schetselaar, E.M., 1998, Fusion by the IHS transform: should we use cylindrical or spherical coordinates? *International Journal of Remote Sensing*, **19**, 759–765.

Schistad, A.H. and Jain, A.K., 1992, Texture analysis in the presence of speckle noise. *Proceedings of the International Symposium on Geoscience and Remote Sensing (IGARSS'92)*, Houston, Texas, May 1992. New York: IEEE Press, 884–886.

Schott, J.R. and Volchok, W.J., 1985, Thematic Mapper thermal infrared calibration. *Photogrammetric Engineering and Remote Sensing*, **51**, 1351–1357.

Schotten, C.G.J., van Rooy, W.W.L. and Janssen, L.L.F., 1995, Assessment of the capabilities of multi-temporal ERS-1 SAR data to discriminate between agricultural crops. *International Journal of Remote Sensing*, **16**, 2619–2637.

Schowengerdt, R.A., 1997, *Remote Sensing: Models and Methods for Image Processing*, 2nd edn. San Diego: Academic Press.

Schreier, G. (ed.), 1993a, *SAR Geocoding: Data and Systems*. Karlsruhe, Germany: Herbert Wichmann.

Schreier, G., 1993b, Geometrical properties of SAR images. In Schreier, G. (ed.) (1993a), 103–134.

Seftor, J.L. and Larch, D., 1995, The use of the genetic algorithm to optimise rule-based classifiers for land cover categorization. *Canadian Journal of Remote Sensing*, **21**, 412–420.

Sellers, P., 1989, Vegetation-canopy reflectance and biophysical properties. In Asrar, G. (ed.) (1989), 297–335.

Settle, J.J. and Campbell, N., 1998, On the errors of two estimators of sub-pixel fractional cover when mixing is linear. *IEEE Transactions on Geoscience and Remote Sensing*, **36**, 163–170.

Settle, J.J. and Drake, N.A., 1993, Linear mixing and the estimation of ground cover proportions. *International Journal of Remote Sensing*, **14**, 1159–1177.

Shandley, J., Franklin, J. and White, T., 1996, Testing the Woodcock–Harward image segmentation algorithm in an area of southern California chaparral and woodland vegetation. *International Journal of Remote Sensing*, **17**, 983–1004.

Sharma, K.M.S. and Sarkar, A., 1998, A modified contextual classification technique for remote sensing data. *Photogrammetric Engineering and Remote Sensing*, **64**, 273–280.

Shaw, R., Sowers, L. and Sanchez, E., 1982, A comparative study of linear and nonlinear edge finding techniques for Landsat multispectral data. In Richason, B.F., Jr (ed.) (1982), *Proceedings of the Pecora VII Symposium*, Sioux Falls, South Dakota. Falls Church, Va.: American Society of Photogrammetry, 529–542.

Shih, T.-Y., 1995, The reversibility of six geometric color spaces. *Photogrammetric Engineering and Remote Sensing*, **61**, 1223–1232.

Shimabukuro, Y.E. and Smith, J.A., 1991, The least-squares mixing models to generate fraction images derived from remote sensing multispectral data. *IEEE Transactions on Geoscience and Remote Sensing*, **29**, 16–21.

Shipman, H. and Adams, J.B., 1987, Detectability of minerals on desert alluvial fans using reflectance spectra. *Journal of*

Geophysical Research, **92**, 10391–10402.

Shlien, S., 1979, Geometric correction, registration and re-sampling of Landsat imagery. *Canadian Journal of Remote Sensing*, **5**, 74–87.

Sietsma, J. and Dow, R.J.F., 1988, Neural net pruning – why and how. *IEEE International Conference on Neural Networks*. New York: IEEE, I-325 to I-333.

Siljeström, P.A. and Moreno, A., 1995, Monitoring burnt areas by principal components analysis of multi-temporal TM data. *International Journal of Remote Sensing*, **16**, 1577–1587.

Siljeström, P.A., Moreno, A., Vikgren, G. and Cáceres, L.M., 1997, The application of selective principal components analysis (SPCA) to a Thematic Mapper (TM) image for the recognition of geomorphologic features configuration. *International Journal of Remote Sensing*, **18**, 3843–3852.

Simonett, D.S. (ed.), 1983, The development and principles of remote sensing. In Colwell, R.N. (ed.) (1983), 1–36.

Singh, A., 1984, Some clarifications about the pairwise divergence method in remote sensing. *International Journal of Remote Sensing*, **5**, 623–627.

Singh, A. and Harrison, A., 1985, Standardised principal components. *International Journal of Remote Sensing*, **6**, 883–896.

Singleton, R.C., 1979a, Mixed radix fast Fourier transform. *Programs for Digital Signal Processing*. New York: IEEE Acoustics, Speech and Signal Processing Society, IEEE Press/Wiley, Section 1.4-1.

Singleton, R.C., 1979b, Two-dimensional mixed radix mass storage Fourier transform. *Programs for Digital Signal Processing*. New York: IEEE Acoustics, Speech and Signal Processing Society, IEEE Press/Wiley, Section 1.9-1.

Skidmore, A.K., Turner, B.J., Brinkhof, W. and Knowles, E., 1997, Performance of a neural network: mapping forests using GIS and remotely sensed data. *Photogrammetric Engineering and Remote Sensing*, **63**, 501–514.

Slater, P.N., 1980, *Remote Sensing: Optics and Optical Systems*. Reading, Mass.: Addison-Wesley.

Slater, P.N., Biggar, S.F., Holm, R.G., Jackson, R.D., Mao, Y., Moran, M.S., Palmer, M. and Yuan, B., 1987, Reflectance-and radiance-based methods for the in-flight calibration of multi-spectral sensors. *Remote Sensing of Environment*, **22**, 11–37.

Smith, A.R., 1978, Color gamut transform pairs. *Computer Graphics*, **12**, 12–19.

Smith, D.M., 1996, Speckle reduction and segmentation of Synthetic Aperture Radar images. *International Journal of Remote Sensing*, **17**, 2043–2057.

Smith, G.M. and Curran, P., 1996, The signal-to-noise ratio (SNR) required for the estimation of foliar biochemical concentrations. *International Journal of Remote Sensing*, **17**, 1031–1058.

Smith, J.A., Lin, T.L., and Ranson, K.J., 1980, The Lambertian assumption and Landsat data. *Photogrammetric Engineering and Remote Sensing*, **46**, 1183–1189.

Smith, M.O., Ustin, S.L., Adams, J.B. and Gillespie, A.R., 1990, Vegetation in deserts: I. Regional measure of abundance from multispectral images. *Remote Sensing of Environment*, **31**, 1–26.

Snyder, J.P., 1982, *Map Projections used by the US Geological Survey*. US Geological Survey Bulletin 1532. Washington DC: US Government Printing Office.

Soares, J.V., Rennó, C.D., Formaggio, A.R., Yanasse, C.C.F. and Frery, A.C., 1997, An investigation into the selection of texture features for crop discrimination using SAR imagery. *Remote Sensing of Environment*, **59**, 234–247.

Sohn, Y. and McCoy, R.M., 1997, Mapping desert shrub rangeland using spectral unmixing and modelling spectral mixtures with TM data. *Photogrammetric Engineering and Remote Sensing*, **63**, 707–716.

Spanner, M.A., Brass, J.A. and Peterson, D.L., 1984, Feature selection and the information content of Thematic Mapper simulator data for forest structural assessment. *IEEE Transactions on Geoscience and Remote Sensing*, **22**, 482–489.

Sparks, D.N., 1985, Half-Normal plotting. In Griffiths, P. and Hill, I.D. (eds) (1985), *Applied Statistical Algorithms*. Chichester: Ellis Horwood, 65–69.

Srinivasana, A. and Richards, J.A., 1990, Knowledge-based techniques for multisource classification. *International Journal of Remote Sensing*, **11**, 505–525.

Steers, J.A., 1962, *An Introduction to the Study of Map Projections*, 13th edn. London: University of London Press.

Stehman, S.V., 1997, Selecting and interpreting measures of thematic classification accuracy. *Remote Sensing of Environment*, **62**, 77–89.

Steigler, S.E., 1978, *Dictionary of Earth Sciences*. London: Pan Books.

Steven, M.D., 1998, The sensitivity of the OSAVI vegetation index to observational parameters. *Remote Sensing of Environment*, **63**, 49–60.

Stimson, A., 1974, *Photometry and Radiometry for Engineers*. New York: Wiley.

Stolz, R. and Mauser, W., 1996, A fuzzy approach for improving landcover classification by integrating remote sensing and GIS. In Parlow, E. (ed.) (1996a), 33–41.

Story, M. and Congalton, R.G., 1986, Accuracy assessment: a user's perspective. *Photogrammetric Engineering and Remote Sensing*, **52**, 397–399.

Strahler, A.H., 1980, The use of prior probabilities in maximum likelihood classification of remotely sensed data. *Remote Sensing of Environment*, **10**, 135–163.

Strahler, A.H., Logan, T.L. and Bryant, N.A., 1978, Improving forest classification from Landsat by incorporating topographic information. *Proceedings of the 12th International Conference on Remote Sensing of Environment*, Ann Arbor, Mich.: Environmental Research Institute of Michigan, 927–942.

Strang, G., 1994, Wavelets. *American Scientist*, **82**, 250–255.

Stromberg, W.D. and Farr, T.G., 1986, A Fourier-based textural feature extraction procedure. *IEEE Transactions on Geoscience and Remote Sensing*, **24**, 722–731.

Suits, G.H., 1983, The nature of electromagnetic radiation. In Colwell, R.N. (ed.) (1983), vol. I, 37–60.

Swain, P.H. and King, R.C., 1973, Two effective feature selection criteria for multispectral remote sensing. *Proceedings of the 1st International Joint Conference on Pattern Recognition*, 73 CHO821-9C. New York: IEEE, 536–540.

Switzer, P., Kowalik, W.S. and Lyon, R.J.P., 1981, Estimation of atmospheric path radiance by the covariance matrix method. *Photogrammetric Engineering and Remote Sensing*, 47, 1469–1476.

Tanré, D., Deroo, C., Duhaut, P., Herman, M., Morcrette, J.J., Perbos, J. and Deschamps, P.Y., 1986, *Simulation of the Satellite Signal in the Solar Spectrum*. Laboratoire d'Optique Atmosphérique, Université des Sciences et Techniques de Lille/Centre Spatiale de Toulouse.

Teillet, P.M. and Fedosejevs, G., 1995, On the dark target approach to atmospheric correction of remotely-sensed data. *Canadian Journal of Remote Sensing*, 21, 374–387.

Teillet, P.M., Guindon, B. and Goodenough, D.G., 1982, On the slope–aspect correction of multispectral data. *Canadian Journal of Remote Sensing*, 8, 84–106.

Terhalle, U. and Bodechtel, J., 1986, Landsat TM data enhancement technique for mapping arid geomorphic features. *Proceedings of the ISPRS/Remote Sensing Society Symposium, Mapping from Modern Imagery*, Edinburgh, September 1986. Nottingham: The Remote Sensing Society, 725–729.

Thomas, G., Hobbs, S.E. and Dufour, M., 1996, Woodland area estimation by spectral mixing: applying a goodness of fit solution method. *International Journal of Remote Sensing*, 17, 291–301.

Thomas, I.L., 1980, Spatial postprocessing of spectrally-classified Landsat data. *Photogrammetric Engineering and Remote Sensing*, 46, 1201–1206.

Thomas, I.L., Howarth, R., Eggers, A. and Fowler, A.D.W., 1981, Textural enhancement of a circular geologic feature. *Photogrammetric Engineering and Remote Sensing*, 47, 89–91.

Thome, K.J., Gellman, D.I., Parada, R.J., Biggar, S.F., Slater, P.N. and Moran, M.S., 1993, Absolute radiometric calibration of Thematic Mapper. *SPIE Proceedings*, 600, 2–8.

Thomson, A.G., Fuller, T.H., Sparks, T.H., Yates, M.G. and Eastwood, J.A., 1998, Ground and airborne radiometry over intertidal surfaces: waveband selection for cover classification. *International Journal of Remote Sensing*, 19, 1189–1205.

Tidemann, J. and Nielsen, A.A., 1997, A simple neural network contextual classifier. In Kanellopoulos, I. *et al.* (eds) (1997), 186–193.

Todd, S.W., Hoffer, R.M. and Milchunas, D.G., 1998, Biomass estimation on grazed and ungrazed rangelands using spectral indices. *International Journal of Remote Sensing*, 19, 427–438.

Tompkins, S., Mustard, J.F., Pieters, C.M. and Forsythe, D.W., 1997, Optimization of end members for spectral mixture analysis. *Remote Sensing of Environment*, 59, 472–489.

Toselli, F. and Bodechtel, J. (eds), 1992, *Imaging Spectroscopy: Fundamentals and Prospective Applications*. Dordrecht: Kluwer Academic.

Tou, J. and Gonzales, R., 1974, *Pattern Recognition Principles*. Reading, Mass.: Addison-Wesley.

Toutin, T., 1995, Multisource data integration with integrated and unified geometric modelling. In Askne, J. (ed.) (1995), 163–174.

Townshend, J.R.G., 1980, *The Spatial Resolving Power of Earth Resources Satellites: A Review*. NASA Technical Memorandum 82020, Goddard Spaceflight Center, Greenbelt, MD. See also: *Progress in Physical Geography*, 5, 33–35.

Townshend, J.R.G., 1984, Agricultural land-cover discrimination using thematic mapper spectral bands. *International Journal of Remote Sensing*, 5, 681–698.

Townshend, J.R.G. and Harrison, A., 1984, Estimation of the spatial resolving power of the Thematic Mapper of Landsat-4. *Proceedings of the Tenth Anniversary International Conference of the Remote Sensing Society*. Nottingham: The Remote Sensing Society, 67–72.

Townshend, J.R.G., Justice, C., Gurney, C. and McManus, J., 1992, The impact of misregistration on change detection. *IEEE Transactions on Geoscience and Remote Sensing*, 30, 1054–1060.

Townshend, J.R.G. and Skole, D.L., 1994, The Global 1 km data set from the Advanced Very High Resolution Radiometer. In Mather, P.M. (ed.) (1994), 75–82.

Tozawa, Y., 1983, Fast geometric correction of NOAA AVHRR. *Proceedings of the Symposium on Machine Processing of Remotely-Sensed Data 1983*, Purdue University, West Lafayette, Ind., 46–53.

Tso, B., 1997, The investigation of alternative strategies for incorporating spectral, textural and contextual information in remote sensing image classification. Ph.D. Thesis, Department of Geography, The University of Nottingham.

Tucker, C.J., 1979, Red and photographic infrared linear combinations for monitoring vegetation. *Remote Sensing of Environment*, 10, 127–150.

Tukey, J.W., 1977, *Exploratory Data Analysis*. Reading, Mass.: Addison-Wesley

Ulaby, F.T. and Elachi, C., 1990, *Radar Polarimetry for Geoscience Applications*. Dedham, Mass.: Artech House.

Ulaby, F.T., Moore, R.K. and Fung, A.K., 1981–1986, *Microwave Remote Sensing: Active and Passive*, 3 vols. Dedham, Mass.: Artech House.

Unwin, D.J. and Wrigley, N., 1987, Towards a general theory of control point distribution effects in trend surface models. *Computers and Geosciences*, 13, 351–355.

Ustin, S.L., Smith, M.O. and Adams, J.B., 1991, Remote sensing of ecological processes – a strategy for developing and testing ecological models through spectral mixture analysis. In Ehleringer, J.R. and Field, C.B. (eds) (1991), 339–357.

Ustin, S.L., Hart, Q.J., Duan, L. and Scheer, G, 1996, Vegetation mapping on hardwood rangelands in California. *International Journal of Remote Sensing*, 17, 3015–3036.

Vanderbrugg, G.J., 1976, Line detection in satellite imagery. *IEEE Transactions on Geoscience Electronics*, 14, 37–44.

van der Meer, F., 1994, Extraction of mineral absorption features from high spectral resolution data using nonparametric geostatistical techniques. *International Journal of Remote Sensing*, 15, 2193–2214.

van der Meer, F., 1996a, Classification of remotely-sensed imagery using an indicator kriging approach: application to the problem of calcite–dolomite mineral mapping. *Interna-

tional Journal of Remote Sensing, **17**, 1233–1249.

van der Meer, F., 1996b, Performance characteristics of the indicator classifier on simulated data. *International Journal of Remote Sensing*, **17**, 621–627.

van der Meer, F., 1996c, Metamorphic facies zonation in the Ronda peridotites: spectroscopic results from field and GER imaging spectrometer data. *International Journal of Remote Sensing*, **17**, 1633–1657.

van der Meer, F., van Dijk, P.M. and Westerhof, A.B., 1995, Digital classification of the contact metamorphic aureole along the Los Pedroches batholith, south-central Spain, using Landsat Thematic Mapper data. *International Journal of Remote Sensing*, **16**, 1043–1062.

Vane, G. (ed.), 1987, *Airborne Visible/Infrared Imaging Spectrometer (AVIRIS): A Description of the Sensor, Ground Data Processing Facility, Laboratory Calibration and First Results*, JPL Publication 87-38. Pasadena, Calif.: NASA Jet Propulsion Laboratory.

Vane, G., Green, R.O., Chrien, T.G., Enmark, H.T., Hansen, E.G. and Porter, W.M., 1993, The airborne visible/infrared imaging spectrometer (AVIRIS). *Remote Sensing of Environment*, **44**, 127–143.

van Gardingen, P.R., Foody, G.M. and Curran, P.J. (eds), 1997, *Scaling-Up*. Cambridge: Cambridge University Press.

van Wie, P. and Stein, M., 1977, A Landsat digital image rectification system. *IEEE Transactions on Geoscience Electronics*, **15**, 130–137.

Varjo, J., 1996, Controlling continuously updated forest data by satellite remote sensing. *International Journal of Remote Sensing*, **17**, 43–67.

Vermote, E. and Kaufman, Y.J., 1995, Absolute calibration of AVHRR visible and near-infrared channels using ocean and cloud views. *International Journal of Remote Sensing*, **16**, 2317–2340.

Verstraete, M.M. and Pinty, B., 1992, Extracting surface properties from satellite data in the visible and near-infrared wavelengths. In Mather, P.M. (ed.) (1992), *TERRA 1: Understanding the Terrestrial Environment – The Role of Earth Observations from Space*. London: Taylor and Francis, 203–209.

Wakabayeshi, H. and Arai, K., 1996, A new method for SAR speckle noise reduction (CST filter). *Canadian Journal of Remote Sensing*, **22**, 190–197.

Wang, F., 1990, Fuzzy supervised classification of remote sensing images. *IEEE Transactions on Geoscience and Remote Sensing*, **28**, 194–201.

Wang, J., 1993, LINDA – A system for automated linear feature detection. *Canadian Journal of Remote Sensing*, **19**, 9–21.

Wang, R.-Y., 1986a, An approach to tree-classifier design based on hierarchical clustering. *International Journal of Remote Sensing*, **7**, 75–88.

Wang, R.-Y., 1986b, An approach to tree-classifier design based on a splitting algorithm. *International Journal of Remote Sensing*, **7**, 89–104.

Wardley, N.W., 1984, Vegetation index variability as a function of viewing geometry. *International Journal of Remote*

Sensing, **5**, 861–870.

Wasserman, G., 1978, *Color Vision: An Historical Introduction.* New York: Wiley.

Wecksung, G.W. and Breedlove, J.R., Jr, 1977, Some techniques for digital preprocessing, display and interpretation of ratio images in multispectral remote sensing. *Proceedings of the Society of Photo-Optical Instrumentation Engineers (SPIE)*, **119**, 47–54.

Wegener, M., 1990, Destriping multiple sensor imagery by improved histogram matching. *International Journal of Remote Sensing*, **11**, 859–975.

Wegman, E., 1990, Hyperdimensional data analysis using parallel coordinates. *Journal of the American Statistical Association*, **85**, 664–675.

Weinand, H.C., 1974, Cosine theta in components analysis. *Annals of the Association of American Geographers*, **64**, 353.

Westin, T., 1990, Precision rectification of SPOT imagery. *Photogrammetric Engineering and Remote Sensing*, **56**, 247–253.

Wezka, J.S., Dyer, C.R. and Rosenfield, A., 1976, A comparative study of texture measures for terrain classification. *IEEE Transactions on Systems, Man and Cybernetics*, **6**, 269–285.

Wharton, S.W., 1980, A contextual classification method for recognising land use patterns in high-resolution remotely-sensed data. *Pattern Recognition*, **15**, 317–324.

White, K., 1993, Image processing of Thematic Mapper data for discriminating piedmont surficial materials in the Tunisian Southern Atlas. *International Journal of Remote Sensing*, **14**, 961–977.

Wilkinson, J.H. and Reinsch, C. (eds), 1971, *Linear Algebra. Handbook for Automatic Computation*, vol. 2. Berlin: Springer-Verlag.

Wilkinson, J.J. and Mégier, J., 1990, Evidential reasoning in a pixel classification hierarchy – a potential method for integrating image classifiers and expert system rules based on geographic context. *International Journal of Remote Sensing*, **11**, 1963–1968.

Williams, C.S. and Becklund, O.A., 1972, *Optics: A Short Course for Scientists and Engineers*. New York: Wiley-Interscience.

Williams, J., 1995, *Geographic Information from Space*. Chichester: Wiley/Praxis.

Wilson, J.D., 1992, A comparison of procedures for classifying remotely-sensed data using simulated data sets incorporating autocorrelations between spectral responses. *International Journal of Remote Sensing*, **13**, 2701–2725.

Wong, F., Orth, R. and Friedmann, D., 1981, The use of digital terrain models in the rectification of satellite-borne imagery. *Proceedings of the 15th International Symposium on Remote Sensing of Environment*. Ann Arbor, Mich.: Environmental Research Institute of Michigan, 653–662.

Woodcock, C. and Harward, V.J., 1992, Nested-hierarchical scene models and image segmentation. *International Journal of Remote Sensing*, **16**, 3167–3187.

Woodcock, C.E. and Strahler, A.H., 1987, The factor of scale in remote sensing. *Remote Sensing of Environment*, **21**, 311–322.

Woodcock, C.E., Strahler, A.H. and Jupp, D.L.B., 1988a, The use of variograms in remote sensing. I: Scene models and simulated images. *Remote Sensing of Environment*, **25**, 323–348.

Woodcock, C.E., Strahler, A.H. and Jupp, D.L.B., 1988b, The use of variograms in remote sensing. II: Real digital images. *Remote Sensing of Environment*, **25**, 349–379.

Woodham, R.J., 1989, Determining intrinsic surface reflectance in rugged terrain and changing illumination. *Proceedings of the International Geoscience and Remote Sensing Symposium (IGARSS'89)*, Vancouver, British Columbia, Canada, 10–14 July 1989. New York: IEEE Press, vol. 1, 1–5.

Wooding, M.G., Zmuda, A.D. and Griffiths, G.H., 1993, Crop discrimination using multi-temporal ERS-1 SAR data. *Proceedings of the Second ERS-1 Symposium on Space at the Service of our Environment*, Hamburg, Germany. ESA SP-361. Paris: European Space Agency, 51–56.

Wu, H.-H.P. and Schowengerdt, R.A., 1993, Improved fraction image estimation using image restoration. *IEEE Transactions on Geoscience and Remote Sensing*, **31**, 771–778.

Yocky, D.A., 1996, Multiresolution wavelet decomposition image merger of Landsat Thematic Mapper and SPOT panchromatic data. *Photogrammetric Engineering and Remote Sensing*, **62**, 1067–1084.

Yool, S.R., Star, J.L., Estes, J.E., Botkin, D.B., Eckhardt, D.W. and Davis, F.W., 1986, Performance analysis of image processing algorithms for classification of natural vegetation in the mountains of Southern California. *International Journal of Remote Sensing*, **7**, 683–702.

Young, T.L. and Kaufman, Y.J., 1986, Non-Lambertian effects on remote sensing of surface reflectance and vegetation index. *IEEE Transactions on Geoscience and Remote Sensing*, **24**, 699–707.

Zhou, J. and Civco, D.L., 1996, Using genetic learning neural networks for spatial decision making in GIS. *Photogrammetric Engineering and Remote Sensing*, **62**, 1287–1295.

Zhou, Z., Civco, D.L. and Silander, J.A., 1998, A wavelet transform method to merge Landsat TM and SPOT panchromatic data. *International Journal of Remote Sensing*, **19**, 743–757.

Zhuang, X., Engel, B.A., Xiong, X. and Johannsen, C.J., 1995, Analysis of classification results of remotely sensed data and evaluation of classification algorithms. *Photogrammetric Engineering and Remote Sensing*, **61**, 427–433.

Index

TERMS OF SERVICE AGREEMENT AND LICENCE

The following provisions govern the use **of Computer Processing of Remotely Sensed Images CD-ROM** (hereinafter referred to as 'the Product'). They consist of four sections: Copyright, Rules of Use, Additional Terms, and Acceptance Procedure. Each section is important and you should read them all carefully before completing the registration process. These terms of service constitute a licence, and you agree to use **the Product** only as set forth in the Rules of Use below. Any statutory rights you may have remain unaffected by your acceptance of this Agreement.

This licence is for single-user access only.

Requests for technical assistance in connection with the Product should be made to
Paul.Mather@Nottingham.ac.uk

Copyright

Copyright © 1999 Paul Mather. All rights reserved.

All material contained within the Product is protected by copyright, whether or not a copyright notice appears on the particular screen where the material is displayed.

Rules of Use

The rights granted to you under this Agreement cannot be transferred, sold or rented to anyone else.

You may download or print out materials from the Product for your own research or study only; you may not transmit, rent, lend, sell, or modify any materials from the Product or modify, or create derivative works based on, materials from the Product, unless this is expressly permitted in a copyright notice or usage statement accompanying the material. Requests for permission to store or reproduce material for any purpose should be addressed to the Permissions Department, John Wiley & Sons, Ltd, Baffins Lane, Chichester, West Sussex, PO19 1UD, UK; telephone +44 (0) 1243 770 347; Email < permissions@wiley.co.uk >.

If you refuse or fail to abide by these rules or violate any other terms or conditions of this Agreement, the Publisher, in addition to any other available remedies, reserves the right in its sole discretion to terminate your licence to use the Product immediately without notice.

Additional Terms

These terms constitute the entire Agreement between you and the Publisher and supersede any prior written or oral representation or Agreement.

The Product is provided on an 'as is' basis, without warranties of any kind, either express or implied, including but not limited to warranties of title, or implied warranties of merchantablility or fitness for a particular purpose. There will be no duty on the author or Wiley to correct any errors or defects in the software.

You agree to indemnify the Publisher, the author and their agents against any liability for any claims and expenses, including reasonable fees, relating to any violation of the terms of this Agreement.

The author and John Wiley & Sons, Ltd make no warranty as to the accuracy and completeness of any information contained in the Product, nor do they accept any responsibility or liability for any inaccuracy or errors and omissions, or for any loss or damage occasioned to any person or property through using the materials, information, methods or ideas contained herein, or acting or refraining from acting as a result of such use. Any statutory rights you may have remain unaffected by your acceptance of this Agreement.

This Agreement shall be governed by English law applicable to Agreements wholly made and performed in England. If you believe that you have been irreparably harmed by any action relating to the Product, you must inform the Publisher in writing and grant the Publisher thirty (30) days to rectify the harm before initiating any legal action. Any legal action relating to this Agreement or a breach of it will take place in a court of competent jurisdiction in England, and notification will be given to you by overnight courier or express mail at your last known address. Any course of legal action initiated by you must be initiated within one (1) year after the reason for it has arisen.

Acceptance Procedure

By opening the CD-ROM you agree with all the terms and conditions of the Agreement. You will then have access to the content of the Product. Your statutory rights remain unaffected by your acceptance of this Agreement.